Engineering Analogies

ENGINEERING ANALOGIES

Glenn Murphy | Iowa State University
David J. Shippy | Graceland College
H. L. Luo | California Institute of Technology

IOWA STATE UNIVERSITY PRESS, *Ames*, IOWA, U.S.A.

ABOUT THE AUTHORS

Dr. GLENN MURPHY is Head of the Department of Nuclear Engineering and in charge of the curriculum in Engineering Science at Iowa State University. Holding the distinction of the first man named Anson Marston Distinguished Professor of Engineering, he has been a member of the Iowa State staff since 1932. He has served as Head of the Department of Aeronautical Engineering and the Department of Theoretical and Applied Mechanics. In 1962, he was elected president of the American Society for Engineering Education and has held membership and offices in a number of national and international educational and professional societies and committees. His writings include a number of papers and bulletins and seven other university-level textbooks.

DAVID J. SHIPPY began his engineering training at Graceland College and continued at Iowa State University, receiving his Ph.D. in 1963. He spent two years in the functional design and analysis of missile autopilots at General Dynamics/Fort Worth. At the present time, he teaches undergraduate engineering and physics at Graceland College.

HUEY-LIN LUO, born in the mainland of China, received his B.S. degree in 1955 from National Taiwan University. He came to the United States, studied under Professor Glenn Murphy at Iowa State University and received his M.S. degree in 1959, following which he moved to the California Institute of Technology in materials science.

© 1963 by the
Iowa State University Press.
Printed in the U.S.A.
All rights reserved.

International Standard Book Number   0-8138-2225-4
Library of Congress Catalog Card Number: 63-16671

Foreword

Introduction— The general concept of utilizing observations on one system to predict how a second system (different in some respects and similar in some respects to the first) will perform or behave has been known for many years, and the technique has been used in the solution of a number of important problems. As the complexity of problems in the several engineering fields becomes greater, increasing use is being made of the methods of analogies whereby problems in one area, such as the vibration of a mechanical system, may be solved by making observations on an apparently different system, such as an electrical network.

The general principles, necessary conditions and operating rules for the use of analogies have been formulated* and numerous applications have been made. Because of the widespread applicability of analogies and the increasing interest in this method, this survey of analogies systems has been prepared. It consists of a compilation of analogies as reported in the standard publications, with a brief description of a number of the applications. An author index is included.

Acknowledgements— Funds for the library search necessary for the compilation were made available through the generosity of the Research Corporation in the form of a Graduate Assistantship. Completion of the project was sponsored by the Iowa Engineering Experiment Station.

---

*Murphy, Glenn, *Similitude in Engineering*, Ronald Press, New York, 1950.

Table of Contents

Section 100. Flow of Fluids in Conduits

110 Open Channels	1
111 Tidal Channels	1
112 Rivers	4
112A Flood Stages	4
112B Intermittently Used Dam	6
113 Ocean	6
120 Closed Conduits	6
121 Unsteady Flow (Pulsations and Waves)	6
121A Liquid	6
121AA Uniform Flow	6
121AAA Frictionless	6
121AAB With Friction	7
121AB Non-Uniform Flow	8
121B Gas	9
121BA General	9
121BAA Frictionless	9
121BAB With Friction	10
121BB Acoustical Systems	13
121BBA Electrical Analogy A	13
121BBB Electrical Analogy B	14
121BBC Electrical Analogy C	14
121BBD Optical Analogy A	14
121BBE Optical Analogy B	14
121BBF Electromagnetic Analogy	15
121BC Other Systems	15
122 Steady Flow (Flow Distribution)	15
122A Liquid	15
122AA Electrical Analogy A (linear resistors)	15
122AB Electrical Analogy B (tungsten filament and vacuum tube resistors)	18
122AC Electrical Analogy C (servo-controlled resistors)	18

CONTENTS

122B	Gas	18
122BA	Pipeline Networks	18
122BAA	Electrical Analogy A	18
122BAB	Electrical Analogy B	19
122BAC	Electrical Analogy C	19
122BB	Mine Ventilation	19

Section 200. Fluid Flow in General

210	Compressible	21
211	Steady State	21
211A	Irrotational	21
211AA	General Potential Flow	21
211AAA	Electrical Analogy A	21
211AAB	Electrical Analogy B	25
211AAC	Electrical Analogy C	26
211AAD	Electrical Analogy D	28
211AAE	Mechanical Analogy	29
211AB	Modified General Potential Flow (The "Hydraulic" Analogy)	29
211AC	"Under Transonic" Flow	34
211AD	Flow With Steady Decline of Density	35
211B	Rotational	36
211C	Other	37
212	Unsteady Flow	37
212A	Electrical Analogy A	37
212B	Electrical Analogy B	39
220	Incompressible	41
221	Non-Viscous	41
221A	Irrotational	41
221AA	Electrical Analogy A	41
221AAA	Two-Dimensional Flow	41
221AAB	Axially Symmetric Flow	53
221AAC	General Three-Dimensional Flow	56
221AB	Electrical Analogy B	58
221AC	Membrane Analogy	60
221AD	Magnetic Analogy A	61
221AE	Magnetic Analogy B	63
221AF	Hydrodynamical Analogy	63
221AG	Hydraulic Analogy (Transient Flow)	64
221B	Rotational	65
221C	Vortices	66
221CA	Electromagnetic Analogy	66
221CB	Electrical Analogy	67
221D	Other	68
222	Viscous	68
222A	Journal Bearing Lubrication	68
222B	Other	70

CONTENTS

Section 300. Heat Transfer

310 Conduction.	71
311 Unsteady Flow (With and Without Sources)	71
311A Electrical Analogy	71
311B Hydraulic Analogy.	78
311C Pneumatic Analogy	79
312 Steady State With Sources and Sinks	80
312A Membrane Analogy	80
312B Hydrodynamical Analogy A	80
312C Hydrodynamical Analogy B	81
312D Electrical Analogy	81
313 Steady State Without Sources and Sinks	81
313A Electrical Analogy	81
313B Membrane Analogy	86
313C Other	86
320 Convection.	87
321 Unsteady Flow	87
321A Electrical Analogy	87
321B Hydraulic Analogy.	87
322 Steady-State.	87
322A General.	87
322AA Electrical Analogy	87
322AB Hydraulic Analogy	91
322B Turbulent Stream Flowing Between Parallel Walls	91
322C Heat Transfer and Fluid Friction - Heat and Momentum Transfer Analogy	92
330 Radiation.	94

Section 400. Mechanics of Materials (Steady-State)

410 Tension and Compression.	95
411 Membrane Analogy	95
412 Hydrodynamical Analogy.	95
413 Electrical Analogy.	96
420 Flexure.	96
421 Columns.	96
421A Electrical Analogy	96
421B Other Electrical Analogies	99
421C Pendulum Analogy.	100
422 Beams.	100
422A The "Column" Analogy.	100
422B Mechanical Analogy.	101
422C The "Moment Area" or "Conjugate Beam" Analogy	102
422D Electrical Analogies	103

CONTENTS

- 430 Transverse Loads 103
  - 431 Membrane Analogy A 103
  - 432 Membrane Analogy B 105
  - 433 Electrical Analogy........................... 107
- 440 Torsional Loads 107
  - 441 Elastic 107
    - 441A Uniform Bars........................ 107
      - 441AA Membrane Analogy A 107
      - 441AB Membrane Analogy B 111
      - 441AC Electrical Analogy A 112
      - 441AD Electrical Analogy B 112
      - 441AE Hydrodynamical Analogy A......... 114
      - 441AF Hydrodynamical Analogy B......... 115
      - 441AG Hydrodynamical Analogy C......... 116
      - 441AH Hydrodynamical Analogy D......... 117
    - 441B Bodies of Revolution 117
      - 441BA Electrical Analogy A............. 117
      - 441BB Electrical Analogy B............. 119
    - 441C Others 120
  - 442 Plastic 120
    - 442A "Sand-Heap" Analogy.................. 120
    - 442B Electrical Analogy 122
- 450 Plane Elasticity..................................... 122
  - 451 The "Slab" Analogy 122
  - 452 Membrane Analogy A 125
  - 453 Membrane Analogy B 126
  - 454 Hydrodynamical Analogies 126
  - 455 Electrical Analogy A 126
  - 456 Electrical Analogy B 127
  - 457 Electrical Analogy C 127
  - 458 Other Electrical Analogies 128
  - 459 Water Diffusion Analogy 128
- 460 Plates and Shells 128
  - 461 Electrical Analogy A 128
  - 462 Other Electrical Analogies 130
  - 463 Membrane Analogy 130
  - 464 Plane Elasticity Analogy.................. 130
  - 465 Mechanical Analogy 131
- 470 Frames and Other Structures......................... 132
  - 471 Pin-connected 132
    - 471A Statically Determinate 132
    - 471B Statically Indeterminate 133
  - 472 Rigidly Connected 134
    - 472A Stress................................. 134
      - 472AA Electrical Analogy 134
      - 472AB Other Electrical Analogies........ 136

CONTENTS

472AC The "Shear and Torsion" Analogy	137
472AD Frame Analogy A	137
472AE Frame Analogy B	139
472AF Other Frame Analogies	139
472B Deflection	139
472BA Conjugate Frame Analogy	139
472BB Other Frame Analogies	141
472BC Electrical Analogy A	141
472BD Other Electrical Analogies	141
480 Reinforced Sheets, Panels, and Shells	142
481 Electrical Analogy A	142
482 Electrical Analogy B	144
483 Electrical Analogy C	146
484 Other Electrical Analogies	147
490 Other	147
491 Visco-Elasticity	147
492 Miscellaneous	148

Section 500. Mechanical Vibrations and Transients

510 Axial Systems	149
511 Concentrated Masses	149
511A Electrical Analogy A	149
511B Electrical Analogy B	152
511C Other Analogies	153
512 Distributed Masses	153
512A Frictionless	153
512B With Friction	154
520 Beams and Plates (Flexure)	155
521 Electrical Analogy A	155
522 Electrical Analogy B	157
523 Electrical Analogy C	158
524 Other Electrical Analogies	159
530 Torsional Members	160
531 Concentrated Masses	160
531A Electrical Analogy A	160
531B Electrical Analogy B	166
532 Distributed Masses	166
532A Electrical Analogy A	166
532B Electrical Analogy B	167
532C Other Electrical Analogies	168
540 Structures	168

Section 600. Electricity and Magnetism

610	Electrostatic Fields	169
	611 Thermal Analogy	169
	612 Electrodynamic Analogy	169
	613 Hydrodynamical Analogy	170
	614 Pendulum Analogy	170
620	Electric Circuits	170
	621 Basic Principles	170
	622 Coupled Circuits	171
	622A Rolling Ball Analogy	171
	622B Pendulum Analogy	173
630	Magnetic Fields	173
	631 Hydrodynamical Analogy	173
	632 Electrical Analogy A. Sources (Scalar Potential)	174
	633 Electrical Analogy B. Vortices (Vector Potential)	174
640	Electromagnetic Fields	175
	641 Electric Network Analogies	175
	642 Other Electrical Analogies	177
	643 Membrane Analogy	178
	644 Acoustical Analogy	179
650	Motion of Charged Particles in Electric and Magnetic Fields	179
	651 General	179
	651A Gyroscope Analogy	179
	651B Rolling Ball Analogy A	181
	651C Rolling Ball Analogy B	181
	651D Electrical Analogy	182
	651E Electromechanical Analogy	182
	652 In Lenses	183
	653 In Gases	183

Section 700. Modern Physics

710	Vibration of Molecules	185
720	Quantum Theory	185
730	Nuclear Reactors	185

Appendix I. Types of Potential Fields	186
Appendix II. Analogs for Potential Fields	187
General References	191
References	194

Section 100.   Flow of Fluids in Conduits

110 OPEN CHANNELS

111 *Tidal Channels*

Glover, Herbert, and Daum have described an electrical analogy for flow in a tidal channel (Fig. 111.1). The analogy is based on the similarity between the equation of continuity and Newton's Law for the fluid and the definitions of inductance and capacitance in a conductor.

Characteristic equations:

           FLUID   ELECTRICAL

$$\frac{\partial Q}{\partial x} + b \frac{\partial y}{\partial t} = 0 \quad (1) \qquad \frac{\partial I}{\partial \xi} + C \frac{\partial E}{\partial \eta} = 0 \quad (2)$$

$$\frac{\partial y}{\partial x} + \frac{1}{gHb}\frac{\partial Q}{\partial t} = 0 \quad (3) \qquad \frac{\partial E}{\partial \xi} + L \frac{\partial I}{\partial \eta} = 0 \quad (4)$$

where:

	FLUID		ELECTRICAL
Q	= discharge	I	= current
y	= height of the free surface above the undisturbed surface	E	= voltage
x	= distance along the stream	ξ	= distance
b	= width of the channel	C	= capacitance per unit length
g	= acceleration of gravity	L	= inductance per unit length
H	= depth of the channel from the undisturbed surface		
t	= time	η	= time

1

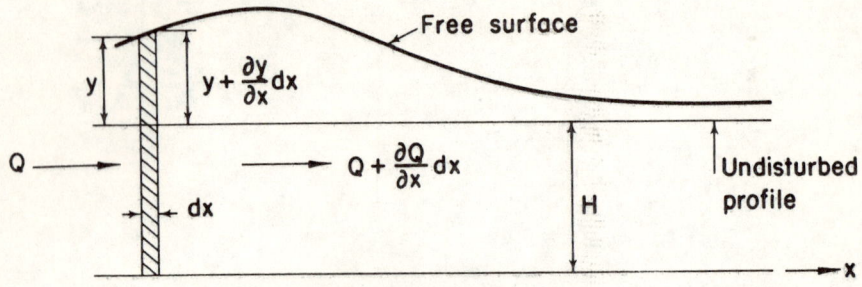

Fig. 111.1. Longitudinal section of a tidal channel.

Thus, the analog consists of a conductor having a capacitance per unit length proportional to the width of the channel at the corresponding station, and an inductance per unit length inversely proportional to the product gH. When a voltage proportional to the height of the disturbance is impressed on the conductor, the resulting current is proportional to the discharge at corresponding times and stations. This analog is based on the assumption that both the friction in the channel and the electrical resistance are negligible.

An analog circuit may be developed by representing a short finite length of the channel with a loop as indicated (Fig. 111.2).

The two pairs of characteristic equations (1) and (3), and (2) and (4) may be combined to give a single characteristic equation for each system.

$$\frac{\partial^2 y}{\partial x^2} = \frac{I}{gH} \frac{\partial^2 y}{\partial t^2} \quad (5) \qquad\qquad \frac{\partial^2 E}{\partial \xi^2} = LC \frac{\partial^2 E}{\partial \eta^2} \quad (6)$$

Van Veen (A) included the effect of friction in the channel by writing equation (3) as

$$\frac{\partial y}{\partial x} + \frac{I}{gHb} \frac{\partial Q}{\partial t} + \frac{kQ}{Hb} = 0 \tag{7}$$

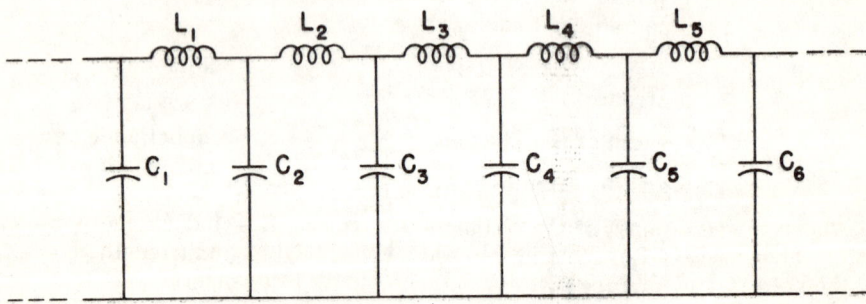

Fig. 111.2. Analog of a frictionless channel.

100. FLOW OF FLUIDS IN CONDUITS

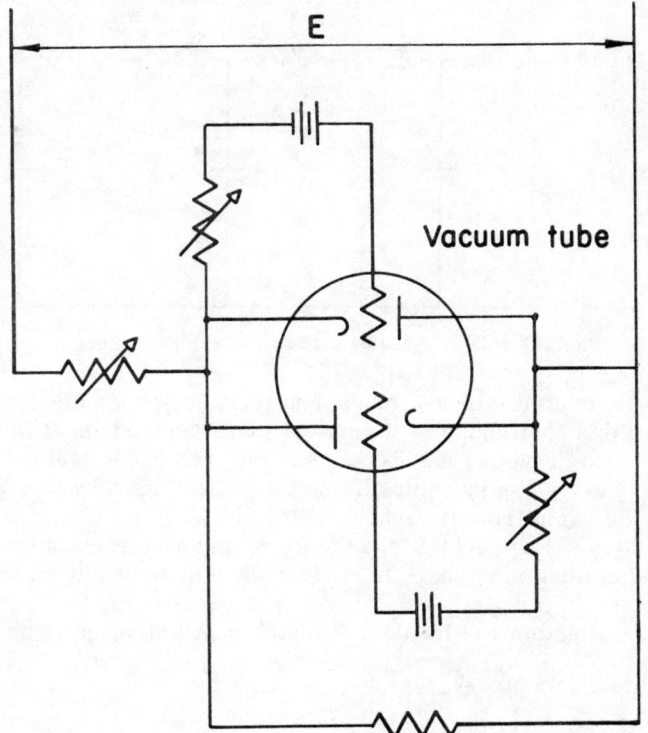

Fig. 111.3. Square-law resistor.

where k = linear resistance coefficient (a constant). A similar term may be included in equation (4) to represent the electrical resistance.

Hydraulic resistance that follows the square law approximately

$$Q = M \sqrt{\frac{\partial y}{\partial x}} \qquad (8) \qquad\qquad I = K \sqrt{\frac{\partial E}{\partial \xi}} \qquad (9)$$

(where M and K are constants), may be represented by special square-law resistors (Fig. 111.3) in the analog circuit.

Low frictional resistance in large channels is nearly linear. For such cases, ordinary resistors governed by the following equation may be used:

$$I = \frac{1}{r} \frac{\partial E}{\partial \xi} \qquad (10)$$

where r = resistance per unit length of circuit.

Then a circuit of the type shown (Fig. 111.4) may be used, with each loop of the circuit representing a finite length of channel.

Van Veen (A) gives descriptions of several tidal phenomena

100. FLOW OF FLUIDS IN CONDUITS

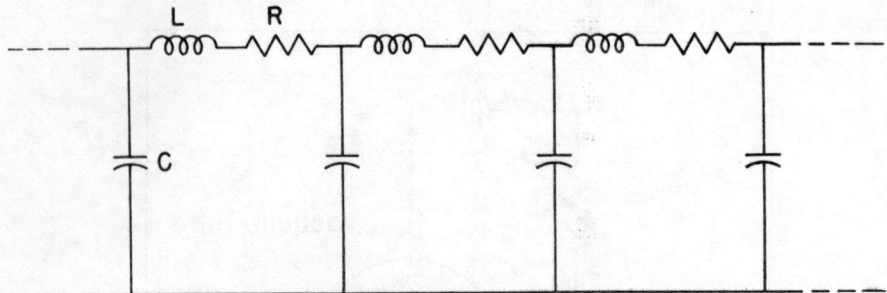

Fig. 111.4. Basic analog circuit for a tidal channel.

analogous to electric-circuit phenomena including phase effects. His report contains oscillograms of vertical tides and horizontal flow. There are also curves of discharge vs. time from data obtained by electrical analogy and by calculation. The circuit used by Van Veen is somewhat different from that shown (Fig. 111.4).

Schoenfeld (C) has applied the theory of a nonhomogeneous electric transmission line to the analysis of tidal wave motion in long, deep canals.

See also Einstein and Harder; Schoenfeld (A); Stroband; Van Veen (B, C).

112 *Rivers*

112A *Flood Stages*

Linsley, Foskett, and Kohler have described a device developed by the U. S. Weather Bureau for routing flow from point to point along a river. In this analogy, the storage within a river reach is assumed to be a function of weighted inflow and outflow.

Characteristic equations:

FLUID	ELECTRICAL
$S = K[xQ_i + (1 - x)Q_o]$ (1)	$q = RCI_i + (RC + 2R'C)I_o$ (2)
$Q_o = Q_i - \dfrac{dS}{dt}$ (3)	$I_o = I_i - \dfrac{dq}{dt}$ (4)

where: Q = discharge (inflow or outflow) I = current

t = time t = time

Δt = period of time for which the flow is routed Δt = period of time during the routing operation

100. FLOW OF FLUIDS IN CONDUITS

- S = storage (volume of water stored in the reach)
- k = storage factor with the dimension of time
- x = weighting factor representing the relative importance of Q_i in controlling storage in the reach
- q = total charge on the capacitors
- R, R′ = resistance
- C = capacitance

Subscripts i and o indicate states at the upstream and downstream ends of the reach, respectively (Fig. 112A.1).

Thus, by choosing values of R, R', and C such that k is proportional to $2(RC + R'C)$ and x is proportional to $\frac{RC}{2(RC + R'C)}$ and by causing I_i to vary with time as Q_i, I_o will vary with time as Q_o.

In operation, the inflow hydrograph (variation of discharge with time) is plotted on the chart of potentiometer P_i, and the inflow current is varied (by adjusting the voltage on the lamp which in turn varies the light intensity falling on the phototube and thus the current) so that the potentiometer pen follows this plotted path. Since the deflection of the pen of potentiometer P_o is directly proportional to I_o, the outflow graph is automatically traced on the chart of potentiometer P_o, which moves with the same speed as the chart on the inflow potentiometer. (Instead of a single circuit containing an inflow potentiometer, several such circuits may be inserted.)

Adjustments of k and x can be determined through successive approximations by running a direct measured problem and comparing observed and computed outflow hydrographs.

Their report includes a plot of outflow hydrographs for varying values of k and x with a given inflow hydrograph.

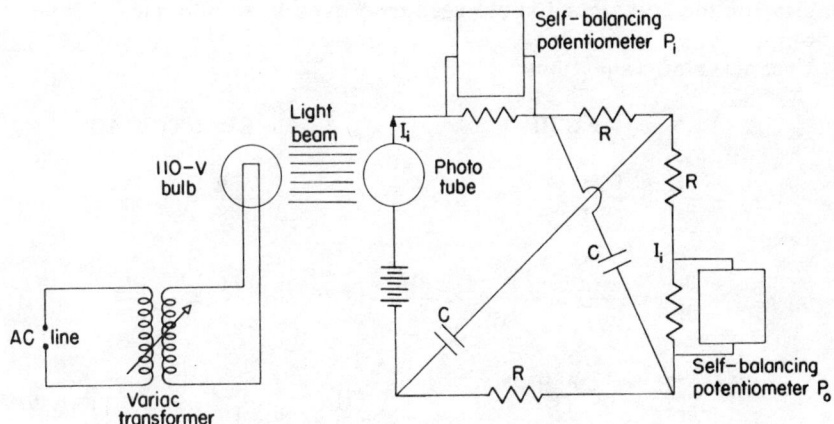

Fig. 112A.1. Analog circuit.

100. FLOW OF FLUIDS IN CONDUITS

M. A. Kohler describes the application of this device to the preparation of river forecasts and to the routing of effective rainfall (runoff) over relatively large basins. He presents plots of typical storage relations and plots of outflow from data obtained by analytical methods, analogy, and actual observation.

See also Linsley, Kohler, and Paulhus; Messerle.

112B *Intermittently Used Dam*

Vartia describes an electric circuit analogy for analyzing the flow of water in an open channel downstream from a dam with an intermittently used power plant where the channel width and depth continually decrease for several miles downstream. The analogous quantities are similar to those in the preceding analogy.

Vartia's report includes curves of elevation and flow vs. time for two-mile intervals up to twelve miles. Analog data are compared with analytical data.

113 *Ocean*

Ishaguro has used electrical analogs similar to those described in section 111 to study long-wave phenomena in the ocean.

120 CLOSED CONDUITS

121 *Unsteady Flow (Pulsations and Waves)*

121A *Liquid*

121AA *Uniform Flow*

121AAA *Frictionless*.

Paynter has described an electric circuit analogy for the transmission of pressure waves in a uniform, frictionless pipeline.

Characteristic equations:

FLUID		ELECTRICAL	
$-\dfrac{\partial H}{\partial x} = \dfrac{1}{g}\dfrac{\partial v}{\partial t}$	(1)	$-\dfrac{\partial E}{\partial s} = L\dfrac{\partial I}{\partial t}$	(2)
$-\dfrac{\partial v}{\partial x} = \dfrac{w}{E_b}\left(1 + \dfrac{E_b}{E_y}\dfrac{D}{e}\right)\dfrac{\partial H}{\partial t}$	(3)	$-\dfrac{\partial I}{\partial s} = C\dfrac{\partial E}{\partial t}$	(4)
$\dfrac{\partial^2 H}{\partial t^2} = c^2 \dfrac{\partial^2 H}{\partial x^2}$	(5)	$\dfrac{\partial^2 E}{\partial t^2} = c^2 \dfrac{\partial^2 E}{\partial s^2}$	(6)

100. FLOW OF FLUIDS IN CONDUITS

$$\frac{\partial^2 v}{\partial t^2} = c^2 \frac{\partial^2 v}{\partial x^2} \quad (7) \qquad \frac{\partial^2 I}{\partial t^2} = c^2 \frac{\partial^2 I}{\partial s^2} \quad (8)$$

The propagation velocities, c, are:

$$c = \sqrt{\frac{E_b g/w}{1 + \frac{E_b}{E_y} \frac{D}{e}}} \quad (9) \qquad c = \sqrt{\frac{1}{LC}} \quad (10)$$

where:
- H = head
- x = distance
- g = acceleration of gravity
- v = velocity
- t = time
- w = specific weight
- E_b = bulk modulus
- E_y = Young's modulus
- D = pipe diameter
- e = pipe-wall thickness

- E = voltage
- s = distance
- L = inductance
- I = current
- t = time
- C = capacitance

These relationships lead to a circuit similar to that shown (Fig. 111.2). Paynter used the analogy to analyze surge, resonance, and water-hammer phenomena. See also Piquemal (A, B).

121AAB *With Friction.* An electric circuit analogy for pulsative liquid flow in a uniform pipeline with friction has been described by Millstone.

Characteristic equations:

FLUID

$$\frac{\partial P}{\partial x} + RQ + \frac{\rho}{A} \left(\frac{\partial Q}{\partial t}\right) = 0 \quad (1)$$

$$\frac{\partial Q}{\partial x} + GP + \beta A \left(\frac{\partial P}{\partial t}\right) = 0 \quad (3)$$

ELECTRICAL

$$\frac{\partial E}{\partial x} + Ri + L \left(\frac{\partial i}{\partial t}\right) = 0 \quad (2)$$

$$\frac{\partial i}{\partial x} + GE + C \left(\frac{\partial E}{\partial t}\right) = 0 \quad (4)$$

where:
- P = pressure
- x = distance
- R = frictional resistance per unit length
- Q = discharge
- $\frac{\rho}{A}$ = equivalent fluid inertial constant per unit length
- t = time
- G = tubing leakage per unit length

- E = voltage
- x = distance
- R = resistance per unit length
- i = current
- L = inductance
- t = time
- G = conductance to ground per unit length

100. FLOW OF FLUIDS IN CONDUITS

βA = fluid compressibility and tube elasticity per unit length

C = capacitance

The Circuit (Fig. 111.4) applies to this analogy.

J. E. Green has discussed the study of a laminar flow pressure surge in an oil pipeline on the basis of an analogy with an electric transmission line.

121AB *Non-Uniform Flow*

Baird and Bechtold have suggested the use of an electrical analogy to analyze the effects of pulsative flow through orifices in pipelines. Electrical terminology and concepts were used to form a theory of the orifice pressure distribution with pulsative flow. The theory was verified by actual measurements on a piping system.

Millstone gives the following as the characteristic equations of this analogy:

$$\Delta P = KQ^2 \quad (1) \qquad \Delta E = (ki^2) = R_n i \quad (2)$$

where:
P = pressure
K = (for the orifice) is a constant dependent upon Reynolds number and orifice geometry
Q = discharge

E = voltage
k = a constant
R_n = non linear resistance
i = current

Since valves are essentially orifices or groups of orifices, they are analogous to networks of non linear electrical resistors.

Millstone also describes an electric circuit analogy for a hydraulic actuator (Fig. 121AB.1).

Characteristic equations:

FLUID ELECTRICAL

$$q_\alpha = \beta V_i \frac{dP_i}{dt} \quad (3) \qquad i_\alpha = C_\alpha \frac{dE_i}{dt} \quad (4)$$

$$q_\beta = \beta V_o \frac{dP_o}{dt} \quad (5) \qquad i = C_\beta \frac{dE_o}{dt} \quad (6)$$

$$q_i = q_\alpha + q_p + q_L \quad (7) \qquad i_i = i_p + i_L + i_\alpha \quad (8)$$

$$q_o = q_p + q_L - q_\beta \quad (9) \qquad i_o = -i_\beta + i_p + i_L \quad (10)$$

$$q_L = \frac{P_i - P_o}{R_L} \quad (11) \qquad i_L = \frac{E_i - E_o}{R_L} \quad (12)$$

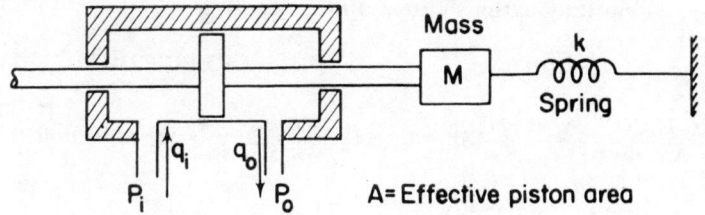

Fig. 121AB.1. Hydraulic actuator.

where: q_p = rate of flow of fluid moved by the actuator

q_L = leakage flow between the piston and the cylinder walls

q_α and q_β = decreases in the flow rates to and from the cylinder

β = fluid compressibility factor

R_L = effective piston-cylinder leakage resistance factor

V_i and V_o = volumes of fluid in the input and output sides of the cylinder

The electrical system symbols are given in figure 121AB.2.

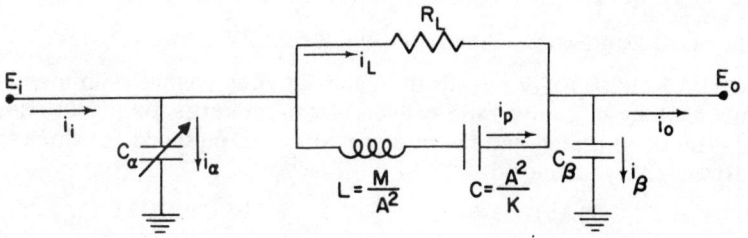

Fig. 121AB.2. Analog circuit for a hydraulic actuator.

Cuenod has applied a similar electrical analogy to the investigation of the influence of water hammer on the speed regulation of hydraulic turbines and to a study of the dynamic characteristics of pressure surges in high head and low head power plants.

See also Gruat.

121B *Gas*

121BA *General*

121BAA *Frictionless.* E. F. Murphy (B) has described an electric circuit analogy for predicting the natural frequencies of gas compressor systems and engine manifolds, resistance being neglected.

100. FLOW OF FLUIDS IN CONDUITS

Characteristic equations (for a single pipe):

FLUID ELECTRICAL

$$\frac{\rho_o L}{A}\frac{dQ}{dt} + \frac{P_o \gamma}{V} \int Q dt = 0 \quad (1) \qquad L\frac{di}{dt} + \frac{1}{C} \int i\, dt = 0 \quad (2)$$

where: ρ_o = average density
L = length of the vibrating mass of gas
A = cross-sectional area of the pipe

L = inductance

Q = discharge
t = time

i = current
t = time

P_o = average absolute pressure
γ = ratio of the specific heats and exponent for adiabatic compression
V = total volume of gas in the pipe

C = capacitance

The analog circuit for a T-shaped pipe system with one open end, one closed end, and one end containing a piston is shown in figure 121BAA.1.

See also Damewood; Sharp and Henderson; Voissel.

121BAB *With Friction.* Chilton and Handley describe an electric circuit analogy to analyze the causes of gas compressor pulsations, their effects on piping and the compressor, and methods by which the pulsations can be reduced.

FLUID ELECTRICAL

Capacitance = $\dfrac{V}{a^2}$ Capacitance = C

Inductance = $\dfrac{L}{a}$ Inductance = L

Resistance = $\dfrac{\Delta P}{Q}$ Resistance = R

in which:
V = tank volume
a = speed of sound
L = length of pipe
ΔP = change in pressure across the valve
Q = discharge

100. FLOW OF FLUIDS IN CONDUITS

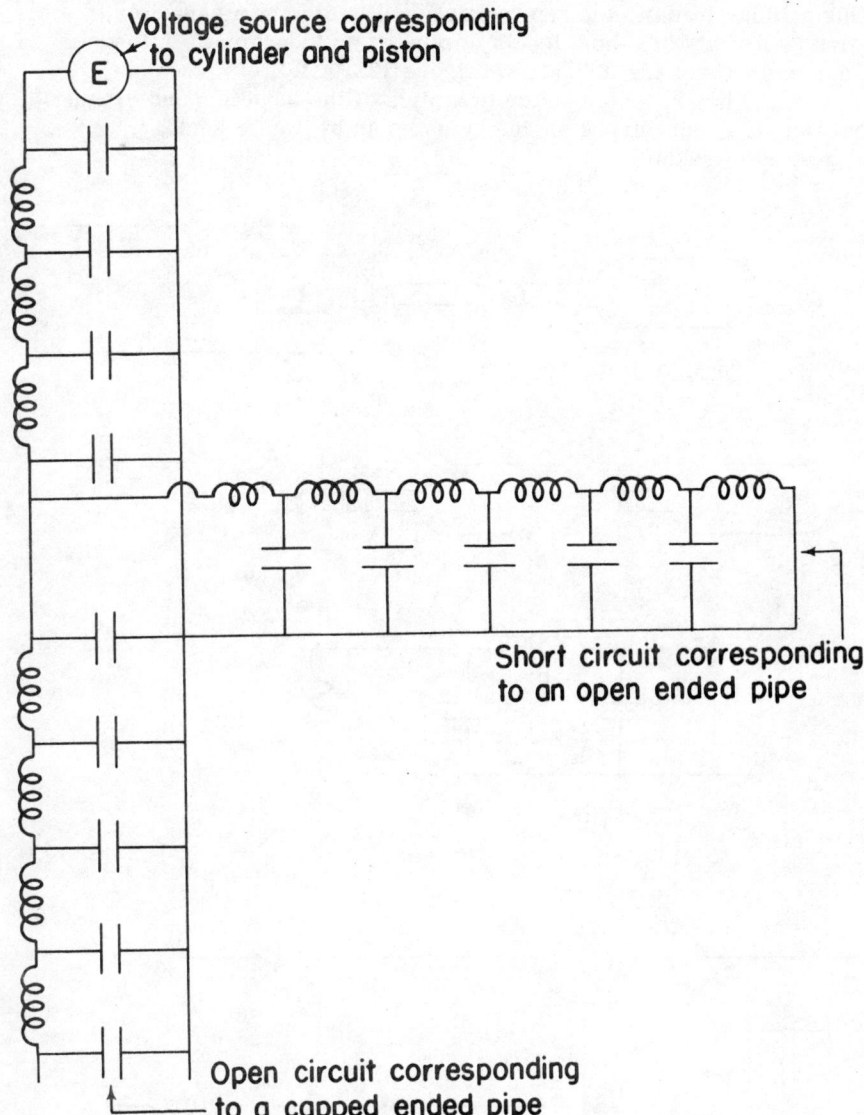

Fig. 121BAA.1. Analog circuit.

100. FLOW OF FLUIDS IN CONDUITS

This analogy leads to the representation of a tank by a capacitor, a valve by a resistor, and a length of pipe by an inductance. A few simple combinations are indicated in figure 121BAB.1.

Loh (C) has reported on the dynamics of the induction and exhaust systems of a four-stroke engine by using an hydraulic analogy.

See also Isakoff.

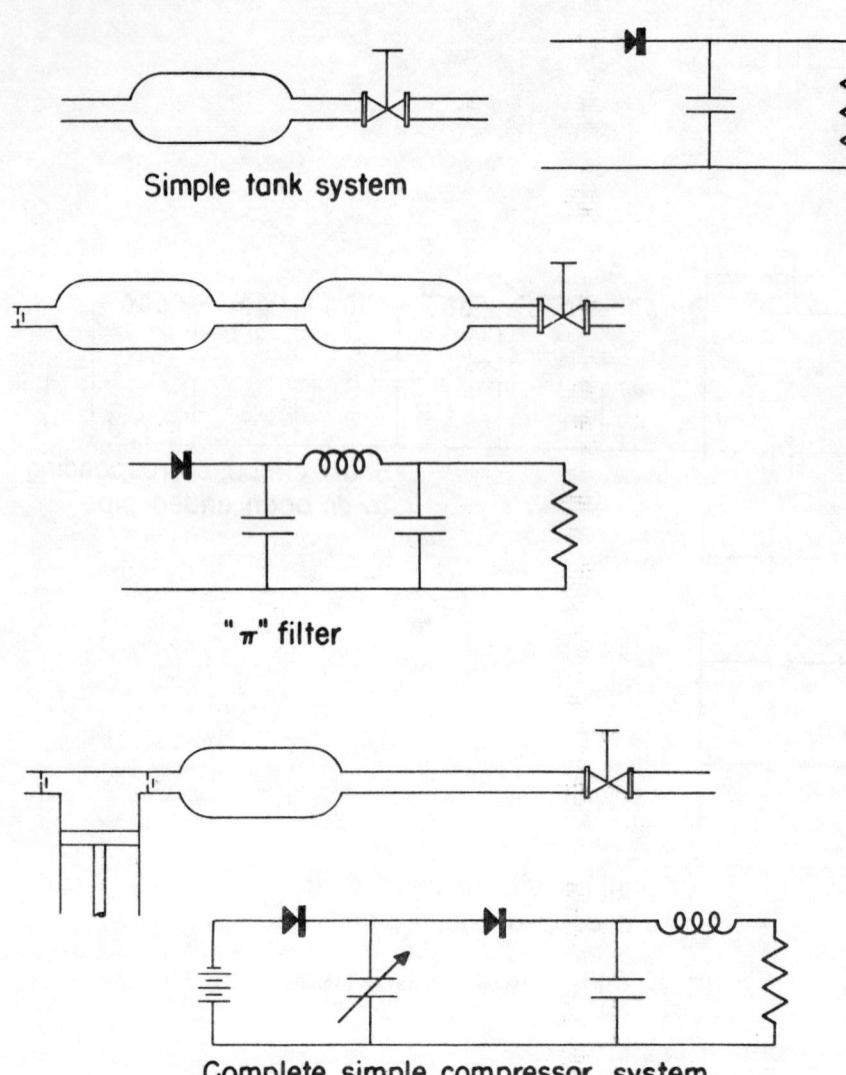

Fig. 121BAB.1. Compressor systems and their analogous circuits.

100. FLOW OF FLUIDS IN CONDUITS

121BB *Acoustical Systems*

121BBA *Electrical Analogy A.* Olson has presented the characteristic equations of an electrical analogy for an acoustical system of volume V.

Characteristic equations:

ACOUSTICAL

$$M\frac{d^2X}{dt^2} + R_A\frac{dX}{dt} + \frac{X}{C_A} = P(t)$$

where:
X = volumetric displacement of the fluid equivalent to the change in volume ΔV.

t = time

M = inertance = $\frac{m}{A^2}$

m = total mass of the fluid

A = cross-sectional area over which the driving pressure acts to drive the mass

R_A = acoustical resistance

C_A = acoustical capacitance = $\frac{V}{\rho c^2}$

ρ = density of the fluid

c = velocity

P(t) = driving pressure

ELECTRICAL

$$L\frac{d^2q}{dt^2} + R\frac{dq}{dt} + \frac{q}{C} = e(t)$$

q = charge stored in capacitor

t = time

L = inductance

R = resistance

C = capacitance

e(t) = applied electrical potential

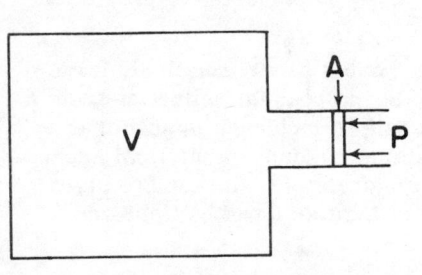

Fig. 121BBA.1. Schematic diagram of a non-flow acoustical system.

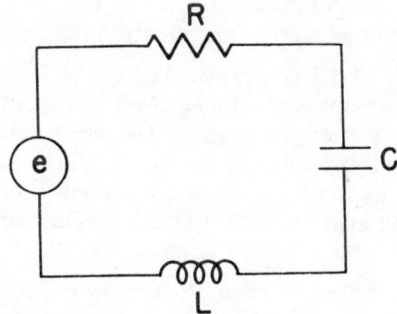

Fig. 121BBA.2. Analog circuit.

See also Bauer (A, B, C).

100. FLOW OF FLUIDS IN CONDUITS

121BBB *Electrical Analogy B.* Saxton has presented another electrical analogy, similar to the previous one, for a non-flow acoustical system.

For further explanation and application of this analogy and the previous one, see Chavasse; Gehlshøj; Sanial; Timmis; Wiggins.

Marchal has developed electric calculating tables which may be used to solve, by analogy, non-electrical problems in mechanics, acoustics and other fields.

See also Van Haver.

121BBC *Electrical Analogy C.* Steven, Kasowski and Fant have developed an electrical analogy for nonsteady flow acoustical systems, with particular application to the analogies of the vocal tract.

The vocal tract is viewed as an acoustic tube of varying cross sectional area, terminated by the vocal cords at one end and by the lip opening at the other. The analog is a lumped-constant electrical transmission line consisting of 35 pi sections. Each section represents a 1/2 - cm. length of the vocal tract and is adjustable to simulate a range of cross sectional areas from 0.17 to 17 cm^2. The electrical line can be excited by a periodic current source representing the vocal cord output or by a random voltage representing the noise from turbulent air flow at a constriction. The electrical analog can synthesize with good quality all English vowels and some consonants. The physical characteristics of the output of the analog for each sound are shown in terms of measured formant frequencies or by conventional spectrograms. For each sound, the vocal tract dimensions that the electrical network simulates are shown in graphical form. Applications of the speech synthesizer to linguistic and engineering research are discussed briefly.

The analog circuit is shown in figure 121BBC.1 and the design equations for the analog are $L = \frac{PI}{KA}$, $C = \frac{KAI}{\rho c^2}$, where ρ = air density; I = tube section length; K = constant determined by impedance level; A = cross sectional area; c = velocity of sound. In the analog circuit, G, the conductance, and R, the resistance, are adjusted to provide the required amount of dissipation.

121BBD *Optical Analogy A.* In Schoch's paper, relatively large lateral displacement of an ultrasonic sound beam at reflection from a sonic surface is indicated when the angle of incidence is such that a Rayleigh wave is excited in the reflecting medium. Analytical expressions for both the displacement and the angle of incidence are given. The analogy with a similar behavior of light as found by Goos and Haenchen is also pointed out.

121BBE *Optical Analogy B.* Kuttruff has investigated with an optical analog the dependence of the diffusivity of a stationary sound field in a room on the geometry and the reflective properties of the walls for the case of very high frequencies. Using a directional photocell, the

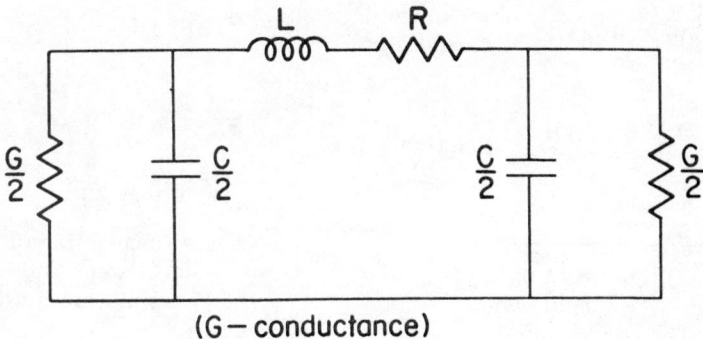

Fig. 121BBC.1. Analog circuit.

directional diffusivity was measured in dependence on the ground plan, the position of the light source and the reflective properties of the walls.

121BBF *Electromagnetic Analogy.* Kraichnan has developed an analogy "between the paths of sound rays in fluids undergoing shear flow and electron trajectories in magnetic fields. If the Mach number of the flow is small, and the characteristic eddy size large compared to the sound wavelength, the ray paths coincide with electron trajectories in a magnetic field everywhere parallel and proportional to the vorticity vector."

121BC *Other Systems*

Key and Lamb have described a non-linear resistance-capacitance circuit used in an electrical analog for simulating air pressure variations due to transient airflow in a system of interconnected chambers.

122 *Steady Flow (Flow Distribution)*

122A *Liquid*

122AA *Electrical Analogy A (linear resistors)*

Electric circuit analogies have been used extensively to predict the flow distributions in complex networks of pipes and other conduits. Possibly the first type of analogy used to analyze such systems involves ordinary adjustable linear resistors. The analogy is based on the equations for head loss, pressure balance around a loop, and continuity for the fluid and on Ohm's Law and Kirchhoff's Laws for the circuit.

Characteristic equations:

FLUID   ELECTRICAL

$\Delta h = (kQ^{n-1})(Q)$ (1)   { for any branch } $\Delta E = (R)(i)$ (2)

100. FLOW OF FLUIDS IN CONDUITS

$$\Sigma \Delta h = 0 \quad (3) \quad \left\{ \begin{array}{c} \text{around any} \\ \text{closed loop} \end{array} \right\} \quad \Sigma \Delta E = 0 \quad (4)$$

$$\Sigma Q = 0 \quad (5) \quad \left\{ \begin{array}{c} \text{at any} \\ \text{junction} \end{array} \right\} \quad \Sigma i = 0 \quad (6)$$

where: h = head

Δh = change in head between ends of a single conduit

Q = discharge

k = frictional resistance

n = empirical exponent (usually about 1.85 for liquids)

E = voltage

ΔE = change in voltage between ends of a single resistor

i = current

R = resistance

Each single branch of the liquid flow network is represented by an electrical resistor and the entire circuit is similar in configuration to the pipe system. Thus, the application of d.c. voltages proportional to the fluid head at reservoirs and/or pump stations will lead to branch currents proportional to the corresponding branch discharges, as in figures 122AA.1 and 122AA.2.

Initially, trial values of R, proportional to the corresponding values of k, are assumed. With an electrical network of resistors whose sizes have been thus calculated, values of Q for each branch are determined.

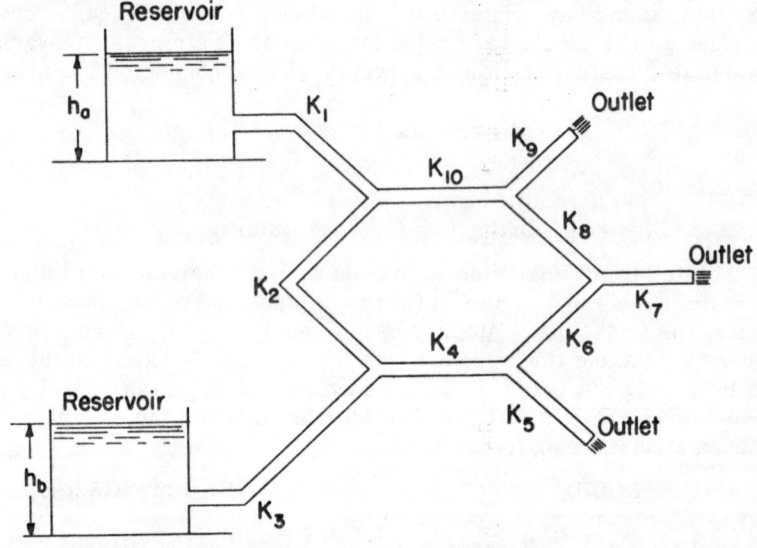

Fig. 122AA.1. Schematic diagram of a simple liquid distribution system.

100. FLOW OF FLUIDS IN CONDUITS

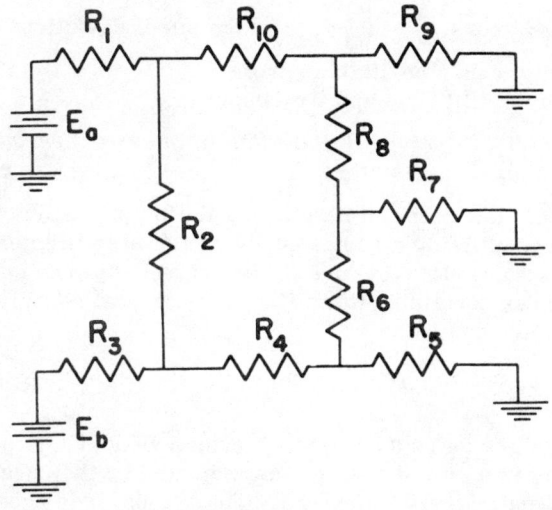

Fig. 122AA.2. Analog circuit for liquid distribution system.

From these values of Q and the corresponding values of k, new values of R, proportional to kQ^{n-1}, are calculated. With new values of resistance, the circuit is used again to determine values of Q. This process of successive approximation is repeated until the desired accuracy is attained (i.e., until the corrections for the values of R become sufficiently negligible).

For further description of the theory of this analogy, see Camp and Hazen; Haupt (A, B); J. L. Jensen; McIlroy (A, C, J); the article "Electronic Brain Reports for Work".

Among the first to apply this analogy were Collins and Jones who, according to Camp, used it to analyze the water distribution system of Warwick, Rhode Island. Camp and Hazen and also Haupt (B) have presented solutions to typical problems solved by this analogy.

Perry, Vierling, and Kohler have modified the foregoing procedure so as to eliminate the many successive approximations. From the values of Q determined from the circuit involving the initial trial values of R, new values of R, proportional to $kQ^{\frac{n-1}{n}}$, are calculated. With a circuit containing resistors whose sizes are thus determined, final values of flow rate and head (sufficiently accurate for the purposes of the inventors of this method) are determined.

Suryaprakasam, Reid, and Geyer (A, B) have indicated another modification of this analogy, and have presented results of the analyses of two complex distribution systems. In this method, the initial approximations of head and volume rate of flow are corrected by determining a correction factor for each loop, based on the ratios of

100. FLOW OF FLUIDS IN CONDUITS

pipe-resistance factors, $\dfrac{k_1}{k_2} = \left(\dfrac{Q_1}{Q_2}\right)^{1.85}$, since small deviations of these factors from unity cause negligible errors.

See also Culver (B); Reid and Wolfenson.

122AB Electrical Analogy B *(tungsten filament and vacuum tube resistors)*

Because of its low cost and its direct relation to pipeline networks, a similar analogy utilizing circuits involving tungsten filaments as non-linear resistors has found wide use in the analysis of such networks. (Commercial tubes containing these filaments are called "Fluistors".)

Characteristic equations:

$$h = kQ^n \quad (1) \qquad\qquad E = Ri^n \quad (2)$$

where quantities are defined as in the previous analogy.

Because this is an exact analogy, as indicated by the equations, successive approximations are not required. Thus, it is necessary to take voltage and current readings only once.

McIlroy (C) and J. L. Jensen have each presented experimental results for typical problems solved by this analogy. The analysis of the water distribution system of Binghamton, New York, by this analogy is also reported by McIlroy (H). Included in the article are a network diagram and results obtained from the analog circuit.

For other similar investigations in which a tungsten filament circuit was used, see Haupt (A); Laurent; McIlroy (A, E, G, J); Standrud.

Another type of circuit used in this analogy involves vacuum tubes as non-linear resistors. According to Camp, Quill succeeded in representing the skeleton water distribution system of Warwick, Rhode Island, with a vacuum tube network analyzer. For further discussion of the use of this type of circuit, see Cook and also Camp.

For other applications of this analogy, see Appleyard; Appleyard and Linaweaver; C. L. Barker; Boyer; Haupt (C); Maytin; McIlroy (B, F, G); the article, "The McIlroy Pipeline Network Analyzer"; McPherson and Radziul; R. C. Moore; Seidling; Suryaprakasam, Reid, and Geyer (A); B. L. Werner and Moore.

122AC Electrical Analogy C *(servo-controlled resistors)*

Kayan and Balmford have studied fluid distribution systems by the use of electric networks whose variable resistances were controlled by servos according to the respective currents corresponding to flow rates.

122B Gas

122BA Pipeline Networks

122BAA *Electrical Analogy A.* Stephenson, Eaton, and Duffy have applied the electric circuit analogy of section 122AA to the analysis of

100. FLOW OF FLUIDS IN CONDUITS

gas distribution systems. Their method, which speeds up the successive-approximations procedure by eliminating most of the calculations, is as follows: A non-linear current-voltage curve (to which the analogous electrical resistance of each branch can be related) is plotted on the scale of an oscilloscope in such a manner that the vertical and horizontal deflections of the beam are proportional to the current and voltage of the resistor. With a d.c. analyzer, the resistance of the element is varied until the small spot on the screen falls on the plotted curve. With an a.c. analyzer, the resistance is varied until the end of a straight line through the origin falls on the curve. Successive adjustments are then made. The solution of a typical problem is also given.

According to Clennon and Dawson, this analogy is valid only for systems operating with less than one psi pressure.

122BAB *Electrical Analogy B.* McIlroy (D) has utilized tungsten filament resistors in applying the analogy of section 122AB to the analysis of gas distribution-systems.

See also Beggs; Chapman; the article, "Columbia Gas System Installs Analyzer To Solve Distribution Problems"; Kroeger; McIlroy and Chow; Trunk.

122BAC *Electrical Analogy C.* For the flow distribution analyses of gas systems operating at pressures above one psi, Clennon and Dawson have presented the following analogy.

Characteristic equations:

FLUID ELECTRICAL

$$h_2^2 - h_1^2 = kQ^2 \quad (1) \qquad E_2 - E_1 = Ri^2 \quad (2)$$

where symbols are defined as in section 122AA. Subscripts 1 and 2 indicate opposite ends of branches.

The analog circuit used by Clennon and Dawson employed linear resistors, necessitating successive approximations to determine the flow distributions. Results are presented for solutions of typical problems by a.c. and d.c. electrical circuits.

For further description of electric circuit analogies for problems of gas distribution in pipeline networks, see the article, "Calculators Solve Flow Problems"; Abbott.

122BB *Mine Ventilation*

Scott and Hinsley used tungsten filament lamps as resistors in an electric circuit analogy designed to analyze the distribution of air flow within the ventilation networks of mines.

Characteristic equations:

FLUID ELECTRICAL

$$\Delta h = kQ^2 \quad (1) \qquad \Delta E = Ri^2 \quad (2)$$

where symbols are defined as in section 122AA.

Results by analogy were compared with analytical results and direct-experimental data.

See also Abramov and Podolsky; Bagrinovskii; Batzel and Schmidt; de Crombrugghe and Remacle; Hiramatsu (A, B); Huebner; Kingery; Maas (A, B); McElroy; Patigny; Scott (A, B, C); Scott and Hudson.

Section 200. Fluid Flow in General

210 COMPRESSIBLE

211 *Steady State*

211A *Irrotational*

211AA *General Potential Flow*

211AAA *Electrical Analogy A.* G. I. Taylor and Sharman have presented an analogy by which an electrolytic tank of variable depth may be used to determine the velocity distribution of the two-dimensional potential flow of a nonviscous compressible fluid. This analogy is based upon the equations of continuity (fluid and electrical, equations (1) and (2)) and definitions of the velocity and electric potentials (Eq. (3), (4), (5), (6)). An electrically conducting plate of variable thickness could be used in place of the tank.

The relationships mentioned above and others pertinent to this analogy are expressed in equation form below.

FLUID ELECTRICAL

$$\frac{\partial(\rho u)}{\partial x} + \frac{\partial(\rho v)}{\partial y} = 0 \quad (1) \qquad \frac{\partial f}{\partial x} + \frac{\partial g}{\partial y} = 0 \quad (2)$$

$$\frac{\partial \phi}{\partial x} = -u \quad (3) \qquad \frac{\partial V}{\partial x} = -E_x \quad (4)$$

$$\frac{\partial \phi}{\partial y} = -v \quad (5) \qquad \frac{\partial V}{\partial y} = -E_y \quad (6)$$

$$\frac{\partial \psi}{\partial x} = \rho v \quad (7) \qquad \frac{\partial W}{\partial x} = g = \left(\frac{t}{t_o \sigma}\right) E_y \quad (8)$$

200. FLUID FLOW IN GENERAL

$$\frac{\partial \psi}{\partial y} = -\rho u \quad (9) \qquad \frac{\partial W}{\partial y} = -f = -\left(\frac{t}{t_o \sigma}\right) E_x \quad (10)$$

where:
- ϕ = velocity potential
- ψ = stream function
- x, y = coordinates
- u, v = x, y components of fluid velocity
- ρ = variable fluid density

- V = electric potential
- W = electric current function
- x, y = coordinates
- E_x, E_y = x, y components of electric field intensity (potential gradient)
- σ = constant resistivity of the electrolyte
- t = variable depth of the electrolyte
- t_o = reference electrolyte depth corresponding to the fluid density at infinity
- f, g = x, y components of electric current density

Equations (1), (3), (5) may then be combined and equations (2), (4), (6), (8), (10) combined.

Characteristic equations:

$$\frac{\partial}{\partial x}\left(\rho \frac{\partial \phi}{\partial x}\right) + \frac{\partial}{\partial y}\left(\rho \frac{\partial \phi}{\partial y}\right) = 0 \quad (11) \qquad \frac{\partial}{\partial x}\left(t \frac{\partial V}{\partial x}\right) + \frac{\partial}{\partial y}\left(t \frac{\partial V}{\partial y}\right) = 0 \quad (12)$$

Thus if the electrolyte depth is made proportional to the fluid density at corresponding points, electric potential will be proportional to fluid velocity potential, electric current density will be proportional to the product of fluid density and velocity, and the electric current function corresponds to the fluid stream function.

Boundary conditions:

1. Along the boundaries of the flow section, lines of constant ϕ (perpendicular to streamlines) are represented by lines of constant V, and lines of constant ψ (streamlines) are represented by lines of constant W.

2. The tank must have the same shape as the flow section. In other words, the end boundaries of the flow section are assumed to be lines of constant ϕ (as at infinity) and are represented by electrodes in the tank. Since there can be no flow through the side boundaries, these boundaries are necessarily lines of constant ψ and are insulated in the tank. To represent flow past a

200. FLUID FLOW IN GENERAL

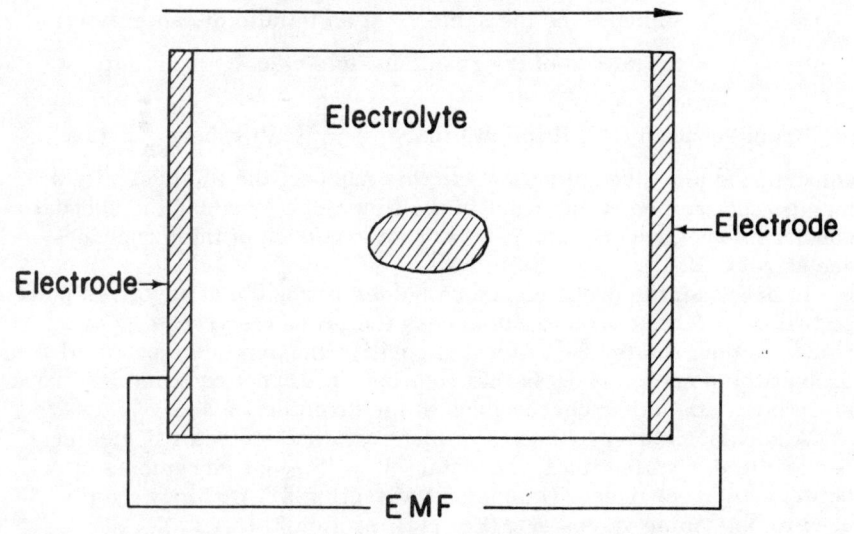

Fig. 211AAA.1. Electrolytic tank analog of a fluid flow section.

solid cylindrical body within the section, an insulated cylinder of the same shape and relative size is inserted in the tank as indicated (Fig. 211AAA.1).

If equation (3) is combined with equation (5) and equation (4) with equation (6), the following relationships result:

$$\frac{\partial \phi}{\partial s} = -q \quad (13) \qquad \frac{\partial V}{\partial s} = -E \quad (14)$$

where: s = displacement in the direction of fluid flow

q = magnitude of the resultant fluid velocity

s = displacement in the direction of electric current flow

E = magnitude of the resultant field intensity

In the analog, a galvanometer may be used to measure the electric potential drop, ΔV, between two probe electrodes, a distance Δs apart. From equations (13) and (14), ΔV is proportional to E, which in turn is proportional to q. From this, it is seen that

$$\frac{\Delta V}{\Delta P} = \frac{q}{U} \quad (15)$$

where: ΔP = electric potential drop between the probe electrodes at the ends of the section, where parallel flow exists (assumed to be the same as at an infinite distance away)

U = magnitude of the resultant fluid velocity at infinity

From equation (15), it follows that $q = \frac{U}{\Delta P}(\Delta V)$, where $\frac{U}{\Delta P}$ is a constant for any given situation. In this manner, the fluid velocity q may be determined at any point in the flow field, provided the boundary conditions are properly applied. For an extension of this procedure, see Miroux (B, C).

In practice, the probe electrode holder is rotated at any given point so that no potential drop exists across the probe electrodes. The holder is then rotated $90°$, in which position the maximum potential drop at the point is read. In this position, the direction from one probe electrode to the other corresponds to the direction of s.

A galvanometer-probe arrangement or a null-detector device may be used to determine lines of constant V, with equal increments of ΔV between adjacent lines. The method of section 211AAC may then be used to determine values of q (Eq. (11), section 211AA).

If equipotential lines, as described above, are plotted with streamlines (determined as in section 211AAB), a flow net is obtained.

The distribution of ρ must be known in order to determine the corresponding distribution of t. This distribution of ρ is represented by the following equation, derived by combining the perfect gas equation with Bernoulli's equation:

$$\frac{t}{t_o} = \frac{\rho}{\rho_o} = \left[1 - \frac{(\gamma - 1)}{2a^2}(q^2)\right]\left[\frac{1}{\gamma - 1}\right] \qquad (16)$$

where: ρ_o = fluid density at infinity
γ = ratio of the specific heats of the fluid
a = velocity of sound in the fluid at infinity

Since q is unknown to begin with, the determination of both q and ρ involves a process of successive approximations. By the use of a tank of constant depth, values of q are determined experimentally. From these values, corresponding values of ρ, and thus t, are computed. A new tank bottom is then constructed to correspond to the new distribution of t. This process is then repeated over and over for closer approximations to the actual situation.

The results of the experiments performed by Taylor and Sharman indicated that for situations in which the maximum fluid velocity is less than the velocity of sound in the fluid, the successive values of ρ form a convergent sequence, and that when the maximum fluid velocity exceeds the velocity of sound (which, in the case of a circular cylinder, occurs when U lies between 0.4 and 0.5 the velocity of sound), the successive values of ρ form a divergent sequence and no solution can be

200. FLUID FLOW IN GENERAL

obtained. Their report includes plots of successive approximations of $\frac{\rho}{\rho_o}$ vs. distance along the center line perpendicular to the direction of flow for the case of flow past a circular cylinder in a large flow section with parallel sides.

For further applications of this analogy, see Germain (C); Malavard (Q); Pocock; Taylor (B).

See also sections 211C, 221AAA, 221D.

211AAB *Electrical Analogy B.* Taylor and Sharman have suggested another analogy similar to the previous one. This analogy is based upon the condition of irrotationality of fluid flow (Eq. (1)), the equation of continuity of electric current (Eq. (2)), and definitions of the fluid stream function and electric potential (Eq. (3), (4), (5), (6)). The relationships mentioned above and others pertinent to this analogy are expressed below in equation form.

FLUID ELECTRICAL

$$\frac{\partial v}{\partial x} - \frac{\partial u}{\partial y} = 0 \quad (1) \qquad \frac{\partial f}{\partial x} + \frac{\partial g}{\partial y} = 0 \quad (2)$$

$$\frac{\partial \psi}{\partial x} = \rho v \quad (3) \qquad \frac{\partial V}{\partial x} = -E_x = -\left(\frac{t_o \sigma}{t}\right) f \quad (4)$$

$$\frac{\partial \psi}{\partial y} = -\rho u \quad (5) \qquad \frac{\partial V}{\partial y} = -E_y = -\left(\frac{t_o \sigma}{t}\right) g \quad (6)$$

$$\frac{\partial \phi}{\partial x} = -u \quad (7) \qquad \frac{\partial W}{\partial x} = g \quad (8)$$

$$\frac{\partial \phi}{\partial y} = -v \quad (9) \qquad \frac{\partial W}{\partial y} = -f \quad (10)$$

where symbols are defined as in the previous analogy.

Equations (1), (3), (5) may then be combined and equations (2), (4), (6) combined.

Characteristic equations:

$$\frac{\partial}{\partial x}\left(\frac{1}{\rho}\frac{\partial \psi}{\partial x}\right) + \frac{\partial}{\partial y}\left(\frac{1}{\rho}\frac{\partial \psi}{\partial y}\right) = 0 \quad (11) \qquad \frac{\partial}{\partial x}\left(t \frac{\partial V}{\partial x}\right) + \frac{\partial}{\partial y}\left(t \frac{\partial V}{\partial y}\right) = 0 \quad (12)$$

Thus if the electrolyte depth is made inversely proportional to the fluid density at corresponding points, electric potential will be proportional to the stream function, electric current density will be proportional to (but perpendicular to) fluid velocity, and the electric current function corresponds to fluid velocity potential.

200. FLUID FLOW IN GENERAL

Boundary conditions:

1. Along the boundaries of the flow section, lines of constant ϕ are represented by lines of constant W and lines of constant ψ are represented by lines of constant V.

2. The tank must have the same shape as the flow section. In other words, the end boundaries (assumed to be lines of constant ϕ) are insulated in the tank, and the side boundaries (lines of constant ψ) are represented by electrodes. A solid cylindrical body within the flow section is represented in the tank by a perfectly conducting cylinder. An interesting feature of this analogy is that irrotational flow with circulation may be represented by sending current into the perfectly conducting cylinder.

Since, in this analogy, electric potential V is analogous to the stream function ψ, lines of constant V, as determined by a galvanometer-probe arrangement, will be streamlines of fluid flow.

The variation in density as determined by the successive approximation procedure applied to the previous analogy can thus be used to determine the variation of electrolyte depth for the analogy under discussion.

See also Taylor (B).
See also sections 211C, 221AAA, 221D.

211AAC *Electrical Analogy C.* A modification of the analogy of section 211AAA has been suggested by Poritsky, Sells, and Danforth. This analogy utilizes electrical networks rather than continuous electrically conducting sheets and is based on the finite-difference equations expressing the fluid and electrical flows.

Characteristic equations:

FLUID

$$\frac{\Delta \left(\rho \frac{\Delta \phi}{\Delta x}\right)}{\Delta x} + \frac{\Delta \left(\rho \frac{\Delta \phi}{\Delta y}\right)}{\Delta y} = 0 \quad (1)$$

ELECTRICAL

$$\frac{\Delta \left(\frac{1}{R} \frac{\Delta V}{\Delta x}\right)}{\Delta x} + \frac{\Delta \left(\frac{1}{R} \frac{\Delta V}{\Delta y}\right)}{\Delta y} = 0 \quad (2)$$

where: R = resistance of a network element; Δx, Δy = equally spaced finite-difference intervals; and other symbols are defined as in section 211AAA.

Boundary conditions:

1. Same as condition 1 in section 211AAA.

2. The network (in schematic diagram) must have the same shape as the flow section (Figs. 211AAC.1 and 211AAC.2).

Corresponding to equations (11) and (12) of section 211AAA are the following relations:

200. FLUID FLOW IN GENERAL

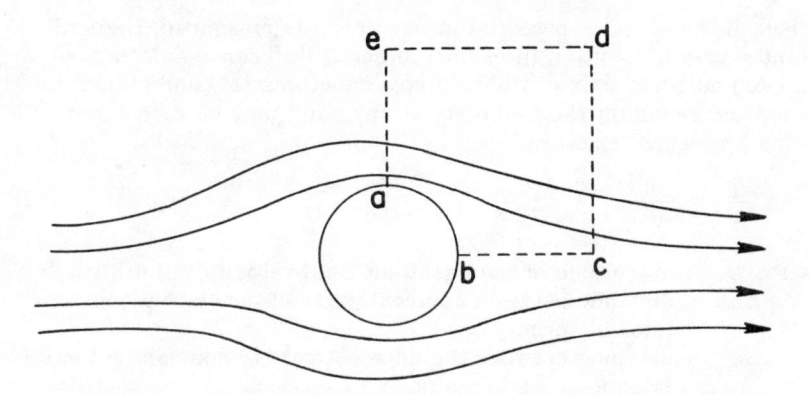

Fig. 211AAC.1. Typical problem: Flow of a compressible fluid past a circular cylinder normal to the direction of flow.

$$\frac{\Delta \phi}{\Delta s} = q \qquad (9) \qquad \frac{\Delta V}{\Delta s} = E \qquad (10)$$

where: q = magnitude of the resultant fluid velocity

E = magnitude of the resultant electric field intensity.

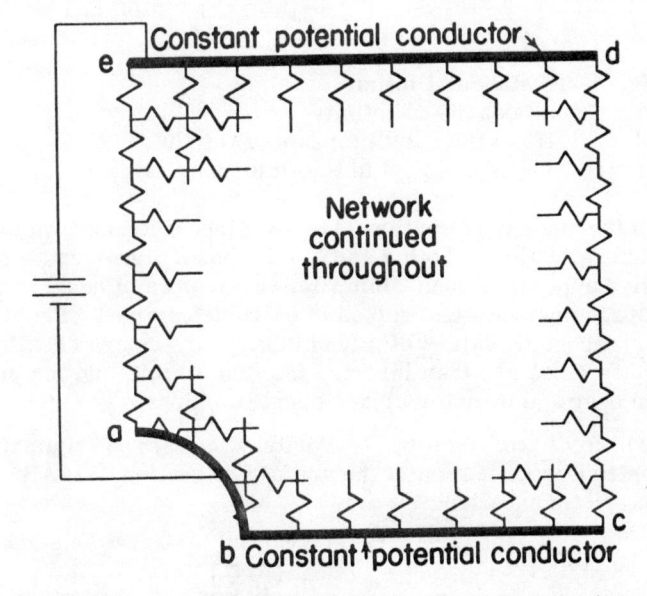

Fig. 211AAC.2. Analog circuit for determining streamlines of one quadrant of the flow region of figure 211AAC.1 (assuming symmetry).

200. FLUID FLOW IN GENERAL

Thus, if the electric potential at each node is measured, lines of constant V may be plotted, the same potential difference existing between each adjacent pair of lines. From equations (11) and (12), it is seen that the resultant fluid velocity at any point may be calculated from the following relation:

$$\frac{q}{U} = \frac{\Delta b}{\Delta s} \qquad (11)$$

where: U = magnitude of the resultant fluid velocity at infinity
Δb = distance between adjacent pairs of constant potential lines at infinity
Δs = distance between the adjacent pair of constant potential lines nearest to the point in question

Thus $q = \frac{U \Delta b}{\Delta s}$, where $U \Delta b$ is a constant for any given situation.

Since R is proportional to $\frac{1}{\rho}$, values of R may be determined from the following relation for ρ:

$$\frac{R}{R_o} = \frac{\rho_o}{\rho} = \frac{1}{\left[1 - \frac{(\gamma-1)}{2a^2} q^2\right]^{\frac{1}{\gamma-1}}}$$

where: R_o = resistance at infinity
ρ_o = fluid density at infinity
γ = ratio of the specific heats of the fluid
a = velocity of sound in the fluid at infinity

As with the analogy of section 211AAA, since q is unknown to begin with, the determination of both q and ρ (and thus R) involves a process of successive approximations. Initially, resistances of equal size are inserted into the network and values of q are determined. From these values, corresponding values of ρ (and thus R) are computed. Resistances of these sizes are then inserted into the circuit, and the process is repeated over and over for closer approximations.

211AAD *Electrical Analogy D.* Poritsky, Sells, and Danforth have also suggested a modification of the analogy of section 211AAB.
Characteristic equations:

$$\frac{\Delta \left(\frac{1}{\rho} \frac{\Delta \psi}{\Delta x}\right)}{\Delta x} + \frac{\Delta \left(\frac{1}{\rho} \frac{\Delta \psi}{\Delta y}\right)}{\Delta y} = 0 \quad (1) \qquad \frac{\Delta \left(\frac{1}{R} \frac{\Delta V}{\Delta x}\right)}{\Delta x} + \frac{\Delta \left(\frac{1}{R} \frac{\Delta V}{\Delta y}\right)}{\Delta y} = 0 \quad (2)$$

where symbols are defined as in sections 211AAA and 211AAC.

200. FLUID FLOW IN GENERAL

Boundary conditions:

1. Same as condition 1 in section 211AAB.

2. Same as condition 2 in section 211AAC (Figs. 211AAC.1 and 211AAC.2).

Using this analogy, Poritsky, Sells, and Danforth analyzed the flow of a compressible fluid, at a Mach number of 0.5, around a circular cylinder placed normal to the stream. Their report includes a plot of streamlines determined by analogy.

Lines of constant electric potential, as determined by measuring the potential at each node, are streamlines of fluid flow.

In this analogy, fluid density is inversely proportional to electrical resistance. Values of density for individual intervals of the flow region may be determined by means of successive approximation. From these, the resistances of the corresponding network elements may be calculated.

See also McCann and Wilts.

211AAE *Mechanical Analogy.* Inoue (C) developed an analogy between a statically determinate, two-dimensional state of stress in a perfectly plastic material and the two-dimensional steady irrotational flow of a compressible fluid.

Stream lines and equipotential lines of the fluid correspond to isostatics of the plastic material. When the flow is supersonic, Mach lines correspond to slip lines. The shock front in supersonic flow of the fluid becomes equivalent to the stress discontinuity surface, and the jump conditions at the shock front yield the continuity of the normal and shearing stresses acting in the stress discontinuity surface when translated into the terminology of plasticity.

See section 454 for more references.

211AB *Modified General Potential Flow (The "Hydraulic" Analogy)*

According to Black and Mediratta, the resemblance between the waves formed by a body moving at a supersonic speed through air, and the surface waves produced by a moving ship, or in water flowing past bridge piers, was first commented on by Mach (B). It was not until 1920 that the analogy between two-dimensional gas flow and the flow of water with a free surface was first developed by Jouquet, and later by Riabouchinsky (B, E). They showed that the mathematical equations governing the two flows were of the same nature if the water were replaced by a fictitious "hydraulic gas" with a ratio of specific heats of 2.0. In 1937, Binnie and Hooker investigated the flow in a channel with a constriction. Preiswerk (A, B), in his application of the methods of gas dynamics to the accurate determination of the water flow in a converging-diverging channel, provided a very complete account of the analogy:

Characteristic equations:

200. FLUID FLOW IN GENERAL

GAS	LIQUID

$$\left(\frac{q}{q_{max}}\right)^2 = \frac{T_o - T}{T_o} \quad (1) \qquad \left(\frac{q}{q_{max}}\right)^2 = \frac{h_o - h}{h_o} \quad (2)$$

$$\frac{\partial(\rho u)}{\partial x} + \frac{\partial(\rho v)}{\partial y} = 0 \quad (3) \qquad \frac{\partial(hu)}{\partial x} + \frac{\partial(hv)}{\partial y} = 0 \quad (4)$$

$$\frac{\partial^2 \phi}{\partial x^2}\left[1 - \frac{\left(\frac{\partial \phi}{\partial x}\right)^2}{a^2}\right] + \frac{\partial^2 \phi}{\partial y^2}\left[1 - \frac{\left(\frac{\partial \phi}{\partial y}\right)^2}{a^2}\right] - 2\frac{\partial^2 \phi}{\partial x \partial y}\frac{\left(\frac{\partial \phi}{\partial x}\right)\left(\frac{\partial \phi}{\partial y}\right)}{a^2} = 0 \quad (5)$$

$$\frac{\partial^2 \phi}{\partial x^2}\left[1 - \frac{\left(\frac{\partial \phi}{\partial x}\right)^2}{gh}\right] + \frac{\partial^2 \phi}{\partial y^2}\left[1 - \frac{\left(\frac{\partial \phi}{\partial y}\right)^2}{gh}\right] - 2\frac{\partial^2 \phi}{\partial x \partial y}\frac{\left(\frac{\partial \phi}{\partial x}\right)\left(\frac{\partial \phi}{\partial y}\right)}{gh} = 0 \quad (6)$$

where:

q	= velocity	q	= velocity
q_{max}	= maximum velocity	q_{max}	= maximum velocity
g	= acceleration of gravity	g	= acceleration of gravity
T	= absolute gas temperature	h	= water depth
ρ	= gas density		
u, v	= x, y components of velocity	u, v	= x, y components of velocity
x	= axial coordinate	x	= axial coordinate
y	= transverse coordinate	y	= horizontal transverse coordinate
a	= velocity of sound	gh	= propagation wave velocity
ϕ	= velocity potential defined as follows:	ϕ	= velocity potential defined as follows:
$u = \frac{\partial \phi}{\partial x}$	$v = \frac{\partial \phi}{\partial y}$	$u = \frac{\partial \phi}{\partial x}$	$v = \frac{\partial \phi}{\partial y}$

The subscript o represents the stagnation state for both systems.

From equations (1), (2), (3), (4), the following relationships can be seen:

$$\frac{T}{T_o} = \frac{\rho}{\rho_o} = \frac{h}{h_o} \quad (7)$$

200. FLUID FLOW IN GENERAL

But since pressure, p equals ρRT,

$$\frac{p}{p_o} = \frac{\rho}{\rho_o}\left(\frac{T}{T_o}\right) = \left(\frac{\rho}{\rho_o}\right)^2 \tag{8}$$

For an isentropic process,

$$\frac{p}{p_o} = \left(\frac{\rho}{\rho_o}\right)^k \tag{9}$$

Thus it is evident from equations (8) and (9) that the ratio of specific heats k of the gas to be investigated must have a value of 2 for the analogy to hold quantitatively.

The analogy consists of a water channel fed by a reservoir of sufficient capacity to provide a stagnation state. The channel has a horizontal bottom with vertical sides the same shape as the boundaries of the two-dimensional gas-flow region.

As seen by equation (9), the gas pressure at any point in the flow region may be determined from a knowledge of the stagnation pressure and from measurements of corresponding depths of water.

The liquid phenomenon analogous to the shock wave is the hydraulic jump. The attempt to thus extend the "Hydraulic" analogy to the analysis of shock waves by means of a hydraulic jump has been described by Black and Mediratta.

Characteristic equations:

GAS   LIQUID

$$\rho_1 V_1 = \rho_2 V_2 \tag{10}$$
$$h_1 V_1 = h_2 V_2 \tag{11}$$

$$p_1 + \rho_1 V_1^2 = p_2 + \rho_2 V_2^2 \tag{12}$$
$$\frac{1}{2}gh^2 + h_1 V_1^2 = \frac{1}{2}gh_2^2 + h_2 V_2^2 \tag{13}$$

$$\frac{1}{2}V_1^2 + \frac{k}{k-1}\frac{p_1}{\rho_1} = \frac{1}{2}V_2^2 + \frac{k}{k-1}\frac{p_2}{\rho_2}$$
$$h_1 + \frac{V_1^2}{2g} = h_2 + \frac{V_2^2}{2g} + e = h_o \tag{15}$$

$$= \frac{k}{k-1}\frac{p_o}{\rho_o} = C_p T_o \tag{14}$$

where: $\dfrac{\rho_2}{\rho_1} = \dfrac{\frac{1}{2}(k+1)M_2^2}{1 + \frac{1}{2}(k+1)M_1^2}$

M = Mach number
V = velocity   V = velocity
C_p = specific heat at constant pressure

200. FLUID FLOW IN GENERAL

The subscripts *1* and *2* represent states just preceding and just following the shock or jump, respectively, and other symbols are as defined for equations (1) to (6).

It is seen that in flow through a shock wave, there are no losses of like nature to the eddying loss involved in the jump. Thus, according to Black and Mediratta, "Quantitative agreement between hydraulic jumps and shock waves was not obtained.... It was found possible, however, to use hydraulic jumps to simulate qualitatively most of the features of gas flow with shock waves, such as flow in a Laval nozzle, or past a supersonic aerofoil; photographs of these jumps are included in the paper. The most valuable aspect of the water channel undoubtedly lies in the ease with which these phenomena may be demonstrated visually, and its construction for this purpose can be recommended."
Others who have used the hydraulic analogy to analyze shock waves are H. A. Einstein and Baird (A); Laitone (C, D); Gilmore, Plesset, and Crossley; Ippen.

The corresponding quantities and conditions of this analogy may be summarized.

TWO-DIMENSIONAL GAS FLOW	LIQUID FLOW WITH FREE SURFACE IN GRAVITATIONAL FIELD
Hypothetical gas with k = 2	Incompressible fluid (water)
Side boundaries geometrically similar	Side boundaries vertical, bottom horizontal
Velocity = $\dfrac{C}{C_{max}}$	Velocity = $\dfrac{C}{C_{max}}$
Temperature ratio = $\dfrac{T}{T_o}$	Water depth ratio = $\dfrac{h}{h_o}$
Density ratio = $\dfrac{\rho}{\rho_o}$	Water depth ratio = $\dfrac{h}{h_o}$
Pressure ratio = $\dfrac{p}{p_o}$	Square of water depth ratio = $\left(\dfrac{h}{h_o}\right)^2$
Pressure on the side boundary walls = $\dfrac{P}{P_o}$	Force on the vertical walls $\dfrac{P}{P_o} = \left(\dfrac{h}{h_o}\right)^2$
Sound velocity = a	Wave velocity = $\sqrt{gh}$
Mach number = $\dfrac{C}{a}$	Mach number = $\dfrac{C}{\sqrt{gh}}$
Subsonic flow	Streaming water
Supersonic flow	Shooting water
Compressive shock	Hydraulic jump

200. FLUID FLOW IN GENERAL

For further discussion of the theory of this analogy, see Laitone (A); Matthews; Orlin, Lindner, and Bitterly; Riabouchinsky (A, C, D).

Black and Mediratta state that "Some attempts to confirm Riabouchinsky's work by quantitative measurements of lift and drag in streaming (subsonic) flow were made in 1939 ['*Some Applications of the Method of Hydraulic Analogy*'] and investigation of the flow through a cascade of turbine blades by this method was made in 1942 ['*Progress in Brown Boveri during 1941*']."

Orlin, Lindner, and Bitterly describe one of the first investigations made with the use of the water channel of the National Advisory Committee for Aeronautics. Their tests were run to analyze the flow about circular cylinders of various diameters at subsonic velocities extending into the supercritical range. Included in their report are data and photographs which indicate reasonably satisfactory agreement of pressure distributions and flow fields between water and air flow about corresponding bodies.

In its usual form, the analogy can be applied strictly only to an imaginary gas whose ratio of specific heats is 2. However, P. Teofilato has developed and established "an analytical transformation which gives the relation between the hodograph of the flow of a real gas and that of the flow of the imaginary gas. This transformation...leads, in the general case, to rather involved expressions relating the velocity, pressure, and density of the flow with the corresponding quantities of the water flow. Two important cases are discussed, however, in which simple results are obtained. In the first of these the transformation is given only for points along a solid boundary rather than for the whole field of flow. In the second, the results are given for the complete field but are based on the assumption that the perturbations from an average velocity are small at all points. The results are applicable to subsonic as well as supersonic regimes." [From APPLIED MECHANICS REVIEW.]

Szebehely and Whicker (A) present a generalization of the analogy to include channels of non-rectangular cross section. In general, most of the quantities involved in the flow within a channel of general shape are found to be functions of a shape factor, n, related to the ratio of specific heats, k, by the following equation: $k = \dfrac{n+2}{n+1}$. A channel of general shape requires a specially shaped model, which must satisfy a constant obstruction ratio of the model area (projected normal to the flow direction) to the channel cross-sectional area. Tests were conducted in a triangular, return-flow water channel. Included in the report are photographs of the equipment, flow patterns, and curves showing the relation between base pressure and Mach number. Szebehely and Whicker (B) have shown that $k = 1.5$ can be obtained with a triangular section and $k = 1.4$ with a parabolic section. For further discussion of such modifications, see Bryant (C).

Riabouchinsky (A) has investigated the effect of resistance upon this analogy.

200. FLUID FLOW IN GENERAL

Oppenheim (B) has simulated a combustion-front discontinuity by an unidimensional source formed by admitting water from the bottom of the channel.

Loh (A) has suggested a modification of this analogy for one-dimensional flow. The modified analogy may be used to simulate flows of gases of any ratio of specific heats.

For reports of other investigations involving this analogy, see Becker; Bömelburg (A, B); Broude; Bruman (A, B); Bryant (A, B, D, E, F, G); Bryant and Grant; Flowers; Harleman and Ippen; Hedgecock; Ippen and Harleman; Isaachsen; Itaya and Tomita; Joukovsky; Laitone (A, B); Laitone and Nielson; Laitone and Stout; Lamb; Legendre; Levin; Loh (B, C); Mackie; Matsunaga (A, D); Matthews; Morris, R. E. and Haythornthwaite; Okamoto; Prandtl (A); Probstein and Hudson; Randels; Shapiro (A, B); Stodola; Stoker; Syogo; Szebehely and Whicker (C); Tearnen; Teofilato, S.; Tomita (A, B); von Karman (B); Whicker and Szebehely; Zhoukovsky.

211AC "Under Transonic" Flow

According to Truesdell, the analogy between a pressureless membrane and the two-dimensional potential flow of a nonviscous compressible fluid was co-discovered by Bateman and Chapylgin.

Szebehely has presented the characteristic equations of this analogy. The fluid flow equation is developed by combining the Euler equation with the continuity equation.

Characteristic equations:

FLUID

$$\frac{\partial^2 \phi}{\partial x^2}\left[a^2 + \left(\frac{\partial \phi}{\partial y}\right)^2\right] + \frac{\partial^2 \phi}{\partial y^2}\left[a^2 + \left(\frac{\partial \phi}{\partial x}\right)^2\right] - 2\frac{\partial^2 \phi}{\partial x \partial y}\frac{\partial \phi}{\partial x}\frac{\partial \phi}{\partial y} = 0 \qquad (1)$$

MEMBRANE

$$\frac{\partial^2 z}{\partial x^2}\left[1 + \left(\frac{\partial z}{\partial y}\right)^2\right] + \frac{\partial^2 z}{\partial y^2}\left[1 + \left(\frac{\partial z}{\partial x}\right)^2\right] - 2\frac{\partial^2 z}{\partial x \partial y}\frac{\partial z}{\partial x}\frac{\partial z}{\partial y} = 0 \qquad (2)$$

where: ϕ = velocity potential defined by the following equations:

$u = \dfrac{\partial \phi}{\partial x}$ (3)

$v = \dfrac{\partial \phi}{\partial y}$ (4)

x, y = coordinates

$a = \sqrt{c^2 - v^2}$

c = velocity of sound

v = fluid velocity

z = transverse deflection of the membrane

x, y = coordinates

200. FLUID FLOW IN GENERAL

These equations show complete analogy for $a^2 = 1$, "which is the case of the 'under transonic' flow, since the Mach number in this case is $M = \sqrt{1-(1/c^2)}$."

See also Behrbohm and Pinl; Sauer.

211AD Flow With Steady Decline of Density

Miles and Stephenson have given a membrane analogy for determining the instantaneous density distribution within a compressible fluid flowing out of a petroleum reservoir through wells with a steady decline of density.

Characteristic equations:

$$\frac{\partial^2 \gamma}{\partial x^2} + \frac{\partial^2 \gamma}{\partial y^2} = \frac{\beta \mu f}{k}\left(\frac{\partial \gamma}{\partial t}\right) \quad (1) \qquad \frac{\partial^2 z}{\partial x^2} + \frac{\partial^2 z}{\partial y^2} = -\frac{p}{2T} \quad (2)$$

where: γ = density $\qquad\qquad$ z = membrane elevation
$\qquad\;\;$ x, y = coordinates $\qquad$ x, y = coordinates

Also, for isothermal flow *(only)*, $\gamma = \gamma_o P$

$$\therefore \frac{\partial^2 P}{\partial x^2} + \frac{\partial^2 P}{\partial y^2} = \frac{\beta \mu f}{k}\left(\frac{\partial P}{\partial t}\right)$$

P = fluid pressure
β = compressibility of the fluid
μ = viscosity of the fluid
f = porosity of the medium
k = permeability of the medium
$\frac{\partial \gamma}{\partial t}$ = constant rate of decline of fluid density
t = time

p = differential pressure across membrane surfaces
T = surface tension in the membrane

Application of the analogy involves establishing a membrane over an opening the shape of which corresponds to the shape of the boundary of the reservoir and increasing the pressure on one side of the membrane by an amount proportional to the term on the right hand side of equation (1). Since the elevation of each point on the membrane boundary is proportional to the fluid density at the corresponding point on the reservoir boundary, the elevation of the membrane at points corresponding to wells must be proportional to bottom hole flowing pressures.

200. FLUID FLOW IN GENERAL

Miles and Stephenson give a comparison between an analytical solution and a solution obtained using the analogy.

211B *Rotational*

Goldsworthy has presented an analogy between two-dimensional rotational flow at high Mach number past thin airfoils and unsteady one-dimensional flow produced by the motion of a piston.

For a uniform stream flowing at zero incidence past an airfoil whose profile is given by $y = y_w(x)$ situated such that the x axis is along the chord of the profile:

$$U \frac{\partial \rho}{\partial x} + v \frac{\partial \rho}{\partial y} + \rho \frac{\partial v}{\partial y} = 0 \tag{1}$$

$$U \frac{\partial v}{\partial x} + v \frac{\partial v}{\partial y} + \frac{1}{\rho} \frac{\partial p}{\partial y} = 0 \tag{2}$$

$$U \frac{\partial S}{\partial x} + v \frac{\partial S}{\partial y} = 0 \tag{3}$$

where: U = uniform velocity of the stream
   $U + u, v$ = x, y components of velocity in the region of disturbed flow
   ρ = density
   p = pressure
   S = entropy

This problem is analytically identical with the unsteady one-dimensional flow problem in y and t, where $t = x/U$, resulting from the motion of a piston with velocity $U \frac{dy_w}{dx}$ at time t.

By making the substitution, $x = Ut$, equations (1)-(3) become

$$\frac{\partial \rho}{\partial t} + v \frac{\partial \rho}{\partial y} + \rho \frac{\partial v}{\partial y} = 0 \tag{4}$$

$$\frac{\partial v}{\partial t} + v \frac{\partial v}{\partial y} + \frac{1}{\rho} \frac{\partial p}{\partial y} = 0 \tag{5}$$

$$\frac{\partial S}{\partial t} + v \frac{\partial S}{\partial y} = 0 \tag{6}$$

Boundary conditions:

1. At the shock, $y = y_s(x)$:

200. FLUID FLOW IN GENERAL

$$\frac{v}{U} = \frac{2}{\gamma+1}\left[\left(\frac{dy_s}{dx}\right)^2 - \frac{1}{M_o^2}\right] \bigg/ \frac{dy_s}{dx}$$

$$\frac{p}{p_o} = \frac{2\gamma}{\gamma+1} M_o^2\left(\frac{dy_s}{dx}\right)^2 - \frac{\gamma-1}{\gamma+1}$$

2. At the upper surface of the airfoil:

$$U\frac{dy_w}{dx} - v = 0$$

where: y_s = equation of the shock profile

$\gamma = \dfrac{c_p}{c_v}$ = ratio of specific heats for the fluid

M_o = undisturbed stream Mach number

p_o = undisturbed stream pressure

211C Other

For further applications of the above analogies and other analogies for the steady flow of a nonviscous compressible fluid, see Aitken; Busemen; Carter and Kron (C); Karapetian; Kron (F); Lazzarino; Pitin, Ponnick, and Farberov; Redshaw (D); Sauer, R.; Vandrey.

212 Unsteady Flow

212A *Electrical Analogy A*

The unsteady flow of a compressible fluid of constant viscosity through a porous medium of constant permeability may be investigated by means of an electrical analogy. This analogy is based upon the equations of continuity (Eq. (1), (2)), the relationship between fluid pressure and fluid velocity (Eq. (3), (5)), and the relationship between electric potential and electric current (Eq. (4), (6)). In two dimensions, these relationships are expressed in equation form as follows:

FLUID ELECTRICAL

$$\frac{\partial u}{\partial x} + \frac{\partial v}{\partial y} = fc\frac{\partial p}{\partial t} \quad (1) \qquad \frac{\partial f}{\partial x} + \frac{\partial g}{\partial y} = C'\frac{\partial V}{\partial t} \quad (2)$$

$$\frac{\partial p}{\partial x} = \frac{\mu}{k} u \quad (3) \qquad \frac{\partial V}{\partial x} = \sigma f \quad (4)$$

$$\frac{\partial p}{\partial y} = \frac{\mu}{k} v \quad (5) \qquad \frac{\partial V}{\partial y} = \sigma g \quad (6)$$

200. FLUID FLOW IN GENERAL

where: p = fluid pressure   V = electric potential
        x, y = coordinates   x, y = coordinates
        u, v = x, y components of   f, g = x, y components of current
               fluid velocity   density
        t = time   t = time
        c = fluid compressi-   C' = electrical capacitance per
            bility   unit volume, when used to
        f = fractional porosity   represent single-phase,
            of the medium   compressible fluid flow.
                                         (When used to represent
                                         gas flow, C' may be ef-
                                         fected by electron tubes.)
        μ = constant fluid viscosity
        k = constant permeability σ = electrical resistivity of the
            of the medium   medium

Equations (1), (3), (5) may then be combined and equations (2), (4), (6) combined.

Characteristic equations:

$$\frac{\partial^2 p}{\partial x^2} + \frac{\partial^2 p}{\partial y^2} = \frac{\mu f c}{k} \frac{\partial p}{\partial t} \quad (1) \qquad \frac{\partial^2 V}{\partial x^2} + \frac{\partial^2 V}{\partial y^2} = \sigma C' \frac{\partial V}{\partial t} \quad (2)$$

Thus, if the distribution of C' corresponds to the distribution of the product of porosity and compressibility, the distribution of electric potential will correspond to the distribution of pressure.

Boundary conditions:

1. The electrically conducting medium must have the same shape as the fluid flow field.
2. The distribution of electric potential along the boundary must correspond to the distribution of fluid pressure along the boundary.

For the case of single phase, compressible fluid flow (where C' is capacitance per unit volume), the continuous medium apparatus of section 311A (Fig. 311A.1) may be used as the analog (for one- or two-dimensional problems).

Bruce has modified this analogy in terms of finite differences as expressed below (for one-dimensional flow):

$$\frac{p_{n+1} + p_{n-1} - 2p_n}{\Delta x^2} = \frac{\mu f c}{k} \frac{\partial p}{\partial t} \quad (1) \qquad \frac{V_{n+1} + V_{n-1} - 2V_n}{\Delta x^2} = RC \frac{\partial V}{\partial t} \quad (2)$$

200. FLUID FLOW IN GENERAL

where: Δx = distance between nodes

R = concentrated resistance (resistor)

C = concentrated C' (capacitor or electron-tube circuit)

Subscripts represent nodes (Figs. 311A.2 and 311A.3). This modification may be extended to two or three dimensions.

Bruce suggests two types of models which may be used for oil zone problems. One involves an electrical network and the other utilizes an electrically conducting liquid to represent the field with electrodes to represent the wells. Diagrams are shown of electrically determined pressure contours for a typical field using a 90 percent glycerol solution as the electrolyte. Plots of pressure vs. time are also given.

See also Aleskerov, Babich, Motyakov, and Chal'yan; Aleskerov and Makhmudov; Boisard.

212B Electrical Analogy B

L. Green and Wilts have applied an electric circuit analogy to the analysis of the unsteady flow of a gas through a porous wall. The equation governing the flow is

$$\frac{\partial}{\partial s}\left(y\,\frac{\partial y}{\partial s}\right) = \frac{\partial y}{\partial \tau} \qquad (1)$$

where: y = pressure ratio = $\dfrac{p}{p_o}$

s = displacement ratio = $\dfrac{x}{L}$

τ = dimensionless time = $\left(\dfrac{p_o}{\epsilon\,\alpha\,\mu\,L^2}\right)t$

p = pressure

p_o = a reference pressure

x = displacement in the direction of wall thickness

L = wall thickness

ϵ = porosity

α = viscous coefficient (inverse of the permeability coefficient defined by Darcy's Law)

μ = viscosity of the gas

t = time

This equation is approximated by the finite difference equation for a mesh element of the fluid (Fig. 212B.1).

200. FLUID FLOW IN GENERAL

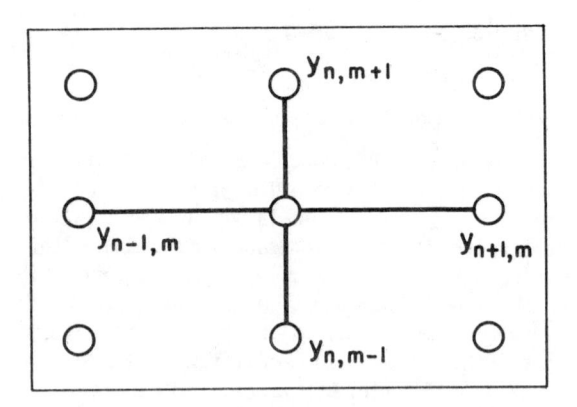

Fig. 212B.1. Typical mesh element for the fluid.

$$\frac{y_{n+\frac{1}{2},m}}{\Delta S_{n+\frac{1}{2}}}(y_{n+1,m} - y_{n,m}) - \frac{y_{n-\frac{1}{2},m}}{\Delta S_{n-\frac{1}{2}}}(y_{n,m} - y_{n-1,m})$$

$$- \frac{\Delta S_n}{2\Delta \tau_{m+\frac{1}{2}}}(y_{n,m+1} - y_{n,m}) - \frac{\Delta S_n}{2\Delta \tau_{m-\frac{1}{2}}}(y_{n,m} - y_{n,m-1}) = 0 \quad (2)$$

with the boundary conditions

$y(0, \tau) = y_1 = N$ (a constant)

$y(L, \tau) = y_0 = 1$

$y(s, 0) = y_0 = 1$

The analogous finite difference equation for a node of the electrical network (Fig. 212B.2) is

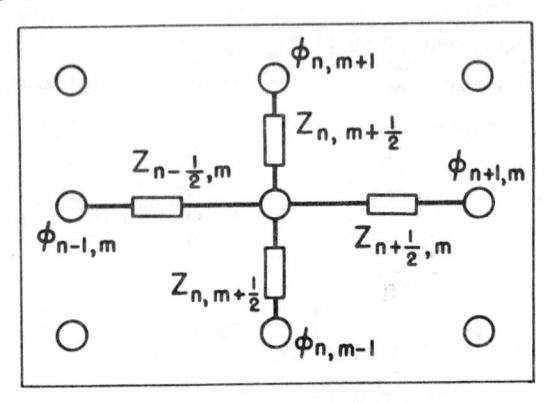

Fig. 212B.2. Typical node of the electrical network.

200. FLUID FLOW IN GENERAL

$$\frac{\phi_{n+1,m}-\phi_{n,m}}{Z_{n+\frac{1}{2},m}} + \frac{\phi_{n-1,m}-\phi_{n,m}}{Z_{n-\frac{1}{2},m}} + \frac{\phi_{n,m+1}-\phi_{n,m}}{Z_{n,m+\frac{1}{2}}} + \frac{\phi_{n,m-1}-\phi_{n,m}}{Z_{n,m-\frac{1}{2}}} = 0 \quad (3)$$

and the analogous quantities are

GAS	ELECTRICAL
$Z_{n+\frac{1}{2},m}$	$\dfrac{\Delta S_{n+\frac{1}{2}}}{y_{n+\frac{1}{2},m}}$
$Z_{n,m+\frac{1}{2}}$	$\dfrac{2\Delta \tau_{m+\frac{1}{2}}}{\Delta S_n}$
$Z_{n-\frac{1}{2},m}$	$\dfrac{\Delta S_{n-\frac{1}{2}}}{y_{n-\frac{1}{2},m}}$
$Z_{n,m-\frac{1}{2}}$	$\dfrac{2\Delta \tau_{m-\frac{1}{2}}}{\Delta S_n}$

The requirement that $Z_{n,m+\frac{1}{2}}$ be opposite in sign to the other three impedances at a given node is satisfied by using alternating current and using inductances and capacitors as impedances as shown in figure 212B.3.

If boundary voltages equal numerically to boundary values of the pressure ratio y are applied, then the voltage at any point (n,m) is equal to the value of y at displacement ratio n and dimensionless time m.

The nonlinear nature of the problem, manifested in the horizontal impedances, which are seen to be functions of the pressure ratio, necessitates an iterative process of solution, in which case the initial values of the horizontal impedances were determined from a rough estimate of y as a function of s for various values of τ.

Curves are shown of y vs. τ for various values of s and for N=2 and N=10.

See also L. Green.

220 INCOMPRESSIBLE

221 *Non-viscous*

221A *Irrotational*

221AA *Electrical Analogy A*

####### 221AAA *Two-Dimensional Flow.* Two electrical analogies for the irrotational flow of an incompressible nonviscous fluid have been

200. FLUID FLOW IN GENERAL

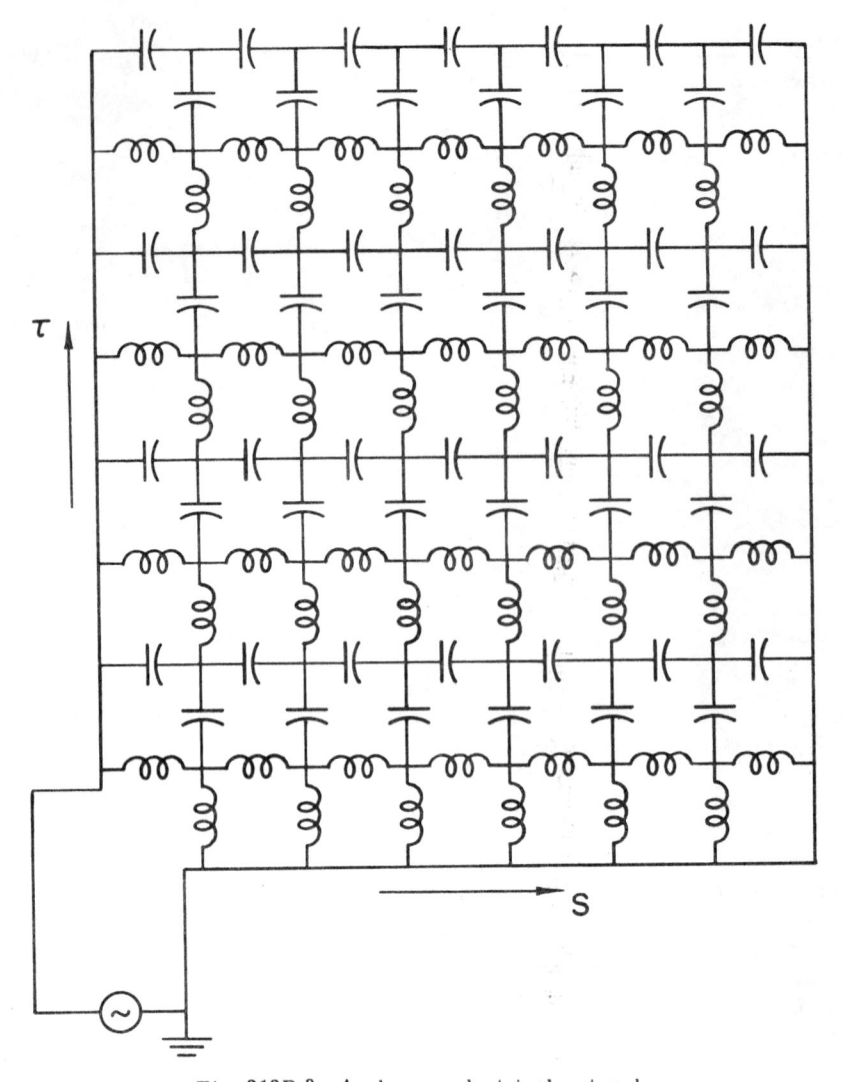

Fig. 212B.3. Analogous electrical network.

described by Hahn (A, B), who was among the first to apply them. The first of these analogies relates fluid velocity potential to electrical potential.

Characteristic equations:

FLUID

$$\frac{\partial^2 \phi}{\partial x^2} + \frac{\partial^2 \phi}{\partial y^2} = 0 \qquad (1)$$

ELECTRICAL

$$\frac{\partial^2 V}{\partial x^2} + \frac{\partial^2 V}{\partial y^2} = 0 \qquad (2)$$

200. FLUID FLOW IN GENERAL

where: x, y = coordinates

ϕ = velocity-potential function defined by the following equations:

$$\frac{\partial \phi}{\partial x} = -u \quad (3)$$

$$\frac{\partial \phi}{\partial y} = -v \quad (5)$$

u, v = x, y components of fluid velocity

x, y = coordinates

V = electric pontential satisfying the following equations:

$$\frac{\partial V}{\partial x} = -E_x \quad (4)$$

$$\frac{\partial V}{\partial y} = -E_y \quad (6)$$

E_x, E_y = x, y components of electric field intensity

Boundary conditions are the same as in section 211AAA.

The methods of sections 211AAA and 211AAC may be used to determine values of fluid velocity and lines of constant potential.

If equipotential lines are plotted with streamlines (determined by the analogy of section 221AAB), a flow net is obtained.

Utilizing an electrolytic tank, Hahn (A, B), who investigated the flow "round the vanes or through the diffusers and wheels of a turbine or pump," applied this analogy to the study of the flow around stationary blades. The metallic periphery of his circular tank (with insulated base) was one electrode, while the other electrode was a metal cylinder in the center of the tank.

One portion of the situation studied involved the direct flow of fluid from the center to the region outside the blades. To simulate this portion, the blades of the turbine or pump were represented in the tank by insulated cylinders of the same cross section as the blades (Fig. 221AAA.1). Hahn's article includes a plot of equipotential lines determined by analogy for this portion of the situation. See also Gerber and Ackert.

Hay and Markland used this analogy to study discharge over weirs.

Malavard (U) has described a French laboratory equipped with electrolytic tanks for studying various aerodynamical problems by the application of the analogies of this section and section 221AB. The general arrangement used is indicated in figure 221AAA.2. The electrolyte was water. Included in his report is a description of a "wing calculator" apparatus used to find lift distributions for wings and to solve other problems relevant to the linear theory of the wing of finite span, such as determining wind tunnel corrections.

Siestrunck (A, B) has described an electrical analogy apparatus called the "propeller calculator" (derived from the "wing calculator" mentioned above) used to study the theory of propellers.

For further application of the analogies of this section and section 221AB to problems of aerodynamics, see Bernard; Borden, Shelton, and Ball; Brower (B); Campbell; Castagno (A, B); Davorin; Duquenne;

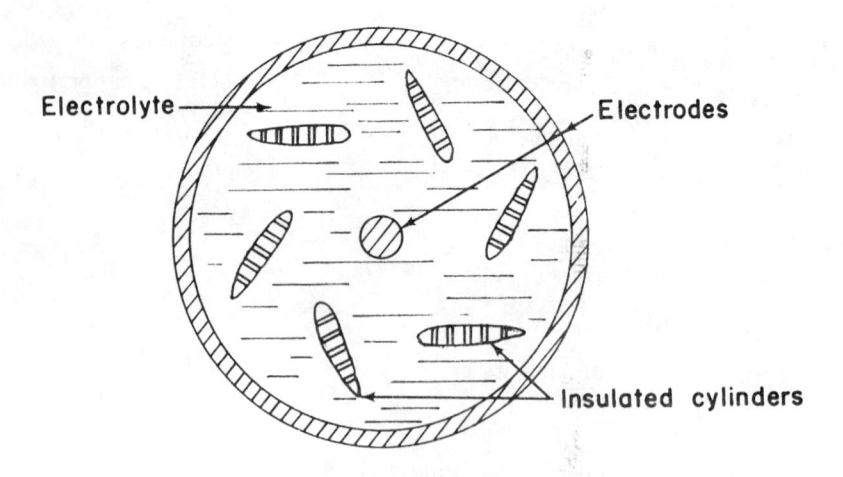

Fig. 221AAA.1. Electrolytic tank analog of a fluid flow section.

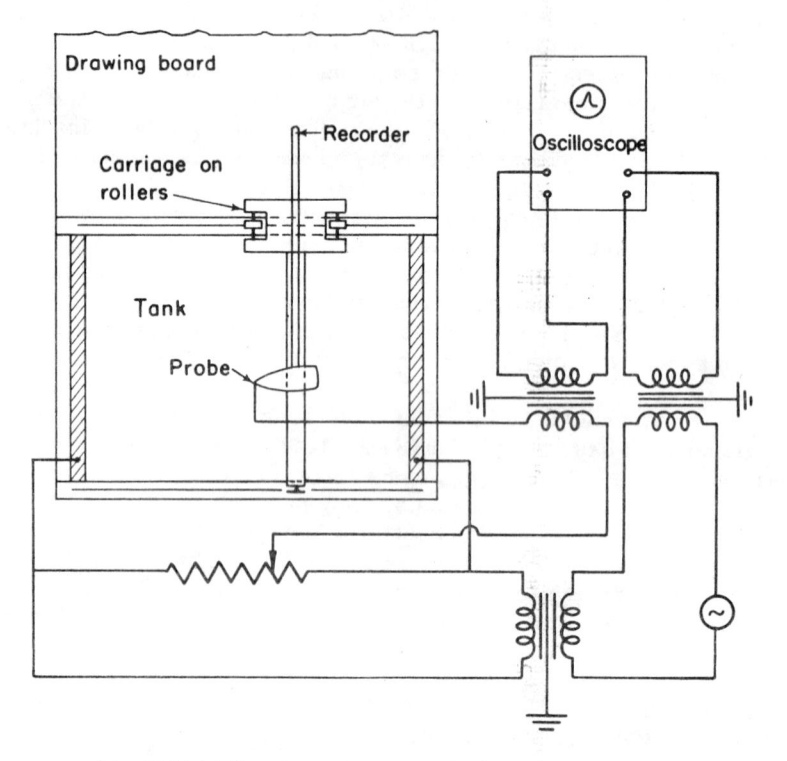

Fig. 221AAA.2. Electrolytic tank described by Malavard.

200. FLUID FLOW IN GENERAL

Duquenne and Grandjean (A, B, C); Enselme (A, B); Falkner and Gandy; Ferrari (B); Hacques (B, C, D); Hargest (A); Kelk, Misener, and d'Arcy; Kucheman and Redshaw; Landahl and Stark; Legras and Malavard; Malavard (A, B, C, D, F, H, J, M, N, O, P, Q, R, S, T); Malavard and Duquenne; Malavard, Duquenne, Enselme, and Grandjean; Malavard, Germain, and Siestrunck; Malavard and Hacques; Malavard and Siestrunck (A, B); Matsunaga (C); Otsuka; Palmer (B); Patraulea and Camarasescu; Peres (A, B); Peres and Malavard (B, C); Peres, Malavard, and Romani; Pizer, Wallis, and Warden; Poisson — Quinton; Revuz (A, B, C); Romani and Revuz; Scanlan (C); Vincent.

For flow through a porous medium, equation (1) may be replaced by the following equation (since velocity potential is proportional to fluid pressure):

$$\frac{\partial^2 p}{\partial x^2} + \frac{\partial^2 p}{\partial y^2} = 0 \qquad (7)$$

where p is fluid pressure.

Pavlovsky was among the first to utilize the analogies of this section and section 221AAB in analyzing the seepage of water under dams. By using equipment with the general arrangement shown in figures 221AAA.3 and 221AAA.4 he determined equipotential and flow patterns for one symmetrical half of the permeable region under a dam with three sheet-pilings of equal length. He also determined the equipotential flow pattern (by analogy) beneath a dam built on non-homogeneous porous ground.

Lane, Campbell, and Price have described an apparatus similar to Pavlovsky's, used to analyze the flow distribution at Hoover Dam.

Ram, Vaidhianathan, and Taylor (A) have described the use of an electrolytic tank to analyze the uplift pressures on dams or floors when the impervious surface is flush with the surface of the porous

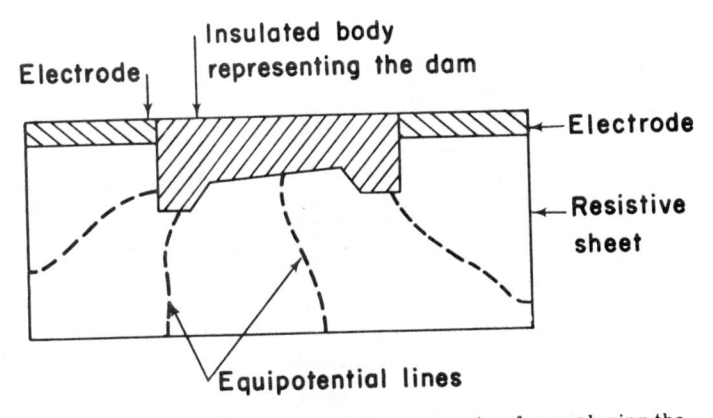

Fig. 221AAA.3. Electrical analogy apparatus for analyzing the seepage of water under dam.

46 200. FLUID FLOW IN GENERAL

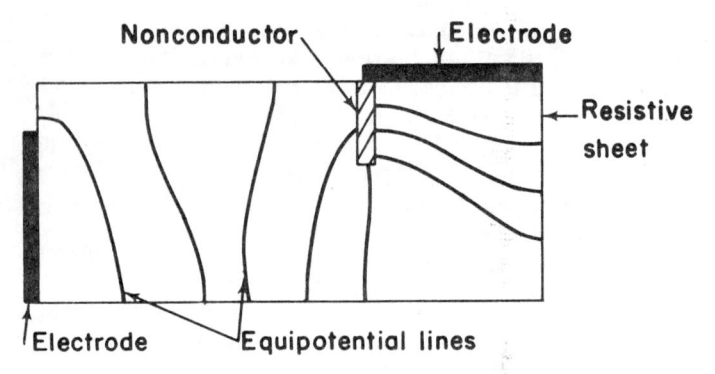

Fig. 221AAA.4. Apparatus for determining equipotential lines for one symmetrical half of the permeable region for a dam with three sheet pilings of equal length.

strata and when sheet piles are present. The authors present curves of percentage pressure distribution vs. fractional distance, obtained from the analogy, from direct pressure measurements, and from analytical methods.

A sketch of their equipment is shown (Fig. 221AAA.5). The electrolyte used was a dilute solution of ammonium chloride, and the probe was made by fusing a platinum wire into a glass tube leaving only the end of the wire exposed.

By using the electrolytic tank method, Gentilini made an investigation of the seepage of water through earth dams.

From electrolytic tank experiments, Selim (A, B) has presented experimental data on the uplift forces acting on a dam built on a porous medium. Flow nets are shown for several combinations of simple horizontal floors and cutoff walls (sheet piles) and for horizontal floors on permeable media of both finite and infinite depth. Also included in the report are tables of loss of head and uplift forces for various cases.

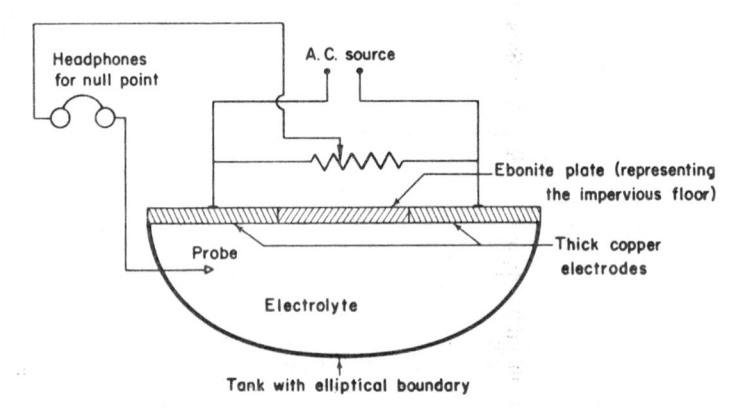

Fig. 221AAA.5. Electrical tank apparatus for analyzing the uplift pressure on dams.

200. FLUID FLOW IN GENERAL

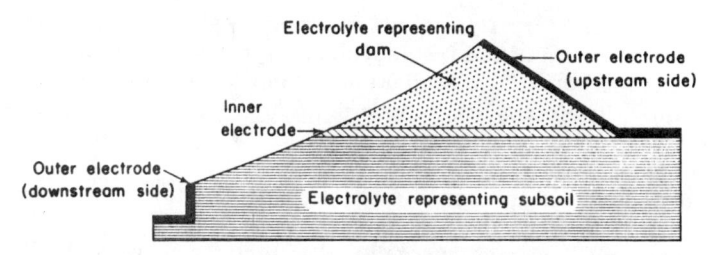

Fig. 221AAA.6. Electrolytic tank representing an earthen dam built upon a subsoil with larger porosity.

He used a tank about 40 in. long by 30 in. wide containing about 3/4 in. of an electrolyte consisting of 50 g of copper sulfate and 25 cc of sulphuric acid per 10 gal. of distilled water. He used a 10 volt ac source and a vacuum tube voltmeter to establish equipotential lines. One feature of his equipment was a pantograph unit attached to the probe so that equipotential lines could be drawn directly on a map of the region near the base of the structure being studied.

Vreedenburgh and Stevens (A) have developed an electrolytic tank representation of two adjoining earth bodies of different coefficients of percolation and have applied their method to the analysis of flow for cases such as an earthen dam of dense material built upon a subsoil with larger porosity. The depth of the salt solution electrolyte is varied to represent the variation in porosity. The general arrangement of their equipment is indicated in figures 221AAA.6 and 221AAA.7. The cascade arrangement of wires at the inner boundary was used to satisfy the continuity equation at the sudden change in depth of electrolyte.

Their report contains the following results obtained by analogy:
a) the flow net within and under an earth dam with smaller permeability than the subsoil and
b) the flow net of the phreatic and capillary flow for a canal in high embankments. They also include a brief discussion of the use of this analogy to investigate two-dimensional flow through soils of homogeneous-anisotropic permeability.

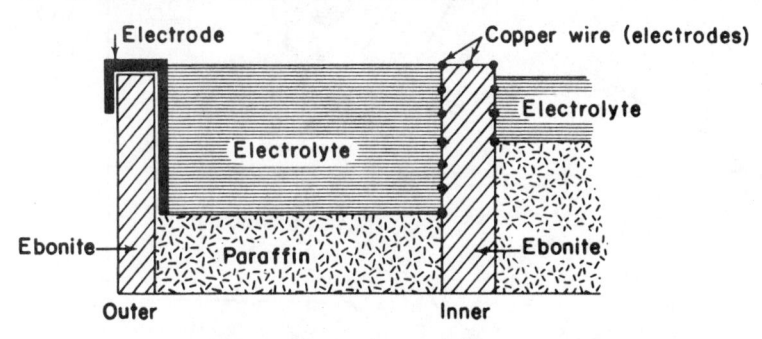

Fig. 221AAA.7. Cross section of electrodes.

200. FLUID FLOW IN GENERAL

By using an electrically conducting graphite sheet Wyckoff and Reed determined the shape of the free surface, the extent of the surface of seepage, and the potential distribution of the flow of water within permeable dams. Their results include plots of velocity distributions along vertical faces, comparative analytical data, and potential plots for a few typical cases.

Reinius described the use of a sheet of electroplated silver to investigate the flow net in the upstream portion of an earth dam of homogeneous material, bounded by a vertical watertight core and built on an impervious foundation. The desired potential variation along the upstream face of the model was obtained by tapping a resistor as indicated in figure 221AAA.8. For other methods of effecting a variation of electric potential along a boundary, see Childs (this section) and sections 221AAB (Babbitt and Caldwell), 221B, and 441AD.

A similar technique, described by Filchakov (A), was used to determine the seepage flow nets through dams made of materials with two different coefficients of permeability.

Boreli (A, B) has used the analogy to study the effect of impermeable cores in dams.

Ram and Vaidhianathan (B) used an electrolytic tank to determine the uplift pressure distributions under weirs with two sheet piles. Their investigation included tests for each of the three following:

a) two sheet piles of equal length at heel and toe of weir,
b) two sheet piles of different length at heel and toe, and
c) two sheet piles of equal length not at heel and toe.

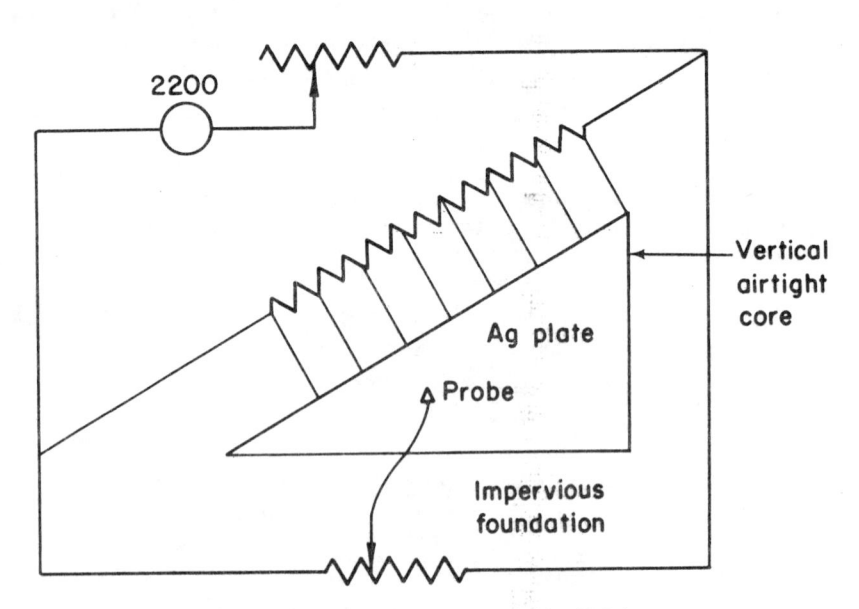

Fig. 221AAA.8. Equipment used by Reinius.

200. FLUID FLOW IN GENERAL

The results from these tests include curves of percentage pressure vs. distance for several pile lengths for each case.

Using a similar apparatus, H. R. Luthra and Vaidhianathan investigated the uplift pressure distributions under weirs with three sheet piles. Results of tests are included in their report.

Ram and Vaidhianathan also used the electrolytic tank to determine the uplift pressure distributions under flush floors with inclined sheet piles. Their report includes plots of percentage pressure vs. distance for various angles and lengths of piles.

An investigation of the uplift pressure distributions under the floors of falls was made by Ram, Vaidhianathan, and Taylor. Two series of experiments were conducted, one in which the length of the downstream apron was varied, with other parameters held constant, and one in which the length of the sheet pile was varied, with other parameters held constant. Included in the article are curves of pressure vs. distance for both series.

Childs has used a graphite electrically conducting sheet in the determination of the pressure potential distribution in drained land in planes perpendicular to the drain lines (Figs. 221AAA.9 and 221AAA.10).

Initially, the equally spaced slits are cut to the same level approximating the water table. The sliding contacts of the potential divider are adjusted until all separate strip currents are equal. The locus of points for which $V = Ah$ (where V = electric potential, h = height of the point above the drain level, and A = constant) is drawn and the slits are extended to this line. Since this will change the potential distribution, another $V = Ah$ line is plotted. By successive approximations, an accurate representation of the water table can be determined. When this has been done, the potential distribution can be plotted over the entire cross section. For other methods of effecting a variation of electric potential along a boundary, see Reinius (this section) and sections 221AAB (Babbitt and Caldwell), 221B, and 441AD.

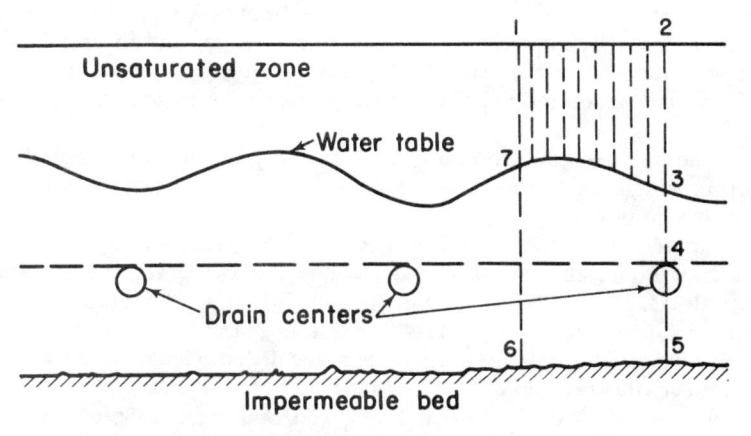

Fig. 221AAA.9. Cross section of drained land.

200. FLUID FLOW IN GENERAL

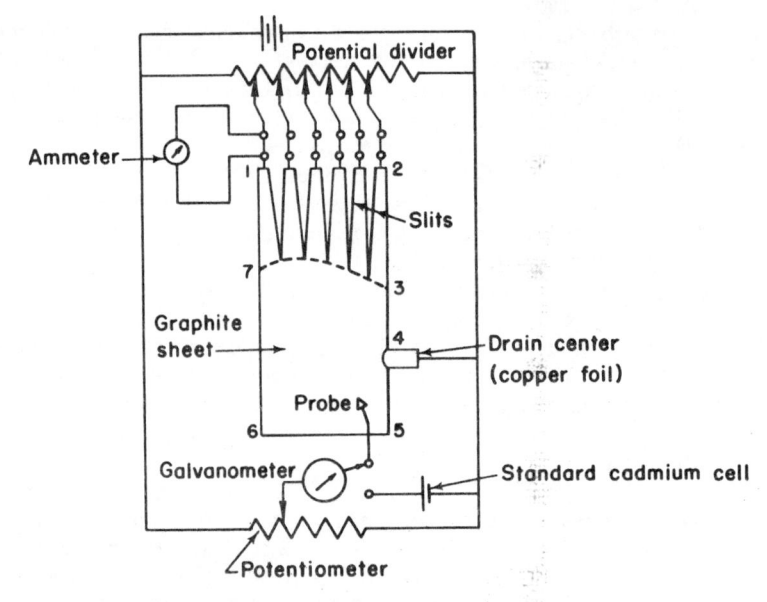

Fig. 221AAA.10. Electrical analogy apparatus.

In part I of his report are found plots of distributions for various drain diameters and heights of drain level above an impermeable bed. Part II includes results of analog experiments to determine

a) the perturbing effect of holes such as are commonly used in attempts to locate the water table in the field,
b) the relative efficacy of flooded and empty drains,
c) the effect of piping and filling in a previously open trench, and
d) the role of the capillary fringe.

In part III are shown results of analog experiments where the permeability is varied above and below the capillary fringe, and in part IV appear results showing the water table and the flow net (determined by analog) in land protected against foreign water by pipe drains and open ditches.

Probine has used an electrical resistance network to investigate the manner in which the permeability of unsaturated porous materials varies with moisture content.

For further references dealing with the use of the analogy of this section in studying the two-dimensional seepage of water, see the article, "Analog Indicates Reservoir Seepage"; Balanin; Chadeisson; Comier; Crausse and Poirier; Dachler; De la Marre; Dolcetta; Druzhinin; Filchakov and Panchishin; Gerber and Pilod; Gruat; Guha and Ram; Hebert; Harza; the article, "How To Trace Ground Water Flow"; H. A. Johnson; H. P. Johnson; Li, Bock, and Beuton; Luthra (A, B); Luthra and Ram (A, B); S. D. L. Luthra (A, B); S. D. L. Luthra and

200. FLUID FLOW IN GENERAL

Ram (A, B); Markland and Hay; "Model Studies of Penstocks and Outlet Works"; Motyakov (A); Nougaro; Peattie; Puppini; Schneebeli; Taylor and Vaidhianathan; Vaidhianathan and Ram (A, B); Vaidhianathan, Ram, and Taylor (A, B); Vreedenburgh and Stevens (B, C); K. R. Wright.

Wyckoff and Botset were among the first to apply this analogy experimentally to flow within petroleum reservoirs. They have described various arrangements of input and output wells used in oil fields and the use of symmetry in designing sections to study flow. Their procedure involved the use of pieces of blotting paper (representing two-dimensional petroleum fields) soaked with a weak solution of potassium sulphate to which was added a generous quantity of a saturated solution of phenolphthalein. Input and output wells were represented by electrodes as shown in figure 221AAA.11. The advance of the hydroxyl ions due to the applied potential difference between electrodes was indicated by the colored front due to the phenolphthalein. In this way, the advance of the fluid front in the actual field could be predicted. Apparatus, diagrams, and photographs of the progression of fluid fronts in various types of fields are included in the article describing this method. See also Muskat and Wyckoff; Fothergill.

Kelton has presented results of an oil-field study made by means of an electrolytic gelatin model using colored ions and has compared them with results from field data. The dry-gas boundaries of the field have been plotted from the electrolytic-model data.

Horner and Bruce have described two types of electrical models

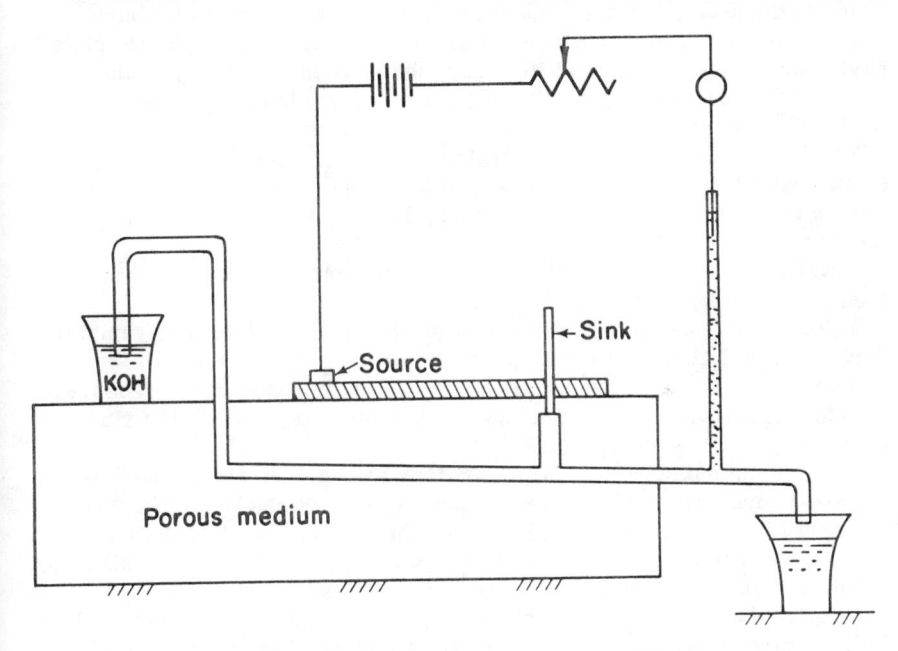

Fig. 221AAA.11. Electrolytic analog of oil field.

200. FLUID FLOW IN GENERAL

used to predict the probable composition-time curves of oil wells and to aid in controlling injection and withdrawal volumes:

a) electrolytic conductors (salt water or salt solutions in gels);
b) electrically-conducting metal plate (Fig. 221AAA.11).

Results are shown for the solution of a theoretical gas-injection problem by analogy.

The article "Electrolytic Model Developed by Gulf Corp.," describes the use of this analogy in determining how much oil can be produced from a field and the most efficient procedure for recovery operations.

In the model is a sensitized mat made of gelatin (representing the field) into the top of which are inserted small plastic tubes with hollow tips. Tubes representing output wells are filled with a colorless liquid. Those representing input wells contain a blue fluid.

Controlled electric current is fed through the intake wells into the mat, turning the gelatin blue as far as it penetrates. A color pattern is thereby formed of the flow, showing how dry gas will force wet gas, or water will force oil toward the producing wells. By controlling input current, adding new input or output "wells," and neutralizing others, the operator can determine the best procedure to follow.

Botset has explained the application of this analogy in predicting recovery from petroleum wells by gas-recycling and water-flooding. His electrolytic model consisted of a glass base on which was poured a 1/16 in. thick layer of a 1% agar gelatin solution containing 0.1 normal zinc-ammonium chloride. The solution was enclosed by a plastic plate having an inside boundary of the same shape as the boundary of the field. The wells (transparent plastic, 1/2 in. inside diameter and 1 1/2 in. long) rested on an opaque white bakelite cover through which the 3/32 in. diameter well tips penetrated into the gelatin. The input (positive potential) wells were filled with a 0.1 molar deep-blue copper-ammonium chloride solution containing 1.5% agar. A camera was mounted underneath the model and photographs were taken at various intervals of time. Typical photographs are shown of the advancing front of the field as indicated by the electrical model.

Muskat (B) has discussed the use of an electrolytic tank of variable depth in simulating flow within petroleum fields of variable porosity and permeability and in which the fluid density depends upon pressure.

Nobles and Janzen have utilized a resistance network in the study of mobility ratio effects.

For further discussion of this analogy as a tool in analyzing flow with petroleum reservoirs, see Aronofsky; Bruce; Burke and Parry; Calhoun (A, B); Hines; Hurst; Hurst and McCarty; Lee; Marshall and Olivery; Morris, W. L.; Motyakov (B); Muskat (A); Ramey and Nabor; Selim (A, B); Swearingen; Wolf; Wyckoff, Botset, and Muskat.

For other applications of this analogy, see Benedict and Meyer; de Haller; Efros; Hargest (B); Larras; Luu; Malavard (E, G, K); Markland and Hay; Peres and Malavard (A, D, F, G); Renard; Ryasanov.

See also sections 221AAB, 221AAC, 221D.

200. FLUID FLOW IN GENERAL

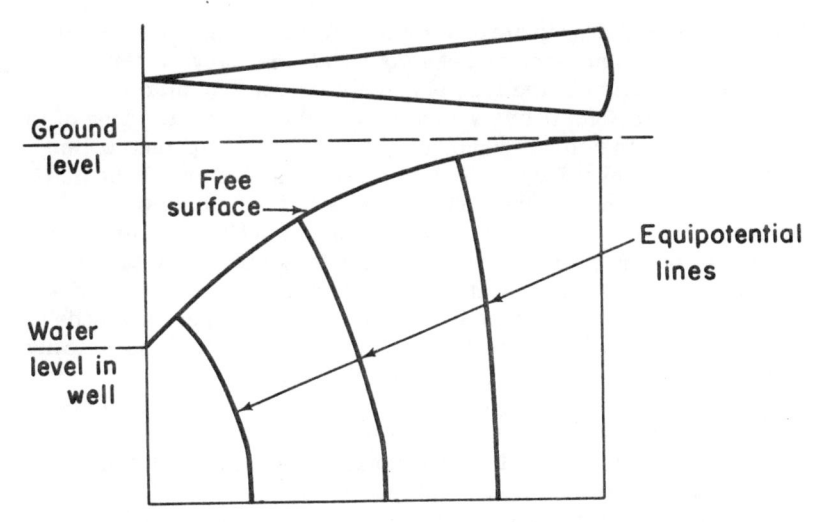

Fig. 221AAB.1. Segment of the volume surrounding well.

221AAB *Axially Symmetric Flow.* Babbitt and Caldwell have described an electrical apparatus to determine, by use of the analogy described in section 221AAA, the free surface of water surrounding a gravity well during pumping. They used a pressed carbon wedge to represent a pie-shaped segment of the volume surrounding the well. As indicated in figures 221AAB.1 and 221AAB.2, vertical distances on the model were proportional to depths in the prototype. The resistance strip was of much lower resistance than the model so that potential drop along the strip was linear. The length of the strip represented the distance from the water surface in the well to the ground surface.

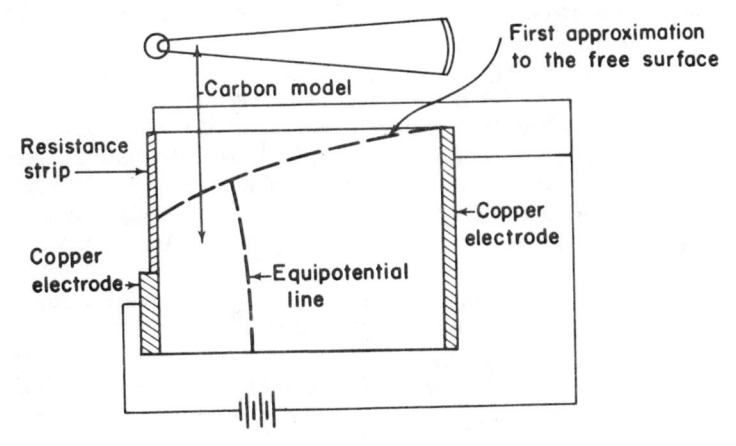

Fig. 221AAB.2. Electrical analogy apparatus for water well during first approximation.

200. FLUID FLOW IN GENERAL

The total potential drop across the model was divided into ten increments, and the corresponding equipotential lines were determined. Then an approximation to the free surface was found by marking the point on each equipotential line which indicated the same fraction of the total drawdown as the corresponding voltage of the equipotential line was of the total voltage drop. The line joining these points indicated the approximated free surface. The model was then cut down to this line and the process was repeated several times until an accurate representation of the free surface was found, after which the actual equipotential lines were plotted. Data from the analog were compared with direct experimental data and analytical results. For other methods of effecting a variation of electric potential along a boundary, see sections 221AAA (Reinius and also Childs), 221B, and 441AD.

On page 82 of "Conference on Control of Underseepage" are given graphical results of a typical test on an electric analogy model of a well, showing the effect of different penetrations of the well point into the pervious layer.

Muskat (A) has analyzed the pressure and velocity distributions around an artesian well partially penetrating a water bearing sand by means of an analytical investigation of the electric potential distributions in large cylindrical disks with partially penetrating electrodes.

Cheers, Raymer, and Fowler and also Ferrari (A) have analyzed axial flow around bodies of revolution by using the analogies of sections 221AAA and 221AB. The analogy consisted of an electrolytic tank representing a thin segment (bounded by two diametrical planes) of the flow section surrounding such a body. The tank contained an insulated member whose profile was the same as that of a symmetrical half of a longitudinal section of the body. Bounding the electrolyte at the ends of the tank were metal electrodes (Fig. 221AAB.3).

The report of Cheers, Raymer, and Fowler includes a diagram of streamlines of the flow around a typical body and a plot of the velocity distribution along the surface of the body.

By applying the analogy of section 221AAA, Hubbard has developed a technique for determining boundary pressures for the flow of incompressible nonviscous fluids confined by surfaces of revolution. The analog apparatus consisted of an electrolytic tank of the same shape as a thin segment (bounded by two diametrical planes) of the flow section. This tank consisted of a glass bottom, end-plate electrodes, and an insulated longitudinal boundary made of a strip of Pyralin (Fig. 221AAB.4).

Along the side (longitudinal boundary) of the tank, narrow electrodes were spaced at equal intervals, Δs (Fig. 221AAB.4), where s is in the direction of flow and thus represents distance along the side boundary since fluid cannot flow across the side boundary. The expression for the magnitude, q, of the resultant fluid velocity at any point is

$$q = \frac{\partial \phi}{\partial s} \approx \frac{\Delta \phi}{\Delta s} \tag{1}$$

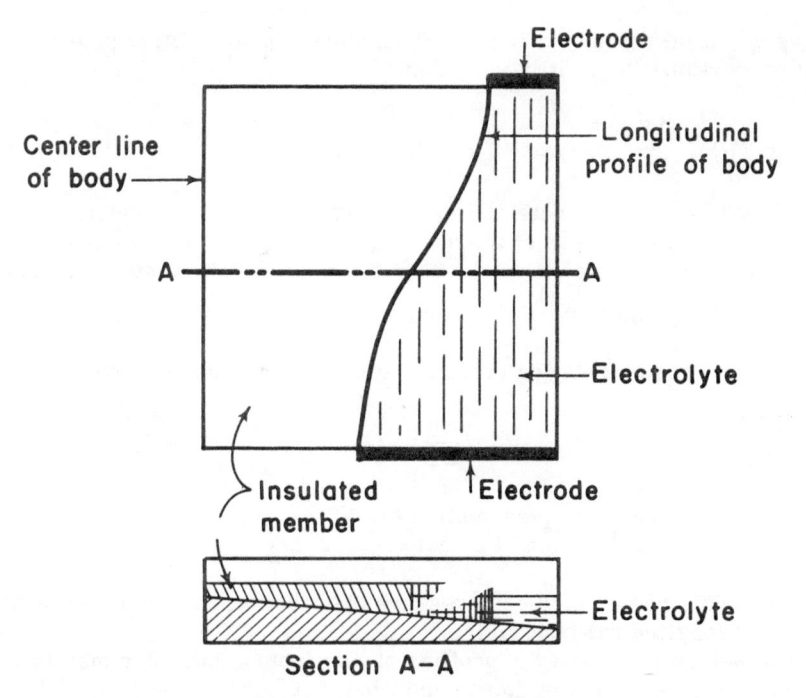

Fig. 221AAB.3. Electrolytic tank analog of the axial flow around a body of revolution.

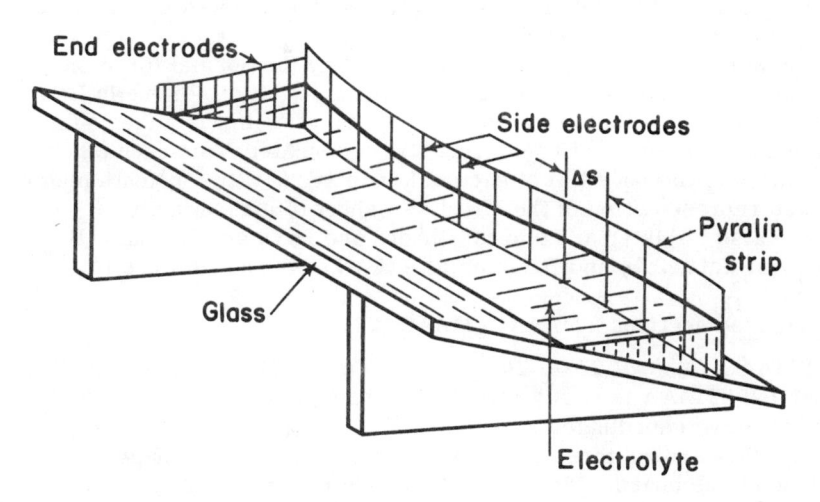

Fig. 221AAB.4. Electrolytic tank analog of flow bounded by a surface of revolution

where ϕ = fluid velocity potential. Therefore, since electric potential V is proportional to ϕ, it is seen that

$$\frac{q}{U} = \frac{\Delta V}{\Delta P} \qquad (2)$$

where: ΔP = electric-potential drop between adjacent side electrodes at the ends of the section, where parallel flow exists (assumed to be the same as at an infinite distance away)

U = magnitude of the resultant fluid velocity at infinity

The expression relating pressure and fluid velocity becomes:

$$\frac{\Delta h}{U^2/2g} = 1 - \left(\frac{q}{U}\right)^2 = 1 - \left(\frac{\Delta V}{\Delta P}\right)^2 \qquad (3)$$

where: Δh = change in piezometric head
  g = acceleration of gravity

Hubbard's report includes pressure distribution curves for several shapes of longitudinal boundaries.

The determination of the profiles of jets from axially symmetric orifices has been made by Rouse and Abul-Fetough, who applied Hubbard's technique. For each situation, the Pyraline sheet boundary strip was adjusted in such a manner as to yield an essentially constant velocity at all points of the simulated free surface of the jet. Each profile was determined by successive approximations.

By this method, Leclerc has investigated the deflection of a liquid jet by a perpendicular boundary.

Brower (A) has determined by this analogy the normal force on slender, closed bodies of revolution inclined slightly to the main flow. In his analog, he simulated a vorticity distribution in the wake region by the use of a series of conducting strips insulated from each other. Regulation of the potential of each wake electrode with a potentiometer allowed representation of the effect of a sheet of discontinuity.

See also Babister and others; Cheers and Raymer; Germain (B); Gibbings; Hahneman and Bammert; Lewis; Lewis and Newman (A, B); Marchet.

See also sections 221AAA, 221AAC, 221D.

221AAC *General Three-Dimensional Flow.* The analogy described in section 221AAA is easily extended to three dimensions by the addition of a third coordinate.

The flow around three-dimensional bodies of general shape was analyzed by Hubbard. Plastic scale models representing the bodies contained built-in electrodes and were immersed in a hexagonal tank as in figure 221AAC.1.

Rouse and Hassan have utilized the technique of Hubbard in

200. FLUID FLOW IN GENERAL   57

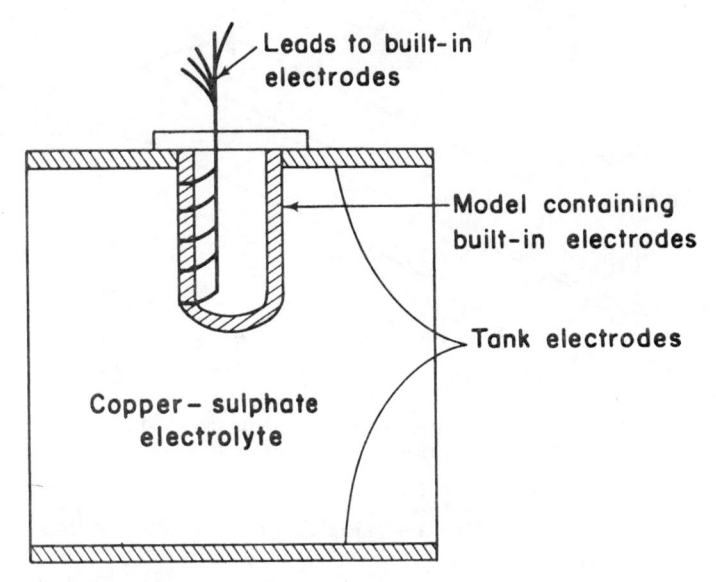

Fig. 221AAC.1. Electrolytic tank analog of the flow around a three-dimensional body of general shape.

designing cavitation and free transition sections of flow passages, cavitation resulting from low fluid pressures. (For the relation between pressure and velocity, see section 221AAB.) Their article includes plots of limiting transition curves for such situations as well as curves of velocity and pressure ratios vs. length. Results from analogy experiments are compared with analytically determined values.

Hubbard and Ling applied the technique in predicting low pressure zones in cylindrical section transitions from a reservoir of fluid to a conduit of square cross section. For this case, the electrolytic tank represented a symmetrical segment of the transition and the conduit (Figs. 221AAC.2 and 221AAC.3). The side electrodes were inserted in an orthogonal pattern. The ratio of the cylindrical section radius R to the width N of the conduit was varied by adjusting the height of the surface of the electrolyte.

A plot of $\frac{\Delta h}{U^2/2g}$ vs. length along the transition at the center line of a side and at the juncture between two sides is superimposed upon a diagram of the developed surface of transition upon which has been constructed isobars. Also included are plots of $\frac{\Delta h}{U^2/2g}$ vs. $\frac{X}{N}$ (where X = length along the conduit) for various values of $\frac{R}{N}$ and plots of $\frac{\Delta h}{U^2/2g}$ vs. $\frac{R}{N}$.

By immersing in an electrolytic tank an insulated model (containing imbedded probe electrodes) of an airplane engine nacelle, Katzoff and

58   200. FLUID FLOW IN GENERAL

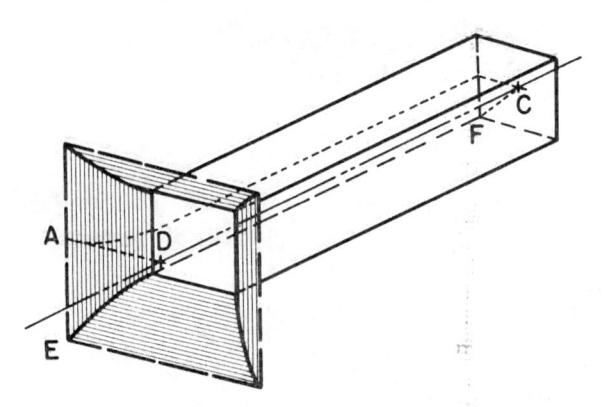

Fig. 221AAC.2. Schematic diagram of a square conduit
and its transition from a reservoir.

Finn determined corrections to the velocities near a cowling flap tip for a typical nacelle.

Using a mixture of graphite and a powdered dielectric, Reltov constructed a three-dimensional model of a dam and its pervious heterogeneous foundation to study the percolation under the dam. Insulated electrodes were imbedded in the model for the measurement of electric potential in the interior. Variations in permeability of the medium were effected by corresponding variations in the model resistivity.

See also Habib and Sabarly; Hassan; Huard de la Marre (A, B); H. A. Johnson; Opsal; Stenstrom.

See also sections 221AAA, 221AAB, 221D.

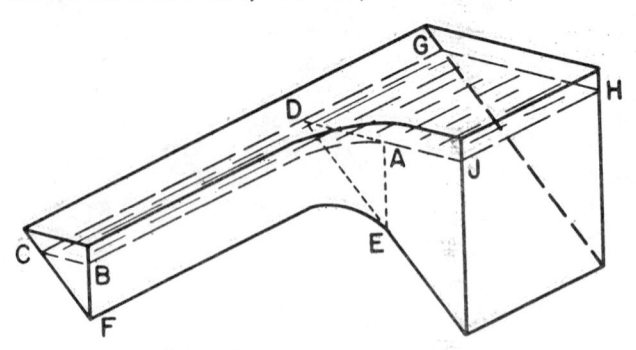

Fig. 221AAC.3. Electrolytic tank analog of a conduit and transition
(ABCDGHJ is electrolyte surface).

221AB *Electrical Analogy B*

The second electrical analogy for the irrotational flow of an incompressible nonviscous fluid relates the fluid stream function to electric potential.

200. FLUID FLOW IN GENERAL

Characteristic equations:

FLUID ELECTRICAL

$$\frac{\partial^2 \psi}{\partial x^2} + \frac{\partial^2 \psi}{\partial y^2} = 0 \quad (1) \qquad \frac{\partial^2 V}{\partial x^2} + \frac{\partial^2 V}{\partial y^2} = 0 \quad (2)$$

where ψ is the fluid stream function defined by the following equations:

$$u = \frac{\partial \psi}{\partial y} \quad (3)$$

$$v = -\frac{\partial \psi}{\partial x} \quad (4)$$

Other symbols are defined in section 221AAA.

Boundary conditions are the same as in section 211AAB.

Lines of constant electric potential, as determined by a probe arrangement, are streamlines of fluid flow.

By means of an electrolytic tank (Sec. 221AAA) this analogy was utilized by Hahn (A, B) in studying the effect of a vortex filament at the center of a set of stationary blades. In this situation, the blades were represented in the tank by cylinders made of metal whose conductivity was large in comparison to that of the electrolyte. In this analogy, the central and peripheral electrodes represent fluid streamlines.

Hahn also investigated the effect of circulation around the blades. For this situation, the following condition must be satisfied:

$$\Gamma_1 - \Gamma_2 = n\,\Gamma_b$$

where: Γ_1 = circulation at the periphery of the flow section
Γ_2 = circulation at the center of the flow section
Γ_b = circulation around each blade
n = number of blades

This condition is satisfied by trial and error. First, an electric potential of assumed value is applied to all of the cylinders representing the blades (see boundary conditions, Sec. 211AAB). From measurements of the lines of constant electric potential (fluid streamlines), the corresponding circulations can be calculated. This process is repeated with different values of electric potential applied to the cylinders until the circulation condition is satisfied. Then the entire flow field is plotted. Included in the report are typical flow patterns.

Relf has used an electrolytic tank in applying this analogy to the investigation of the flow around a flat plate lying oblique to the direction of flow. He has presented a diagram of analytically and experimentally determined streamlines for such a flow. Also included in his report is a plot of experimentally determined values of the distribution of the

200. FLUID FLOW IN GENERAL

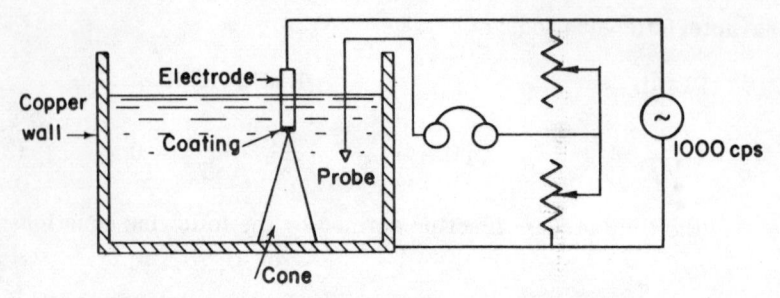

Fig. 221AB.1. Equipment for locating streamlines.

streamline function along an axis bisecting the width of such a plate at right angles (due to circulation only), as well as comparable analytically determined values for both an infinite and a finite fluid.

Frevert used an electrolytic tank to establish the streamlines around a tube in a pervious bed of soil above an impervious layer. A cross section of the tank, which was 68 in. in diameter and 20 in. deep, is shown (Fig. 221AB.1) together with the circuit employed. The end of the brass electrode was coated with glyptal and a copper cone was placed in the tank to simulate the limiting position of the streamlines.

Frevert also used his equipment to assist in the analysis of soil permeability.

See also Walker.

221AC *Membrane Analogy*

Brahtz used a pressureless rubber membrane to determine pressures and flow nets in plane sections of concrete dams.

Characteristic equations:

FLUID

$$\frac{\partial^2 p}{\partial x^2} + \frac{\partial^2 p}{\partial y^2} = 0 \quad (1)$$

$$\frac{\partial^2 \phi}{\partial x^2} + \frac{\partial^2 \phi}{\partial y^2} = 0 \quad (2)$$

MEMBRANE

$$\frac{\partial^2 z}{\partial x^2} + \frac{\partial^2 z}{\partial y^2} = 0 \quad (3)$$

where: p = pore pressure
ϕ = velocity potential
x, y = coordinates of the plane being considered

z = membrane elevation
x, y = coordinates

Boundary conditions:

1. The boundary of the membrane must have the same shape as the boundary of the plane section of the dam.

200. FLUID FLOW IN GENERAL

2. The distribution of z along the boundary must be the same as the distribution of ϕ or p along the boundary. (For the construction of such membrane boundaries, see section 432.)

Thus lines of constant elevation of the membrane correspond to lines of constant pressure or potential.

Brahtz' paper shows photographs of the apparatus used and typical diagrams of equipotential and flow lines.

Hansen has applied this analogy to the study of steady-state flow to well systems.

A combination of this analogy and the analogy of section 221AAA has been utilized by Zee to obtain the free surface for unconfined flow to a well.

221AD *Magnetic Analogy A*

Gardner and LaHatte have utilized a magnetic analogy in the investigation of the induced velocity in front of an inclined propeller (assumed to be a actuator disk) in a horizontal stream.

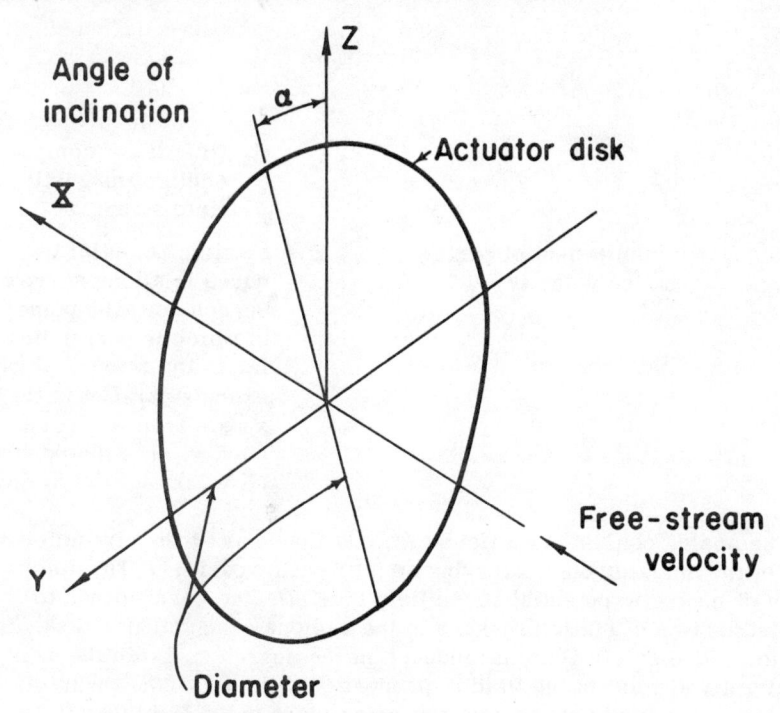

Fig. 221AD.1. Coordinate system for inclined propeller.

Characteristic equations:

$$\frac{\partial^2 p}{\partial x^2} + \frac{\partial^2 p}{\partial y^2} + \frac{\partial^2 p}{\partial z^2} = 0 \quad (1) \qquad \frac{\partial^2 \phi}{\partial x^2} + \frac{\partial^2 \phi}{\partial y^2} + \frac{\partial^2 \phi}{\partial z^2} = 0 \quad (2)$$

200. FLUID FLOW IN GENERAL

$$W = K_1 \int_{-\infty}^{x} \frac{\partial p}{\partial z} dx \quad (3) \qquad E = K_2 \int_{-\infty}^{x} \frac{\partial \phi}{\partial z} dx \quad (4)$$

where: x = horizontal coordinate in direction of free stream

y = horizontal coordinate perpendicular to x and z

z = vertical coordinate

p = local static pressure

x, y, z = coordinates

ϕ = magnetic potential defined by the following equations:

$$H_x = -\frac{\partial \phi}{\partial x}$$

$$H_y = -\frac{\partial \phi}{\partial y}$$

$$H_z = -\frac{\partial \phi}{\partial z}$$

H_x, H_y, H_z = components of magnetic field strength

w = z component of perturbation velocity

E = electric potential induced in a long narrow search coil, the plane of which is perpendicular to the z-axis, which extends parallel to the x-axis from the point $(-\infty, y, z)$ to the point (x, y, z).

The analog consists of a field coil, positioned as at the circumference of the actuator disk, carrying an alternating current. The distribution of magnetic potential in the field thus effected corresponds to the distribution of static pressure in the region of the actuator disk. In addition, the electric potential induced in the search coil with the near end at a given point in the field is proportional to the z-component of perturbation velocity at the corresponding point in the fluid flow field. (The infinitely long search coil may be closely approximated by a finite coil that is long compared with the diameter of the actuator disk.)

The analogy was thus applied experimentally to the determination of values of *w* and was also applied purely analytically to the determination of values of *u* (the *x* component of velocity, proportional to the pressure).

200. FLUID FLOW IN GENERAL

The report of Gardner and LaHatte includes plots of lines of constant u and lines of constant w for various values of α.

221AE *Magnetic Analogy B*

Hurst and Bey have described the use of a magnet to aid in plotting equipotential lines for fluid flow under a dam. It is indicated that the base of the dam is represented by a piece of sheet iron. A horizontal sheet of paper is placed over the region through which flow takes place and iron filings are sprinkled on the paper. A magnet below the paper will cause the filings to align themselves to represent equipotential lines. By using blueprint paper and exposing the assembly a print of the lines may be obtained.

221AF *Hydrodynamical Analogy*

Stokes has presented an analogy between the two-dimensional irrotational flow of an incompressible nonviscous fluid and the two-dimensional irrotational flow on a very thin sheet of an incompressible viscous fluid.

Characteristic equation (same for both situations):

$$\frac{\partial^2 \psi}{\partial x^2} + \frac{\partial^2 \psi}{\partial y^2} = 0 \tag{1}$$

where: x, y = coordinates

ψ = stream function defined by the following equations:

$$\frac{\partial \psi}{\partial x} = -v \tag{2}$$

$$\frac{\partial \psi}{\partial y} = u \tag{3}$$

u, v = x, y components of fluid velocity

Thus the streamlines of both situations are identical to each other. (The analogy holds exactly only for the limiting case of zero thickness of the viscous fluid sheet.)

Boundary conditions:

1. The lateral boundaries of the flow section in both must be streamlines.
2. The flow regions of the two must be geometrically similar in the plane of flow.

Probably the first to apply this analogy was Hele-Shaw (A, B). His analog consisted of two closely spaced glass plates, separated near their lateral edges by a thickness of some other solid substance, its shape defining the boundary of the flow region (Fig. 221AF.1). A

200. FLUID FLOW IN GENERAL

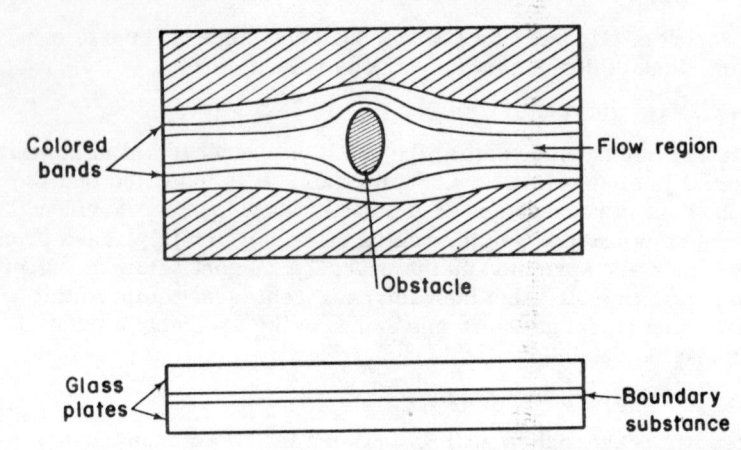

Fig. 221AF.1. Hydrodynamical analog for two-dimensional fluid flow.

viscous liquid, such as glycerine, was forced under pressure to flow through the confine thus formed. Narrow streams of colored glycerine were introduced into the flow at regular intervals near the entrance to the flow region, thereby indicating the streamlines within the region. The effect of a uniform obstacle in the flow of a nonviscous fluid was simulated by placing within the glycerine channel a thickness of the boundary substance, cut to the shape of the obstacle cross section.

The advantage of using a viscous liquid to determine the streamlines, rather than using a liquid of low viscosity, such as water, is that the less viscous liquid exhibits a degree of instability of motion. The flow patterns formed in the Hele-Shaw apparatus are remarkably distinct and correspond very accurately to the computed patterns of a perfect (incompressible, nonviscous) fluid.

Using this type of apparatus, Balloffet has investigated flow through elliptical contractions. See also Fossa.

A. D. Moore (A, C, E) devised a modification of the Hele-Shaw type of apparatus which is less expensive and more flexible than the original. In Moore's "mappers," the lower glass plate is replaced by a plaster or cast stone slab. In addition, flow lines are made visible by streaks of dissolved colored crystals placed within the flow space rather than by streams of dyed fluid. Further, plain water or water treated with a viscosity-increasing substance is used in place of glycerine as the flow medium. For further modifications made by Moore, see sections 312B and 313C.

Meyer and Searcy have utilized this analogy in making a study of water coning in petroleum fields. See also Matthews and Lefkovits.

221AG *Hydraulic Analogy (Transient Flow)*

Barron has used an hydraulic device, similar to that described in section 311B, to study the time and position variation of hydrostatic excess head of the pore water in a consolidating soil mass.

See also Güenther.

200. FLUID FLOW IN GENERAL

221B *Rotational*

Hahn (A, B) has also described an electrolytic tank analogy for the rotational flow of an incompressible nonviscous fluid.

The characteristic equation for the fluid is

$$\frac{\partial^2 \psi}{\partial x^2} + \frac{\partial^2 \psi}{\partial y^2} = -2\omega \tag{1}$$

where ω = angular velocity of the fluid. An electrolytic tank may be used if a new function, ψ_1, is defined as follows:

$$\psi = \psi_1 - \frac{\omega \rho^2}{2} \tag{2}$$

where $\rho = \sqrt{x^2 + y^2}$ = distance of any point from the origin.

Characteristic equations:

FLUID

$$\frac{\partial^2 \psi_1}{\partial x^2} + \frac{\partial^2 \psi_1}{\partial y^2} = 0 \tag{3}$$

ELECTRICAL

$$\frac{\partial^2 V}{\partial x^2} + \frac{\partial^2 V}{\partial y^2} = 0 \tag{4}$$

where symbols are defined as above and in section 221AA.

When $\rho = \infty$, ψ_1 vanishes. Along a boundary contour G,

$$(\psi_1)_g = \psi_g + \frac{\omega \rho^2}{2}$$

where: $(\psi_1)_g$ = distribution of ψ_1 along G
ψ_g = distribution of ψ along G

If ψ_g is arbitrarily set equal to zero, then $(\psi_1)_g$ will vary as ρ^2. Thus, V must vary as ρ^2 along each boundary contour. This may be effected approximately by fixing along each contour a number of electrodes, the potentials of which satisfy the above condition.

By using an electrolytic tank (Sec. 221AAA), Hahn applied this analogy to the study of the effect of a vortex filament at the center of a turbine pump.

As indicated above, ψ_1 must vary as ρ^2 along a boundary. In the tank, this condition is approximately satisfied at the blades by representing them with several discrete electrodes. The electric potential (analogous to ψ_1) at each of these electrodes is made proportional to ρ^2 (ψ is arbitrarily assigned a value of zero at the blades) by appropriate connections to a potentiometer (Fig. 221B.1). The center and periphery of the tank are electrodes also, as before. For other methods of effecting a variation of electric potential along a boundary, see sections 221AAA (Reinius and also Childs), 221AAB (Babbitt and Caldwell), and 441AD.

200. FLUID FLOW IN GENERAL

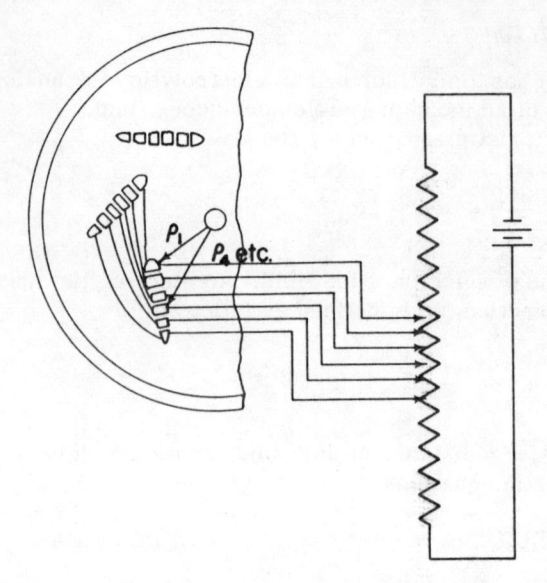

Fig. 221B.1. Electrolytic tank representation of blades of a turbine pump when a vortex filament exists at the center.

Fluid streamlines (lines of constant electric potential) for the situation described are easily determined by means of this analog.

221C *Vortices*

221CA *Electromagnetic Analogy*

Swanson and Crandall used the magnetic field around an electrical conductor to study the induced velocity field around a vortex and applied the analogy in making lifting-surface calculations.

Characteristic equations:

FLUID	ELECTROMAGNETIC
$d\bar{v} = \dfrac{\Gamma}{4\pi} \dfrac{d\bar{L} \times \bar{r}}{\lvert\bar{r}\rvert^3}$ (1)	$d\bar{H} = i \dfrac{d\bar{L} \times \bar{r}}{\lvert\bar{r}\rvert^3}$ (2)
where: v = induced velocity	H = magnetic field strength
r = distance from an element of the vortex to the point in question	r = distance from an element of the conductor to the point in question
L = length of vortex	L = length of conductor
Γ = vortex strength	i = current in the conductor

and *d* denotes "differential," and the bar over a symbol denotes a vector quantity.

200. FLUID FLOW IN GENERAL

Electrical conductors were arranged to represent the vortex sheet. The magnetic field strength was then determined by measuring with an electronic voltmeter the voltage induced in a small search coil by the alternating current in the wires representing the vortex sheet.

A comparison was made of the downwash determined by means of a preliminary electromagnetic analogy model with the downwash obtained by calculation for an elliptic wing having an aspect ratio of 3. In this investigation, "a set of 50 circular electric wires carrying the same current (connected in series) representing 50 columnar vortices of equal strength were distributed over the wing and wake. The arrangement of the wires was determined as follows: fifty contour lines of Γ were calculated.... Each wire was placed halfway between two adjacent Γ contour lines and thus represented $\frac{\Delta\Gamma}{\Gamma_{max.}} = 0.02$." In another set-up aluminum strips were used instead of copper wires to represent the vortex sheet.

In another investigation involving this analogy, Swanson and Priddy studied the induced velocity field around a thin elliptic wing of aspect ratio 6 in a steady roll. From the results, aspect ratio corrections for the damping in roll and aileron hinge moments for a wing in steady roll were obtained. The magnetic field strength was measured at 4 to 5 vertical heights, 15 spanwise stations, and 25 to 50 chordwise stations. Photographs of the apparatus and plots of the results are included.

Crandall applied this analogy to obtain solutions for two vortex patterns representing unswept elliptic wings having aspect ratios of 3. The induced magnetic field strength was measured at about 50 chordwise points for 5 vertical heights and 20 spanwise positions for both models.

Swanson, Crandall, and Miller experimented with an electromagnetic analogy model of the vortex load, estimated from lifting line theory with an arbitrary fairing near the elevator tip, on a thin elliptical tail surface of aspect ratio 3 with 0.5 chord and 0.85 span elevator. The aspect ratio corrections to the lift and hinge moments were calculated from the measured results. They show comparative analytical solutions and wind-tunnel data.

See also Castles, Durham, and Kevorkian; Gray; W. P. Jones; McCormick.

221CB *Electrical Analogy*

By the use of an electrical analogy for the aerodynamic field behind a propeller, Theodorsen (A, B) has determined values of the mass coefficient K and the potential distribution circulation function K*(x)*.

The electric field for determining K is simulated by perfect helices (fabricated from celluloid strips) in a large cylindrical tank (made of insulating materials) with copper ends and filled with tap water. K is calculated from the following equation:

$$\frac{KF}{S} = \frac{\Delta I}{I_o} \tag{1}$$

200. FLUID FLOW IN GENERAL

where: F = projected area of the helix

S = area of the end plates of the cylindrical tank

I_o = current through the test tank with the helix removed

ΔI = reduction in current due to the presence of the helix

For single rotation wake models, K*(x)* is measured by the use of apparatus similar to that used to determine K. For dual rotation wake models, K*(x)* is measured by the use of a rubber membrane helix stretched across the ends of an open V-shaped tray.

Included in Theodorsen's report are plots of K vs. $\frac{Vw}{nD}$ for single and dual rotating multi-blade propellers where:

V = advance velocity of the propeller

w = rearward displacement velocity of the helical vortex surface (at infinity)

n = rotational speed of the propeller

D = diameter of the vortex sheet

His report also contains plots of fractional potential along the unit cell (helix surface between two successive lines of intersection) vs. fractional radius of the helix and plots of K*(x)* vs. the fractional radius.

For other vortex analogies, see Hahn (A, B) (Secs. 221AB and 221B).

221D *Other*

For further references to analogies for the flow of a nonviscous incompressible fluid, see Carter and Kron (C); Concordia and Carter; Filchakov (B); K. H. Grossman; Karplus and Allder; Katzoff, Gardner, Diesendruck, and Eisenstadt; Kron (F, L); Lazzarino; McMahan; Merbt and Hansson; Molnar; Nechayev; Zangar and Haefeli.

222 *Viscous*

222A *Journal Bearing Lubrication*

The electrical analogy for the determination of the pressure distribution in the oil film of a journal bearing was first utilized by Kingsbury, who employed an electrolytic tank.

Characteristic equations:

FLUID ELECTRICAL

$$\frac{\partial}{\partial x}\left(h^3 \frac{\partial p}{\partial x}\right) + \frac{\partial}{\partial z}\left(h^3 \frac{\partial p}{\partial z}\right) = f(x,y) \quad (1) \qquad \frac{\partial}{\partial x}\left(\frac{t}{t_o \sigma} \frac{\partial V}{\partial x}\right) + \frac{\partial}{\partial z}\left(\frac{t}{t_o \sigma} \frac{\partial V}{\partial z}\right) = i(x,y) \quad (2)$$

200. FLUID FLOW IN GENERAL

where:
- x = distance along the bearing surface in the direction of motion
- z = distance along the bearing surface perpendicular to the direction of motion
- p = oil pressure
- h = oil film thickness
- f(x, y) = volume displacement independent of pressure

- x = coordinate
- z = coordinate
- V = electric potential
- t = depth of electrolyte
- t_o = reference electrolyte depth
- σ = constant resistivity of the electrolyte
- i(x, y) = distributed current source strength (current density)

Thus if the depth of the electrolyte is made proportional to oil film thickness at corresponding points, and if the distribution of current source strength corresponds to the distribution of volume displacement independent of pressure, the electric potential at any point in the tank will be proportional to oil pressure at the corresponding point of the oil film.

The distributed current source strength is approximated by introducing discrete amounts of electric current in each of small uniform sections of tank area by means of vertical electrodes which extend the depth of the electrolyte. The total current flowing into these electrodes is proportional to the total volume flow of fluid in the film.

Boundary conditions:

1. In the x-z plane, the tank must be geometrically similar to the oil film.
2. Along the boundary, the distribution of electric potential must correspond to the distribution of oil pressure. This is effected by placing a series of individual electrodes along the boundary and applying appropriate potential.

Kingsbury obtained solutions for a number of cases of plane bearing surfaces and journal bearings. Included are comparable analytical solutions.

By the same method, Needs investigated 120 degree centrally supported journal bearings.

Applying the same basic method but with improved techniques, Morgan, Muskat, and Reed analyzed journal bearings of finite width,

200. FLUID FLOW IN GENERAL

bearings with circumferential grooves, and bearings with sources of lubricant.

D. S. Carter has applied finite difference equations in modifying this analogy to utilize electric circuits in place of electrolytic tanks. See also Kettleborough (B).

According to Kingsbury, Michell applied an elastic analogy experimentally to the study of lubricating films between plane surfaces.

Loeb has modified this analogy and used a conducting-paper analog in a study of hydrostatic bearings. See also Loeb and Rippel.

Light has extended and modified this analogy for the investigation, by means of conducting-paper analogs, of externally pressurized thrust and journal guide bearings when the lubricant is a gas.

See also Belgaumkar (A).

222B Other

Roscoe has shown that "in certain cases of slow viscous flow round thin plates the solutions of Stokes' equations can be reduced to a solution of LaPlace's equation with simple boundary conditions, the flow being thus relatable to the potential distribution in an electrostatic position." For example, a region of uniform electrostatic potential disturbed by the presence of a grounded conducting lamina is analogous to a region of flow of a viscous fluid around a thin plate of the same form as the conducting lamina.

Janssen has used an electric network analog to study flow past a flat plate at low Reynolds numbers.

Shearer has studied viscous flow, normal to parallel, evenly spaced cylinders by the use of an electric network for solving the Navier-Stokes equations.

Expressing the Navier-Stokes equations in biharmonic form, Goodier suggested an analogy between slow viscous flow and plane elasticity. In Goodier's analogy, lines of constant thickness (isopachics) in the stress problem will correspond to lines of constant velocity in the flow field and the lines of constant rotations, being orthogonal trajectories of the isopachics, will be equivalent to the lines of constant fluid pressure in the flow problem. It also follows that the stress trajectories and isochromatics will represent in the field of flow the orientation of maximum distortion and the lines of constant maximum distortion respectively.

Rayleigh suggested a similar analogy between slow viscous flow and the flexure of an elastic plate. Southwell (B) utilized this analogy in a theoretical manner to resolve Stokes' paradox of slow flow around a cylinder. By means of a special photographic technique, Richards has determined the geometry of flexed plates and has thereby predicted the flow characteristics in analogous flow situations. (See section 451.)

See also section 121BBF.

Section 300. Heat Transfer

310 CONDUCTION

311 *Unsteady Flow (With and Without Sources)*

311A *Electrical Analogy*

The electrical analogy for unsteady flow heat conduction has found many varied applications. This analogy is based upon the equations of continuity (Eq. (1), (2)), the relationship between temperature and heat flow (Eq. (3), (5)), and the relationship between electric potential and electric current (Eq. (4), (6)). In two dimensions, these relationships are expressed in equation form as follows:

HEAT		ELECTRICAL	
$\frac{\partial q_x}{\partial x} + \frac{\partial q_y}{\partial y} = c\rho \frac{\partial \theta}{\partial t}$	(1)	$\frac{\partial f}{\partial x} + \frac{\partial g}{\partial y} = C' \frac{\partial V}{\partial t}$	(2)
$\frac{\partial \theta}{\partial x} = \frac{1}{K} q_x$	(3)	$\frac{\partial V}{\partial x} = \sigma f$	(4)
$\frac{\partial \theta}{\partial y} = \frac{1}{K} q_y$	(5)	$\frac{\partial V}{\partial y} = \sigma g$	(6)

where: θ = temperature
$\quad$ x, y = coordinates
$\quad$ q_x, q_y = x, y components of rate of heat flow per unit area
$\quad$ t = time

V = electric potential
x, y = coordinates
f, g = x, y components of current density
t = time

300. HEAT TRANSFER

c = specific heat of the heat-conducting medium

ρ = density of the heat-conducting medium

K = coefficient of thermal conductivity of the medium

C' = electrical capacitance per unit volume of the electrically-conducting medium

σ = electrical resistivity of the medium

Equations (1), (3), (5) may then be combined and equations (2), (4), (6) combined.

Characteristic equations:

$$\frac{\partial}{\partial x}\left(K \frac{\partial \theta}{\partial x}\right) + \frac{\partial}{\partial y}\left(K \frac{\partial \theta}{\partial y}\right) = c\rho \frac{\partial \theta}{\partial t} \quad (7) \qquad \frac{\partial}{\partial x}\left(\frac{1}{\sigma}\frac{\partial V}{\partial x}\right) + \frac{\partial}{\partial y}\left(\frac{1}{\sigma}\frac{\partial V}{\partial y}\right) = C' \frac{\partial V}{\partial t} \quad (8)$$

Thus if electrical resistivity is made inversely proportional to thermal conductivity at corresponding points and electrical capacitance is made directly proportional to the product of specific heat and density, the distribution of electric potential will correspond with the distribution of temperature.

If the thermal conductivity of the medium to be investigated is constant, equations (7) and (8) reduce to the following:

$$\frac{\partial^2 \theta}{\partial x^2} + \frac{\partial^2 \theta}{\partial y^2} = \frac{c\rho}{K} \frac{\partial \theta}{\partial t} \quad (9) \qquad \frac{\partial^2 V}{\partial x^2} + \frac{\partial^2 V}{\partial y^2} = \sigma C' \frac{\partial V}{\partial t} \quad (10)$$

Boundary conditions:

1. The electrically-conducting medium must have the same shape as the heat-conducting medium.

2. The distribution of electric potential along the boundary must correspond to the distribution of temperature along the boundary at all times. Thus portions of the boundary at constant temperature would be represented by lines of constant electric potential, and portions of the boundary that are thermally insulated would be represented by portions of boundary that are electrically insulated. For methods of effecting a variation of electric potential along a boundary, see sections 221AAA (Reinius and also Childs), 221AAB (Babbitt and Caldwell), 221B, and 441AD.

One of the principal difficulties involved in the direct application of the analogy is in effecting continuously distributed capacitance of sufficient magnitude to permit a feasible time scale in the analog.

Shippy has suggested an experimental technique whereby such

300. HEAT TRANSFER

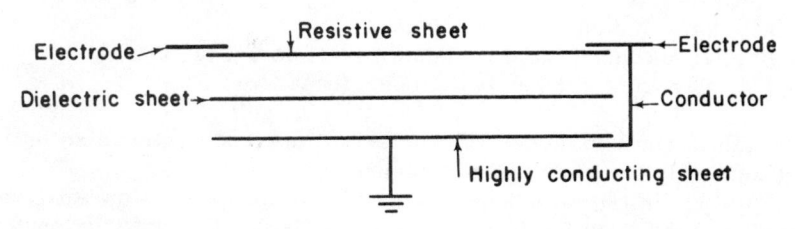

Fig. 311A.1. Electrical analog for heat flow along a uniform rod.

capacitance can be effected for one- or two-dimensional problems. His analog consists primarily of layers of a resistive medium and a highly conducting medium, with a dielectric sheet separating them. These are all cut to the shape of the thermal medium to be investigated, and appropriate electrodes are attached. For the one-dimensional flow of heat along a uniform homogeneous rod with an insulated lateral surface, the analog in basic form is shown in figure 311A.1 (the ideal analog in this case being a uniform, homogeneous, insulated, inductionless electric cable).

The effect of a variation of temperature with time at one end of the rod may be determined by a similar variation of electric potential at the ungrounded electrode of the analog and measuring the variation of potential with time at points along the resistive strip.

Variations of specific heat, density, and thermal conductivity are simulated by corresponding variations in capacitance and resistivity.

By the application of finite difference equations, this analogy has been modified to allow the use of standard resistors and capacitors instead of an electrical conductor geometrically similar to the thermal member. For illustration, the uniform homogeneous rod described above may be represented by a series of equal size elements (called heat reservoirs) connected by resistive conducting rods (Fig. 311A.2).

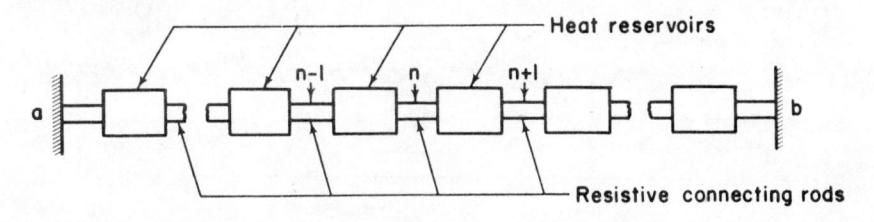

Fig. 311A.2. Finite difference representation of the uniform homogeneous rod.

Expressing equations (9) and (10) in terms of finite differences for one-dimensional flow,

$$\left(\frac{\theta_{n+1} + \theta_{n-1} - 2\theta_n}{\Delta x^2}\right) = \frac{c\rho}{K}\frac{\partial \theta_n}{\partial t} \quad (11) \quad \left(\frac{V_{n+1} + V_{n-1} - 2V_n}{\Delta x^2}\right) = RC\frac{\partial V_n}{\partial t} \quad (12)$$

where: Δx = distance between nodes
R = concentrated resistance (resistor) (Fig. 311A.3)
C = concentrated capacitance (capacitor)

Subscripts represent nodes as indicated. This modification also may be extended to two and three dimensions.

Paschkis (E, G, M) has developed a "Heat and Mass Flow Analyzer" based upon the modified (network) analogy above. For descriptions of this analyzer, see "Complex Heat Transfer Solved by Electrical Analogy"; "Electrical Analogy Method for Investigation of Transient Heat-Flow Problems"; "Electrical Analyzer Solves Heat-Flow Problems"; "Heat Transfer Problems Solved with Roomful of r-c Networks"; "Heat Transfer Problems Solved by Electrical Analyzer"; Paschkis and Heisler (B); Seymour; Swain.

For an extension of the network analogy to include sources, see Ackroyd and others.

Within homogeneous panels, slabs, and walls of uniform thickness, the flow of heat normal to the surfaces is almost the same as the flow of heat along an insulated, uniform, homogeneous cylinder (the greater the extent of the surfaces, the more exact the relationship). Similarly, the flow of heat parallel to the surfaces is a simple illustration of the corresponding two-dimensional case (assuming insulated surfaces). If the heat flow is within a panel, slab, or wall, the flow of heat is three dimensional.

An investigation of the heat losses through a wall in intermittent operation has been made by Bradley, Ernst, and Paschkis. By means of the electric circuit analogy, curves have been developed which permit the ready determination of the heat losses for any temperature, any single material wall, any intermittency, any length of period, and any film conductance. These curves permit the determination of the economic thickness for intermittent operation. The application of the

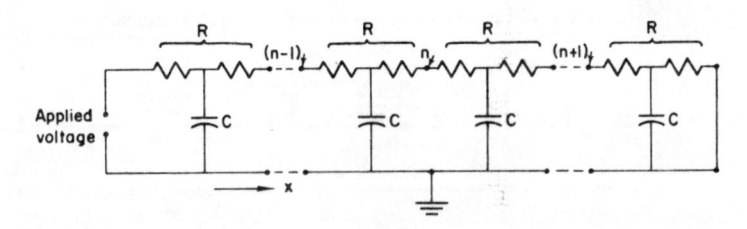

Fig. 311A.3. Electrical circuit analog for a uniform homogeneous rod.

300. HEAT TRANSFER

curves is illustrated by an example. The influence of intermittency on the economic thickness is discussed for one specific case.

Paschkis (N) has used the electric circuit analogy for determining the variation with time of the temperature distribution within a building wall consisting of layers of different materials. Included in the report are plots of temperature vs. time for various depths and surfaces of boundary-air layers. See also Paschkis (J).

Bradley and Ernst have presented a comparison of direct measurement solutions of problems involving heat flow through walls of furnaces in cyclic operation with solutions from the Heat and Mass Flow Analyzer. The performance and results of an actual problem solved by both methods are given.

See also Beuken and Hamaker.

Burnand utilized an electronic circuit in studying the heating of buildings. His results, recorded on an oscilloscope, include plots of inside air temperature and rate of heat supply vs. time. Diagrams of equipment are shown.

Linvill and Hess also used an electronic circuit in their investigation of the heat flow through walls of insulated houses. Included in their report are typical circuit diagrams. See also Paschkis (L).

An electrical network thermal analyzer described by Tribus has been used to investigate intermittent heat transfer during aircraft icing conditions with application to propellers and jet engines. The analyzer is used to demonstrate how the protection afforded for propeller deicing may be increased without an increase in the energy required from the electrical supply of the airplane and without major redesign of the propeller heaters. Results from the analyzer are compared with measured time-temperature data. See also F. Weiner (A).

Otis has used an electrical network with servo controlled resistances to study transient heat conduction in a medium whose thermal properties are arbitrary functions of temperature. For another method of simulating such functions, see Friedmann (A).

Using an electrical network, Neel (B) has made a comparison of the relative advantages of tubular heaters and internal shoe type heaters for hollow steel propellers. This study was limited to the 48-inch radial station of the propeller (Figs. 311A.4, 311A.5).

"A number of special circuits were employed to simulate the boundary and input functions representative of a cyclically heated propeller in icing conditions. A constant current source connected to a switching system provided the electrical arrangement representative of the cyclic application of heat. Additional constant current circuits were arranged to simulate the application of heat at the blade surface resulting from aerodynamic heating and release of the heat of fusion of the supercooled drops impinging on the blade surface as they solidified to ice. Relay circuits switched in resistors and condensers to typify the continuous growth of ice on the blade after each removal process. Other circuits, termed 'heat of fusion' circuits, were provided to simulate the melting of those regions of ice formation which reached the

300. HEAT TRANSFER

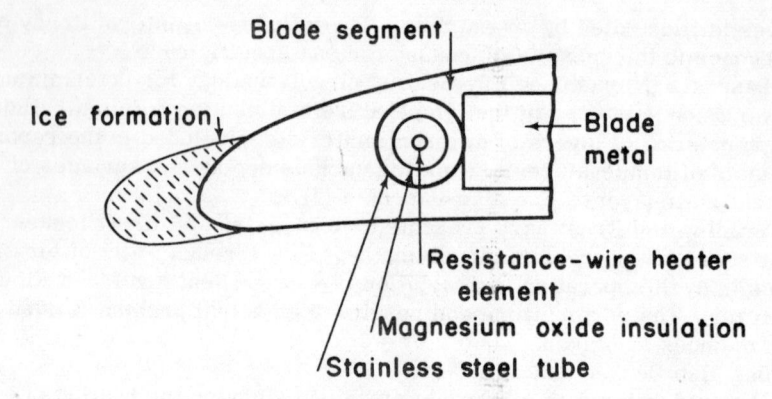

Fig. 311A.4. Cross section of an airplane propeller with a tubular heater.

melting point before release of the ice. This was accomplished by retaining these regions at a constant voltage representative of 32°F until simulated removal of the ice was attained."

Included in the report are temperature vs. time curves for various parts of the investigated sections. There are also curves of heating intensity vs. time.

"The study showed the impracticability of using an internal tubular heater and illustrated the advantages of employing an internal shoe type heater, which distributes the heat more evenly to the blade surface." See also Hauger.

Jackson and others have utilized electrical circuits to determine temperature distributions in steel masses undergoing heating or cooling. Typical applications involved solidification of steel in an ingot mold and heating of an ingot in a furnace.

The use of the Heat and Mass Flow Analyzer in solving common thermal problems encountered in foundry work has been described by Paschkis (H). Application to one specific foundry problem, that of optimum size of opening in the sand core between the casting and the feeding head of the casting, is discussed.

In another investigation, Paschkis (Q) has analyzed (a) the influence

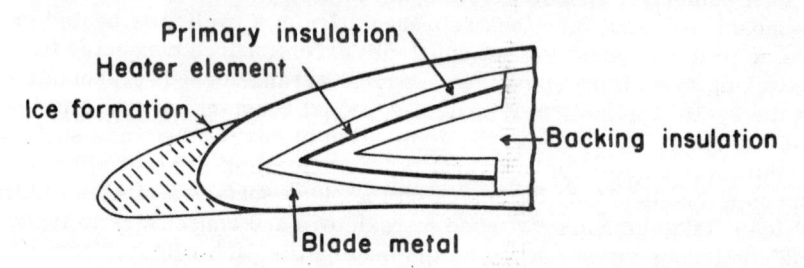

Fig. 311A.5. Cross section of an airplane propeller with an internal shoetype heater.

of the material of the bottom of an ingot on cooling, comparing a cast-iron bottom with a carbon insert, (b) the solidification at midheight of the ingot, and (c) a comparison of two hot tops.

The quenching of steel spheres and rings has also been studied by use of the Heat and Mass Flow Analyzer and reported by Paschkis (O). For spheres, general curves are presented in which the delaying effect of the heat of transformation in the range from 250 to 150° C has been included. In addition, a large number of investigations have been carried out in which the change of thermal properties (conductivity and specific heat) with temperature has been considered. For steel rings, charts have been developed which show the complete temperature-time-space relationships in rings of any size and material of constant thermal properties.

Paschkis (F) also described the determination by electrical analogy of general curves of the cooling rates of welds for any welding conditions for the top pass of a 1-1/2 inch plate.

Vogel and Krueger have used electrical circuits to study heat conduction from a moving, cylindrical heat source in the earth.

Woodcock, Thwaites, and Breckenridge have studied the dynamics of a thermal system composed of the human body, its clothing, and its environment.

A determination of heat losses from insulated steam pipes has been made by Paschkis and Baker.

The analysis by electrical analogy of the heat flow through an axially connected rod and cylinder has been described by Willey.

Examples of applications of the electrical circuit analogy in the glass and ceramics industries are given by Paschkis (M). Included are a description and results of analyses of the heat flow within airplane windows composed of heated wires imbedded between layers of glass and plastic.

For further discussion of the theory and application of this analogy, see Aleskerov, Babich, Motyakov, and Chal'yan; Avrami and Paschkis; Benedict; Billington (A); Buchberg (A, B, C); Clarke; J. de Laclemandiere; L. de Laclemandiere; Ellerbrock et al; "Experimental Determination of Fluctuating Heat Flow"; "Experimental Determination of Non-Permanent Heat Flow"; Gelissen; Goldenberg; Guile; Kayan and McCague; Klein, Touloukian and Eaton; Lawson and McGuire; McKeon and Eschenbrenner; Miroux (A); Muncey; "New Electrical Method for Mass and Heat Transfer Problems...."; "New Heat Transfer Research Tool"; Paschkis (B, C); Paschkis and Heisler (A); Potseluiko and Trofimenko; Potter; Price and Sarjant; Provasnik; Przemieniecki; Robertson and Gross; Vyalov and Kogan; J. C. Williams.

See also section 313A.

A. V. Clark has devised a combined geometric and network analog for transient heat flow. In this analog, electrically conductive paper provides a continuous resistive medium used in conjunction with a network of lumped capacitances. Paschkis (A) has used a similar analog employing an electrolytic tank rather than conductive paper. See also Simmons (A).

Liebmann (C) has developed and applied a finite difference form of this analogy which requires only resistors in the network analog. This form of the analogy involves finite intervals not only of space but also of time. The network analog simulating a uniform homogeneous rod (the finite difference representation of which is shown in figure 311A.2) is illustrated in figure 311A.3. The resistances R_o and R_x are related by the following equation:

$$R_o = \frac{K}{c\rho} \frac{\Delta t}{\overline{\Delta x}^2} R_x \qquad (13)$$

where Δt = finite time interval and other quantities are as defined previously.

The characteristic equations then become

$$\left(\frac{\theta_{m,n+1} + \theta_{m,n-1} - 2\theta_{m,n}}{\overline{\Delta x}^2}\right) = \frac{c\rho}{K} \frac{\theta_{m,n} - \theta_{m-1,n}}{\Delta t} \qquad (14)$$

$$\left(\frac{V_{m,n+1} + V_{m,n-1} - 2V_{m,n}}{\overline{\Delta x}^2}\right) = RC \frac{V_{m,n} - V_{m-1,n}}{\Delta t} \qquad (15)$$

where $\theta_{m,n}$ is the temperature at a distance $x = n\Delta x$ from the boundary $x = 0$ at the time $t = m\Delta t$ since the beginning ($t = 0$).

Voltages representing the initial temperatures at the ends and nodes of the thermal rod are applied to the corresponding ends and nodes of the electric network at the *lower* ends of the resistances R_o. The voltages then at the *upper* ends of the resistances R_o represent the temperatures at the corresponding nodes at the time $t = \Delta t$. These new voltages are then applied to the lower ends of the resistances R_o and the same process is performed repeatedly.

Liebmann (M) has extended this procedure to problems involving heat transfer across a surface, or generation or absorption of heat in the interior of the system.

Newman has used this procedure to determine the temperature distribution within a calorimeter for measuring the heat generation in powdered coal and coke.

See also Karplus (B).

311B *Hydraulic Analogy*

In place of an electric circuit to determine temperature distributions, Moore (D) devised an hydraulic apparatus consisting of transparent standpipes connected at their lower ends by small bore flow tubes which provide resistance to flow. In this analogy, the mass flow of water corresponds to the flow of heat.

Initially, the valve is closed and the tubes are drained, representing

300. HEAT TRANSFER

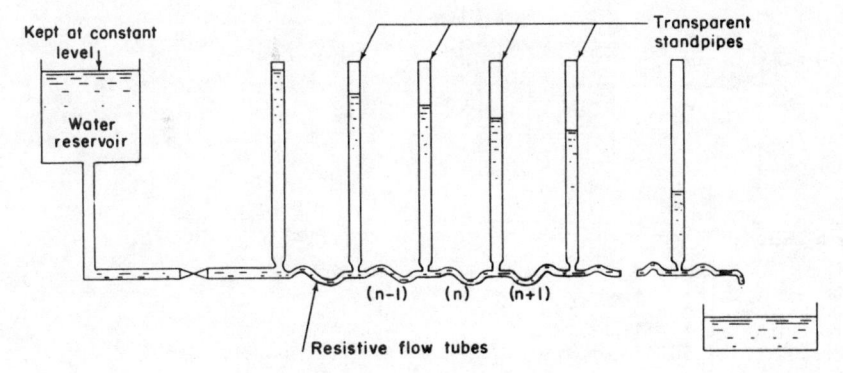

Fig. 311B.1. Hydraulic apparatus.

a constant, low level temperature along a uniform rod (Fig. 311B.1). When the valve is opened, water from the reservoir flows through the tubes and up into the standpipes, the water level in each standpipe (a measure of the fluid pressure at the bottom of the standpipe) representing the temperature at the corresponding position in the uniform rod. The water level of the reservoir is kept constant and is indicated by the level in the first standpipe.

This analogy also may be extended to more than one dimension.

A device similar to Moore's has been developed by Leopold, whose system uses oil rather than water, thereby eliminating some of the serious difficulties encountered in Moore's system. Leopold applied his hydraulic apparatus to finding cooling requirements and air and structure temperature distributions for the cooling of an infinite interior zone in a multi-story building. He considered convection and radiation as well as conduction. His report included plots of heat transfer and temperature vs. time.

Aldrich and Scott have utilized a refined apparatus of the type to solve complex one-dimensional freezing and thawing problems in soil. See also R. F. Scott and the report "Design and Operation of an Hydraulic Analog Computer for Studies of Freezing and Thawing in Soils."

For other reports dealing with this analogy, see Backstrom (A, B); Budrin; Ericksson; "Fluid Analog Studies Air-Conditioning Loads"; Knuth and Kumm; Lukyanov; Molchanov.

311C *Pneumatic Analogy*

A pneumatic analogy similar to the hydraulic analogy in the previous section has been developed by Coyle (A, B). This analogy is applicable to the analysis of the temperature distribution in a material having specific heat and thermal conductivity which vary appreciably with temperature. In this analogy, the mass flow of air corresponds to heat flow.

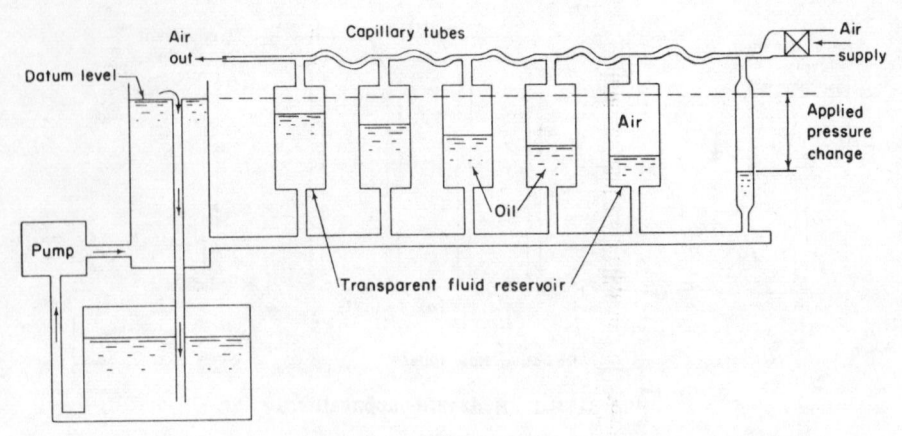

Fig. 311C.1. Pneumatic apparatus.

The valve on the air supply tube is initially closed, and, at the same time, the oil level in each fluid reservoir is the same as the datum level of the overflow tank. When the valve is opened, air flows through the capillary tubes and down into the fluid reservoirs, the level in each reservoir (a measure of the air pressure in the reservoir) representing the temperature at the corresponding position in a uniform rod (Fig. 311A.2). The oil level of the overflow tank is kept constant by the pump system as shown.

Variations of specific heat and thermal conductivity in the uniform rod can be represented in the apparatus by inserting variable-thickness cores into the reservoirs, thereby causing the reservoir cross sectional area to vary with height and thus causing a change in the relationship between the mass of air within a reservoir and the air pressure in the same reservoir.

The analogy may be extended to more than one dimension.

Coyle (B) presents plots of the temperature distribution in a mild steel slab, data for which were obtained by this analogy and by mathematical methods.

312 *Steady State With Sources or Sinks*

312A *Membrane Analogy*

Schneider (B) has described a membrane analogy for the study of steady state, heat conduction fields with uniformly distributed sources or sinks. See also Schneider and Cambel (B).

312B *Hydrodynamical Analogy A*

Moore (A, E) has described the use of fluid flow apparatus, similar to that described in section 221AF, for the simulation of steady state heat flow with sources or sinks. The sources and sinks are simulated by holes in the base place of the apparatus through which fluid enters or leaves the flow space. Distributed sources and sinks are simulated with large holes above which is placed granular material resting on a screen.

312C *Hydrodynamical Analogy B*

Schneider (A) has suggested that the two-dimensional, steady state heat flow in a body with sources or sinks may be simulated by the rotational flow of a nonviscous, incompressible fluid within a container of the same shape as the body.

312D *Electrical Analogy*

Simmons (B) has described an electrically-conducting paper apparatus for simulating two-dimensional heat conduction fields with uniformly distributed sources.

McNall and Janssen (A, B) have described an electrolytic tank analog used in the determination of the temperature field within a transistor. Special features which made possible the simulation of heat flow through several different materials connected at nonisothermal interfaces, controlled distribution of heat generation at a boundary, and interface thermal resistance were discussed.

Gilbert and Gilbert utilized a capacitively coupled, conductive paper analog to study two-dimensional fields with distributed sources.

See also Eckert, Hartnett, Irvine, and Birkebak; Mashovets and Korobov; Simmons (A).

313 *Steady State Without Sources or Sinks*

313A *Electrical Analogy*

For steady state conditions, temperature (and therefore electric potential also) becomes a function of space coordinates only. (See section 311A.)

Characteristic equations:

$$\frac{\partial}{\partial x}\left(K\frac{\partial \theta}{\partial x}\right) + \frac{\partial}{\partial y}\left(K\frac{\partial \theta}{\partial y}\right) = 0 \quad (1) \qquad \frac{\partial}{\partial x}\left(\frac{1}{\sigma}\frac{\partial V}{\partial x}\right) + \frac{\partial}{\partial y}\left(\frac{1}{\sigma}\frac{\partial V}{\partial y}\right) = 0 \quad (2)$$

where symbols are defined as in section 311A.

If thermal conductivity is constant, equations (1) and (2) reduce to the following:

$$\frac{\partial^2 \theta}{\partial x^2} + \frac{\partial^2 \theta}{\partial y^2} = 0 \qquad (3) \qquad \qquad \frac{\partial^2 V}{\partial x^2} + \frac{\partial^2 V}{\partial y^2} = 0 \qquad (4)$$

Boundary conditions are the same as in section 311A.

Possibly the first to apply this analogy were Langmuir, Adams, and Mickle, who used an electrolytic tank to investigate the effect of shape

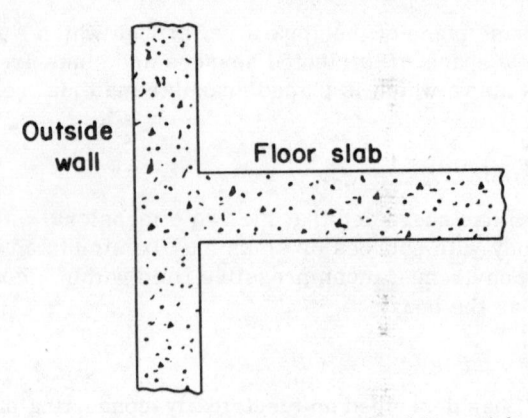

Fig. 313A.1. Outside wall with attached floor slab.

upon heat flow. Finzi-Contini and Beuken (A, B), apparently independently of each other, replaced the continuous electrode described by I. Langmuir, Adams, and Mickle with a number of individual electrodes insulated from each other, according to Paschkis (B). In this manner, a temperature variation with distance along a boundary may be simulated approximately.

Applying this analogy, Kayan (J. O) has used an electrically conducting sheet to determine the temperature distribution in an outside uniformly thick concrete wall and an attached floor slab. Included are diagrams of structures with isotherms determined by analogy for the case of all concrete construction and for the case of insulation within the floor and covering the wall. See also Kayan (A).

Davey and Spooner have described the use of electrically conducting sheets of metal (representing portions of a cross section) to analyze the heat flow through the doors of domestic refrigerators (Fig. 313A.2). The base of the analog model is a 0.0025 inch thick stainless steel sheet. On this sheet are soldered stainless steel conducting wires corresponding with the edges of the structural sheet metal. The insulation material in the wall and the door is represented by a $2(10^{-7})$ inch thick plating of copper. The bare stainless steel plate represents the air film, assumed to be 1/4 inch thick. At the outer boundaries of the plate (inside and outside of the representation of the door and the wall) are soldered 1/4 inch diameter copper bus wires, one for the inside and one for the outside. The electric potential difference corresponding to the temperature difference between the inside and the outside of the refrigerator was applied across these bus wires. The potential at any point was measured by the use of probes and a self balancing potentiometer. Temperature distributions determined by analogy are compared with those directly measured.

Bruckmayer (A) has used an electrolytic tank to investigate the heat flow in cold storage plants.

Pierre applied an electrolytic tank using water as the conducting

300. HEAT TRANSFER

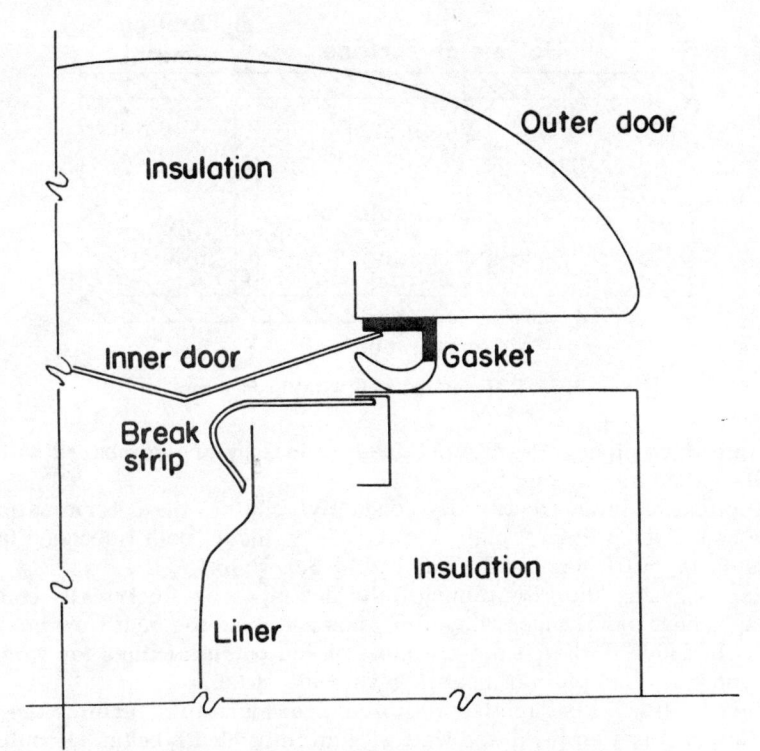

(All lines represent edge views of sheet metal, except for gasket representation and break lines.)

Fig. 313A.2. Cross section of a standard refrigerator door-gap.

medium in determining the influence of frames on the insulation of cold storage chambers on board ships. The effects of the thermal resistances of the bearing structures forming part of the wall of the cold storage chamber, the film between the air in the cold storage chamber and the wall of the chamber, and the film between the ambient fluid and the ship's side were accounted for by computing equivalent thicknesses of insulation for each. Measurements of the flow through the insulation with and without the frames were made and values of the ratios between the corresponding values for the two situations were tabulated and plotted against size of frame for various spacings and shapes of frames. In addition, temperature distributions in the insulation were determined. Diagrams of equipotentials for typical frames are presented. Pierre reported that graphite gave unsatisfactory results when used as the conducting medium for the analog.

Humphreys, Nottage, and Franks and others have described the use of an electrolytic apparatus to determine the heat flow through slabs

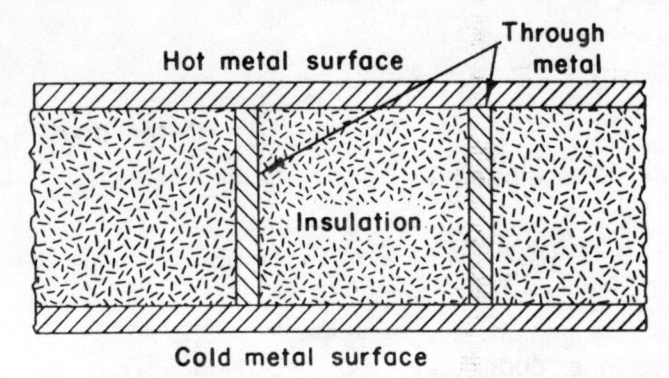

Fig. 313A.3. Insulated wall with through metal.

with imbedded pipes. Results obtained by analogy are compared with results calculated by theory.

Application of an electrically conductive plate to the determination of the heat losses from a plate and from a cylinder, both imbedded in an insulating wall, has been described by Schofield (A, B).

Representing slabs containing imbedded pipes by electrically conducting, sheet metal mesh, Kayan (Q) has analyzed the heat flow in such slabs. Included in the report are plots of equipotential lines for various conditions and plots of heat flow vs. tube spacing.

Kayan (P) has also applied a similar procedure to determine the heat flow through an insulated wall with an imbedded I-beam. Results include plots of equipotential lines.

In another investigation, Kayan (B) applied his electrical analogy technique in predicting the steady state heat transfer and internal temperature distributions for a radiant heat panel with various tube spacings (for both idle and active tubes). Results are in the form of isopotential patterns for the various spacings and curves of panel surface temperature variation and heat flow per unit area for various sections.

An electric circuit modification (Sec. 311A) of this analogy has been utilized by Paschkis and Heisler (C) in an analysis of the heat flow patterns within insulated walls with through metal (Fig. 313A.3).

Mumford and Powell have used electrically conducting sheets of metal to examine the heat flux patterns in fin tubes under radiation in coal fired furnaces.

Bruckmayer (B) has described electrolytic model tests to solve such problems as the determination of heat losses and temperature curves for machinery parts, fittings, and flanges.

Kayan (L) has used both an electrically conducting sheet and an electrical network to analyze the heat flow within a thick two-dimensional corner composed of two different materials of unequal thickness with fluid boundary conductance on each side and with a temperature difference established between the two boundary fluids (Fig. 313A.4). Boundaries were kept at constant potentials. Included in the report are

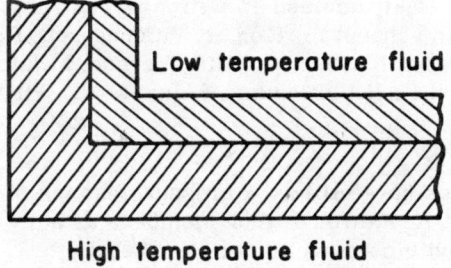

Fig. 313A.4. Thick two dimensional corner.

tabulated values of potential for each junction point of the network as determined by experimental and analytical methods.

An electrically conducting metal sheet was used by Kayan (G) in his investigation of the heat flow within fins (Fig. 313A.5). Results are given in the form of isopotentials on a model diagram and plots of potential ratio vs. distance.

$$\text{Potential ratio} = \frac{\text{potential difference from the given point to the edge}}{\text{over-all potential difference across the section}}$$

For further discussion of this analogy and the analogy of section 311A, see "Analog Points to Optimum Design"; Andreevskii; R. V. Andrews; Awberry and Schofield; Baehr and Schubert; Baer, Schlinger, Berry, and Sage; Baumann; Billington (A, B, C); Blass and Wesser; Brokmeier; Bromberg and Martin; Bruckmayer (B); Bublikov; Dean and Shurley; Ellerbrock, Schum and Nachtigall; Kayan (F, H); Kayan and Gates (A, B); Kettleborough (A); Kostiuk and Sokolov; Kourim; Kozdoba; Landis and Zupnik; Lang and Petrick; Litvinov; Malavard and Miroux; Matsunaga (B); Mattarolo; McCann and Wilts; Moeller (A, B, C);

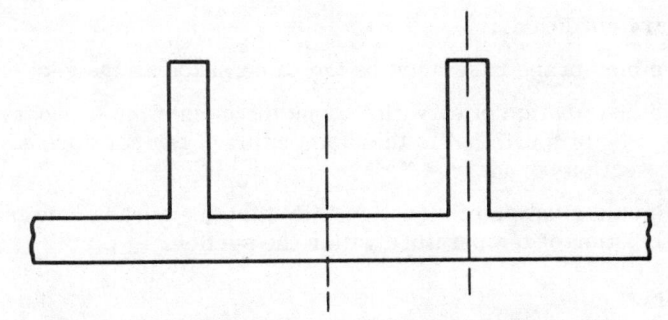

The dashed lines enclose the symmetrical sections represented by the metal sheet.

Fig. 313A.5. Typical fin section.

Mueller; Northrup; Nukiyama and Tanasawara; Paschkis (K, P); Paschkis and Beuken; Ramachandran; Ranger; Reichenbach; Reiniger; Schlitt; R. O. Smith; Torgeson, Kitchar, and Hill; Vernon; Vidal; Volterra; F. Weiner (B); Werner, K. H.; Wiles and Grave; M. L. Williams.

313B *Membrane Analogy*

Wilson and Miles have utilized a pressureless membrane analogy in the analysis of heat flow through such members as hollow circular and square sections, and ribbed sections (Fig. 313B.1).

Hollow circular   Hollow square   Ribbed

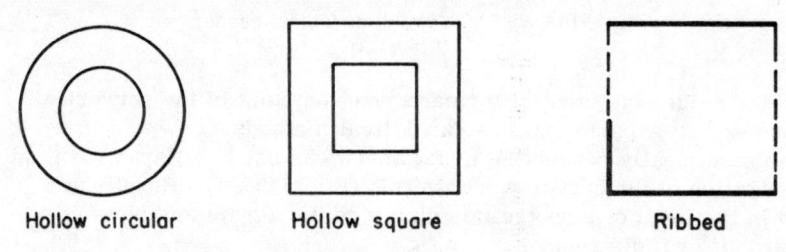

Fig. 313B.1. Typical sections studied in heat flow investigation by membrane analogy.

Characteristic equations:

THERMAL

$$\frac{\partial^2 \theta}{\partial x^2} + \frac{\partial^2 \theta}{\partial y^2} = 0 \quad (1)$$

MEMBRANE

$$\frac{\partial^2 z}{\partial x^2} + \frac{\partial^2 z}{\partial y^2} = 0 \quad (2)$$

where: θ = temperature

x, y = length coordinates of the section

z = membrane elevation

x, y = length coordinates of the membrane base

Boundary conditions:

1. The membrane base must be the same shape as the section.
2. The distribution of elevation along the membrane boundary must be proportional to the distribution of temperature along the section boundary.

Thus the distribution of elevation within the membrane corresponds to the distribution of temperature within the section.

313C *Other*

The analogs of sections 311B and 311C may be used also for steady state problems.

Moore (A, C, E) has suggested the use of an apparatus similar to that described in section 221AF for the study of steady-state heat flow.

Variations in the thickness of the flow space can be used to simulate regions of varying thermal conductivity.

320 CONVECTION

321 *Unsteady Flow*

321A *Electrical Analogy*

Heindlhofer and Larsen have developed an electric circuit analogy for the heat flow in regenerator systems. (Combustion gases flow in one direction over a network of checker brick that absorbs heat, the flow then being reversed, when cold air flows in the opposite direction to reabsorb the heat added to the checker brick in the first half of the cycle, the path of flow being the same in both directions.) Brick cells, as well as quantities of gas and air, are represented by capacitors. Temperature differences are represented by voltage drops across the capacitors. Charge is distributed to each of the "wall" condensers in succession by rotating contacts of the commutator type. The article contains a comparison of actual measurements of temperature with results of analogy experiments.

Developing a method similar to that of Heindlhofer and Larsen, Tipler has described an electric circuit analogy to determine the effectiveness of heat regenerators in gas turbine cycles. The regenerator matrix is represented by a bank of condensers which is scanned first by an initially charged condenser representing the hot gas, in series with a resistance representing the rate of heat transfer between the gas and the matrix, and then in the reverse direction by a second initially charged condenser-resistance unit representing the air "blow."

Ford has described electric circuit analogs for heat exchangers with a view to their application in the solution of automatic control problems relating to the latter.

See also Cima and London; Mozley; Woodcock, Thwaites, and Breckenridge.

321B *Hydraulic Analogy*

Juhasz has investigated crossflow heat exchangers by means of an hydraulic analog. See also Juhasz and Clark.

For similar applications of this analogy, see Mackey and Gay (A, B).

322 *Steady-State*

322A *General*

322AA *Electrical Analogy*

Applying the electrical circuit modification of the analogy of section 311A, Kayan (K) has analyzed the heat flow within fluid-flow heat exchangers (Figs. 322AA.1, 322AA.2).

300. HEAT TRANSFER

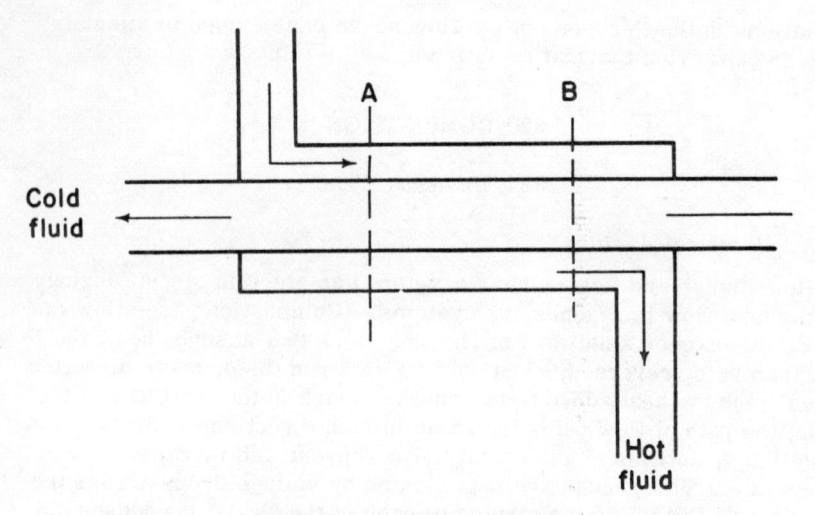

Fig. 322AA.1. Cross section of a counter flow fluid heat exchanger.

The heat-transfer area is divided into a finite number of sections longitudinally. The resistance of each section to the flow of heat is represented by a lumped thermal resistance. In the analog circuit, each thermal resistance r is represented by a corresponding electrical resistance R.

Thermal resistances are defined as follows:

r_{UA} = net heat-transfer resistance for area A of a longitudinal section

r_c = cold-fluid heat-exchange resistance

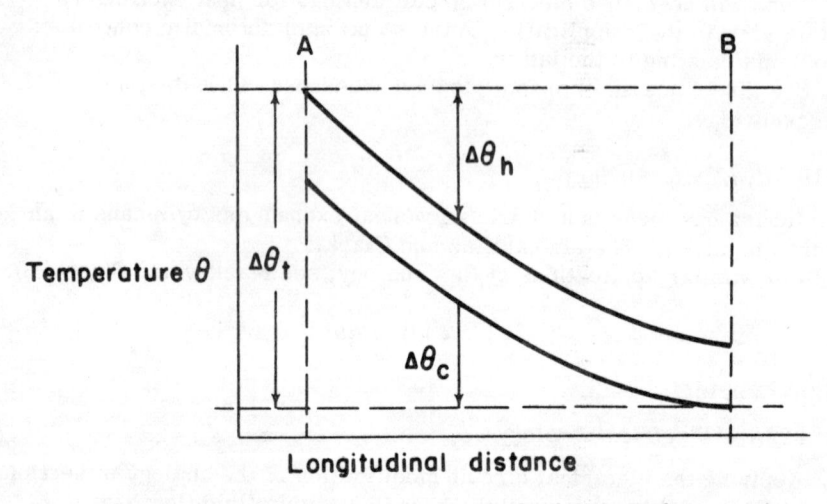

Fig. 322AA.2. Plot of temperature vs. longitudinal distance for a counter flow fluid heat exchanger.

300. HEAT TRANSFER

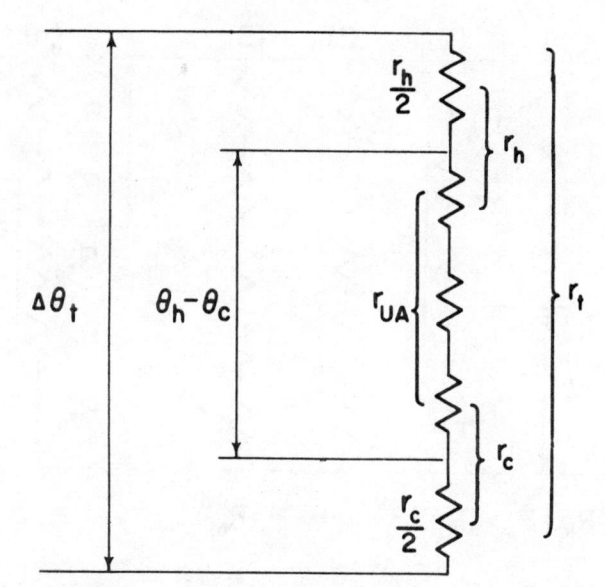

Fig. 322AA.3. Lumped thermal resistance for one section.

r_h = hot-fluid heat-exchange resistance
r_t = total resistance of one section

Ballast resistances, r_x and r_y, are calculated such that the voltages (representing fluid temperatures) of each section coincide with the corresponding voltages of the adjacent sections:

$$R_{y_1} = 0 \qquad R_{x_n} = 0$$
$$R_{y_2} = R_h (P) \qquad R_{x_{n-1}} = R_c (Q)$$
$$R_{y_3} = R_h (P + P^2) \qquad R_{x_{n-2}} = R_c (Q + Q^2)$$

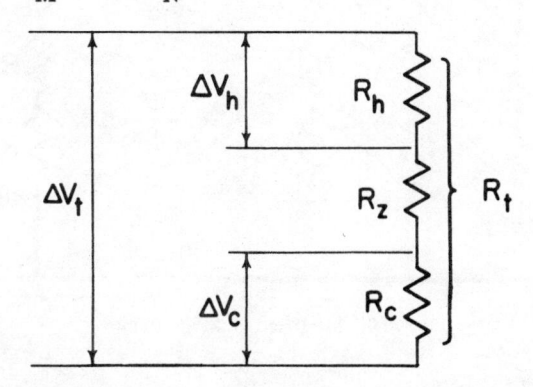

where: $P = \dfrac{N}{M}$; $Q = \dfrac{M}{N}$; $M = r_t - r_h$; $N = r_t - r_c$

Fig. 322AA.4. Equivalent electrical resistance for one section.

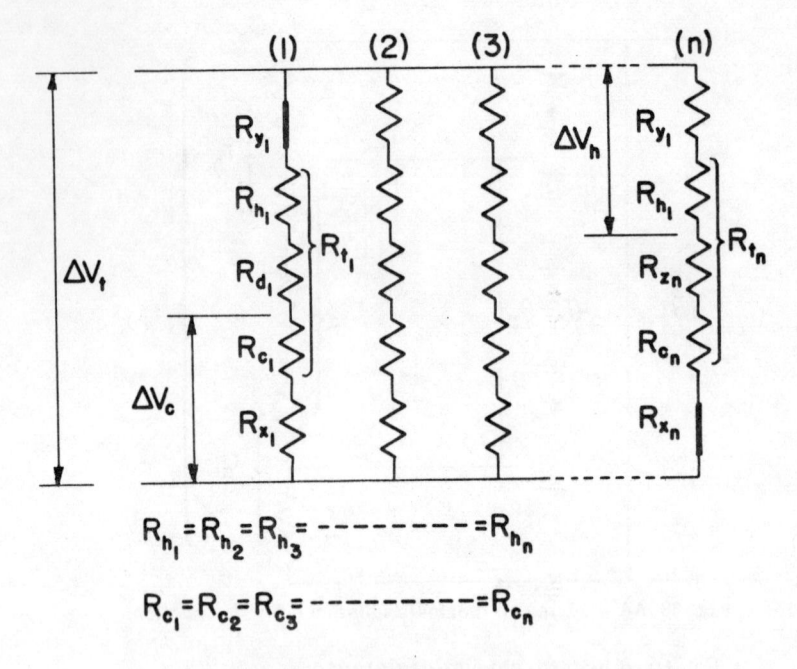

Fig. 322AA.5. Analog circuit of n sections.

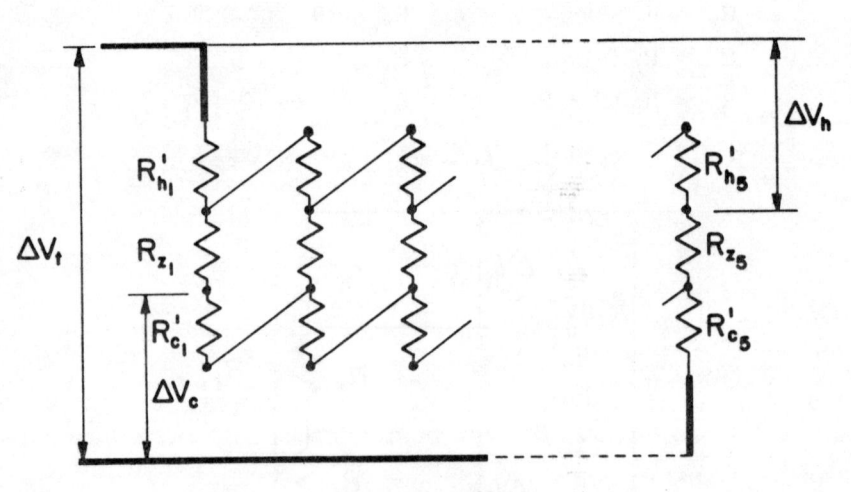

Fig. 322AA.6. Modified analog circuit.

300. HEAT TRANSFER

The resistances of the modified circuit are calculated as follows:

$$R'_{c_1} = R_c \qquad\qquad R'_{h_n} = R_h$$

$$R'_{c_2} = \frac{R_c}{1 + P} \qquad\qquad R'_{h_{n-1}} = \frac{R_h}{(1 + Q)}$$

$$R'_{c_3} = \frac{R_c}{(1 + P + P^2)} \qquad\qquad R'_{h_{n-2}} = \frac{R_h}{(1 + Q + Q^2)}$$

Analytical results are compared with results by analogy for a typical case involving a counter current heat exchanger.

Neel (A) has used a similar analogy in the design of air heated, ice prevention systems for aircraft.

See also Kayan (C, D, E, J, M, N); Weissenbach.

322AB *Hydraulic Analogy*

For discussions of the utilization of hydraulic analogs in solving steady state problems involving fluid flow heat exchangers, see Backstrom and others; Juhasz and Hooper (A, B, C); Prochazka, Landau, and Standart.

322B *Turbulent Stream Flowing Between Parallel Walls*

Schlinger, Berry, Mason, and Sage have described an electric circuit analogy for predicting the temperature distribution within a turbulent stream flowing between two parallel walls (Fig. 322B.1).

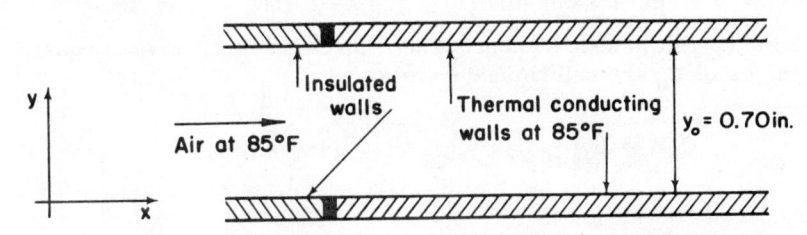

Fig. 322B.1. Turbulent stream flowing between two parallel walls.

If the thickness of the channel is divided into N parts, the temperature distribution may be expressed by a finite difference equation.

Characteristic equations:

THERMAL ELECTRICAL

$$(\epsilon_c + k)_n (u_x)_n \frac{\partial \theta}{\partial x} = \frac{\theta_{n-1} - 2\theta_n + \theta_{n+1}}{\Delta\phi^2} \quad (1) \qquad C_n \frac{\partial V}{\partial t} = \frac{V_{n-1} - 2V_n + V_{n+1}}{\Delta R^2} \quad (2)$$

where: θ = temperature $\qquad$ V = voltage

$\quad$ x = longitudinal distance $\qquad$ t = time

300. HEAT TRANSFER

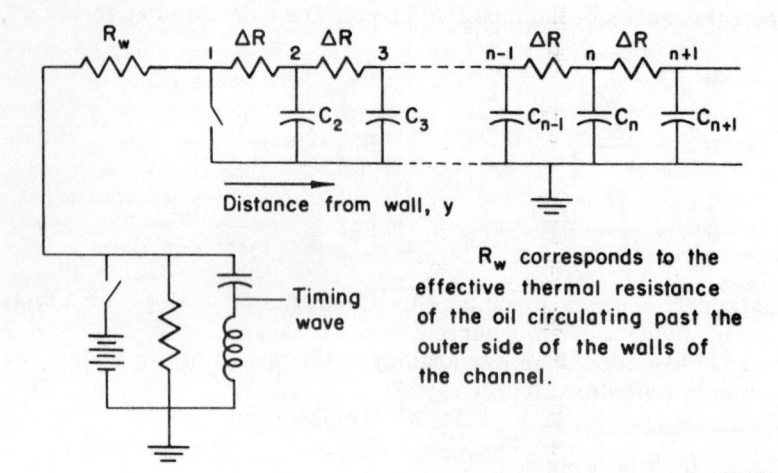

Fig. 322B.2. Analog circuit.

R_w corresponds to the effective thermal resistance of the oil circulating past the outer side of the walls of the channel.

ϵ_c = eddy conductivity
k = thermometric conductivity
u_x = velocity in the direction of flow

C = capacitance

$\phi = \int_0^y \dfrac{1}{(\epsilon_c + k)}\,dy$

R = resistance

y = transverse distance

and *n* indicates the node number. The *N* increments $\Delta\phi$ are all equal. Values of C_n are determined as follows:

$$C_n R = \beta \left[\dfrac{1}{N} \int_0^{y_0} \dfrac{1}{\epsilon_c + k}\,dy \right]^2 \left[(\epsilon_c + k)_n (u_x)_n \right]$$

where β is a conversion factor.

Values of voltage corresponding to values of temperature at a given transverse node are recorded on an oscilloscope, the time axis corresponding to the longitudinal distance. The temperature vs. time curve is accompanied by a timing wave adjusted so that one cycle corresponds to 30 inches along the channel. Included in the report are curves of transverse temperature distribution at three downstream positions from data obtained by analogy and by direct experiment.

322C *Heat Transfer and Fluid Friction - Heat and Momentum Transfer Analogy*

The analogy between heat transfer and fluid friction is primarily analytical. Such analogous characteristics were first proposed by Reynolds. Prandtl (B), Taylor (A), and von Karman (A) have attempted

300. HEAT TRANSFER

to deduce the form of the function ϕ [Nu = ϕ (Re, Pr)] analytically for turbulent flow by comparing the exchange of momentum and heat.

Originally, Reynolds dealt with a purely turbulent system in which exact similarity between temperature and velocity fields was postulated (i.e., Pr = 1). Prandtl (B) and Taylor (A) extended the analogy to other values of the Prandtl modulus Pr by introducing a concept which included a laminar sublayer and a turbulent core. Th. von Karman (A) defined a buffer layer (of particular characteristics) which was located between the laminar layer and the turbulent core.

In von Karman's ideal system, the resistance to heat transfer from a solid boundary to a fluid consists of the following:

1. A laminar sublayer in which viscous forces predominate. The heat is transferred through this sublayer by conduction only.
2. A "buffer" in which both viscous and eddy forces are important and through which heat is transferred by both thermal conduction and eddy diffusion.
3. A core of fluid in which eddy forces predominate. The heat is transferred in the core by eddy motion only.

The rate at which heat is transferred per unit area and the unit shear at any point in the system may be expressed as follows:

	UNIT SHEAR	RATE OF HEAT TRANSFER PER UNIT AREA
Laminar sublayer	$\dfrac{\tau}{\rho} = \nu \dfrac{du}{dy}$	$\dfrac{q}{A} = -K \dfrac{dt}{dy}$
Buffer layer	$\dfrac{\tau}{\rho} = (\nu+\epsilon) \dfrac{du}{dy}$	$\dfrac{q}{AC_p \gamma} = -(\dfrac{\nu}{Pr} + \epsilon) \dfrac{dt}{dy}$
Turbulent core	$\dfrac{\tau}{\rho} = \epsilon \dfrac{du}{dy}$	$\dfrac{q}{AC_p \gamma} = -\epsilon \dfrac{dt}{dy}$

Where: Nu = Nusselt's modulus
Pr = Prandtl's modulus = $\dfrac{\mu C_p g}{K}$
Re = Reynolds' modulus
μ = viscosity of fluid
C_p = unit heat capacity of fluid
K = thermal conductivity of fluid
g = gravity
τ = unit shear at any point, y
ρ = density of fluid
ν = kinematic viscosity of fluid
u = average axial velocity of flow at any point, y
q = radial rate of heat transfer

A = area perpendicular to heat flow
t = temperature at any point, y
y = distance from wall
ϵ = eddy diffusivity
γ = weight density of fluid

Colburn made a comparison by analytical development and experimental data of several men, of forced convection heat transfer (for flow parallel to plane surfaces and for fully turbulent flow inside tubes) and fluid friction. He also indicated a correlation in the transition region between streamline and turbulent flow in tubes from the similarity in the curve "dips."

For further explanation and application of this analogy, see Bedingfield and Drew; Boelter, Martinelli, and Johassen; Callaghan; Garner, Jensen, and Keey; Jenkins, Brough, and Sage; Knudsen and Katz; Martinelli, R. C. (A, B); McAdams; Metzner and Friend; Mizushina; Seban.

330 RADIATION

Mackey and Gay (A, B) have used an hydraulic analog to study the relation between cooling load and the instantaneous rate of heat gain from solar radiation.

See also Adrianov; Oppenheim (A); Woodcock, Thwaites, and Breckenridge.

Section 400. Mechanics of Materials (Steady-State)

410 TENSION AND COMPRESSION

411 *Membrane Analogy*

McGivern and Supper have used the method of Weibel (membrane analogy in conjunction with photoelasticity) (see section 452) to analyze the stresses in a simple phenolite tension member with a circular hole equal to one-half of its width centrally located in it. Their report includes stress distribution diagrams, three-dimensional drawings of the membrane surface and two-dimensional diagrams of the isoclinic lines and the stress trajectories.

412 *Hydrodynamical Analogy*

The irrotational flow of an ideal incompressible fluid has been used by J. Smith to analyze the stress distributions in two-dimensional tension members with fixed edges (effecting the condition of irrotational strain).

Characteristic equations:

STRESS

$$\frac{\partial^2 \phi}{\partial x^2} + \frac{\partial^2 \phi}{\partial y^2} = 0 \quad (1)$$

FLUID

$$\frac{\partial^2 \psi}{\partial x^2} + \frac{\partial^2 \psi}{\partial y^2} = 0 \quad (2)$$

where: ϕ = potential function defined by the following equations:

$$e_x = \frac{\partial \phi}{\partial y} \quad (3)$$

$$e_y = -\frac{\partial \phi}{\partial x} \quad (4)$$

ψ = stream function defined by the following equations:

$$u = \frac{\partial \psi}{\partial y} \quad (5)$$

$$v = -\frac{\partial \psi}{\partial x} \quad (6)$$

where: x, y = coordinates   x, y = coordinates
e_x, e_y = x, y components   u, v = x, y components of
      of strain   fluid velocity

Boundary conditions:

1. The flow channel must have the same shape as the tension member.
2. There is no stress normal to the boundary of the tension member (condition of fixed edges). This is equivalent to the condition of no fluid flow across the boundaries.

The streamlines for perfect flow are then analogous to the lines of strain in the member and are therefore analogous to the lines of longitudinal stress in a similarly shaped member with either fixed or free edges. Lines of transverse stress are orthogonal to the lines of longitudinal stress. In general, the longitudinal stress is directly proportional to the transverse concentration of lines.

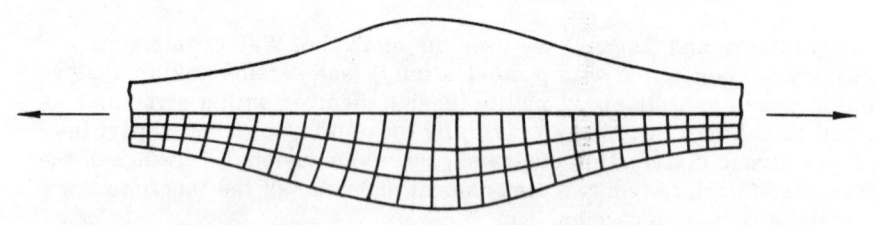

Fig. 412.1. Stress lines for a symmetrical half of a typical tension member.

The author has described the use of the Hele-Shaw apparatus (see section 221AF) in determining lines of fluid flow and thus lines of longitudinal stress in structural members of ships.

413 *Electrical Analogy*

Theocaris (B, C) has used a conductive paper analog to determine stress trajectories (isostatics and isopachics) for members in tension and compression.

420 FLEXURE

421 *Columns*

421A *Electrical Analogy*

Swenson (A) has presented an electric circuit analogy for the determination of the lateral deflections of non-uniform beams or beam columns and for determining the critical loads of non-uniform columns.

400. MECHANICS OF MATERIALS (STEADY-STATE)

For a slender beam-column:

$$\frac{d^2y}{dx^2} + \left(\frac{P}{EI}\right)y = \frac{M}{EI} \tag{1}$$

where: y = lateral deflection
x = axial coordinate
P = axial load
E = modulus of elasticity
I = moment of inertia of the cross section (a function of x)
M = bending moment (a function of x)

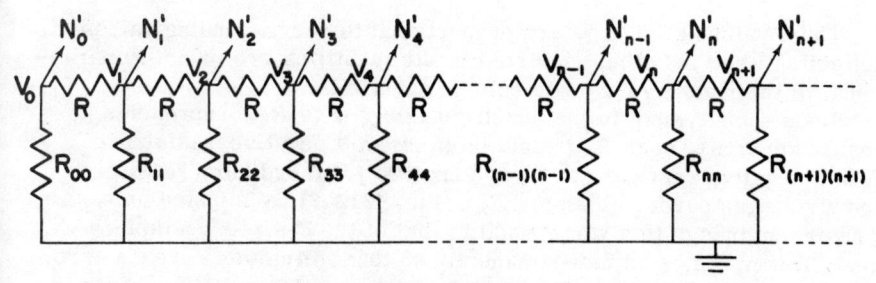

Fig. 421A.1. Analog circuit for a beam column.

This equation may be approximated by a finite difference equation which is developed by dividing the beam into equal intervals and numbering the stations.

Characteristic equations:

BEAM COLUMN ELECTRIC CIRCUIT (Fig. 421A.1)

$$y_{n-1} + y_{n+1} + y_n\left[\frac{h^2 P}{(EI)_n} - 2\right] = h^2\left(\frac{M}{EI}\right)_n \qquad V_{n-1} + V_{n+1} + V_n\left[-\frac{R}{R_{nn}} - 2\right] = i_n R$$

(2) (3)

where: V = voltage
R = resistance between adjacent stations
R_{nn} = resistance from station n to ground
i_n = isolated current leaving station n
n = station number
h = distance between adjacent stations

400. MECHANICS OF MATERIALS (STEADY-STATE)

The analogous quantities are as follows:

BEAM COLUMN	ELECTRIC CIRCUIT
y_n	V_n
$\left(\dfrac{M}{EI}\right)_n$	i_n
h^2	R
$\dfrac{(EI)_n}{P}$ (except at n = 0 where it is assumed y = 0)	R_{nn}
$\dfrac{h^2}{1 - h^2 P/(EI)_0}$	R_{00}

Thus, values of voltage are proportional to corresponding values of deflection when the other electric circuit quantities are made proportional to their analogous beam column quantities.

Since each symbol in the quantity analogous to R_{nn} represents a positive quantity, each R_{nn} must be in effect a negative resistance. Such a negative resistor can be synthesized from ordinary resistors and a voltage source. Resistor R_n (Fig. 421A.2) is adjusted on a Wheatstone bridge to a value equal to that of the desired resistance R_{nn}. If now, R_S is adjusted manually so that voltmeter V reads zero, a current V_n/R_n flows in R_n toward station n. Accordingly, when connected to the analyzer, this circuit comprises the desired negative resistor.

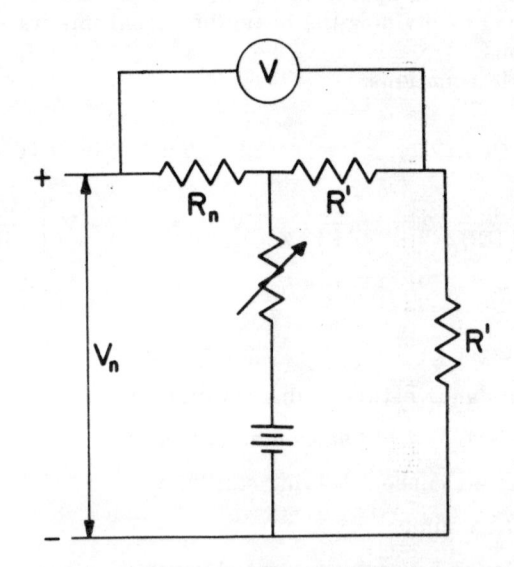

Fig. 421A.2. Negative resistor for beam column circuit.

To set up the circuit, the desired value of h is chosen, and values of R_{nn} are calculated. The node-to-ground resistors are adjusted on a Wheatstone bridge accordingly. Then each potentiometer R_S is adjusted (in turn) until its corresponding voltage V becomes zero. This procedure is repeated until all the V's in the network are zero simultaneously. The iterative process involved in this method converges for all axial loads less than, and also for a small range just above, critical load.

When the circuit is set up to represent a slender column with no lateral loads (the value of P being arbitrary), and when the negative resistors are systematically adjusted, a trivial solution is obtained. The remedy is to apply a battery between some one node and ground to force the network to assume a solution at that point conforming to the applied voltage. When the adjustment process is completed, there is a current in the battery circuit. This current represents the lateral load at that point required to cause a deflection corresponding to the battery voltage. This is true for P either less than or greater than P_{cr} (critical load); the analyzer does not concern itself with the fact that the deflection is unstable. The current is positive (toward the node) for P less than P_{cr} and negative for P greater than P_{cr}. For P equal to P_{cr}, a finite deflection requires no lateral force, so that the battery current is zero. Thus, the critical load is determined by graphing the current required to produce a certain node-to-ground voltage at an arbitrary station in the network and by interpolating to the value of P which corresponds to zero current.

Comparisons of typical values of critical loads determined by analogy and by analytical methods are included in Swenson's report.

421B *Other Electrical Analogies*

For the description of other electrical analogies for the study of columns, see Berry; Chegolin (A); Culver (A); Scanlan (A, B); Shields and MacNeal.

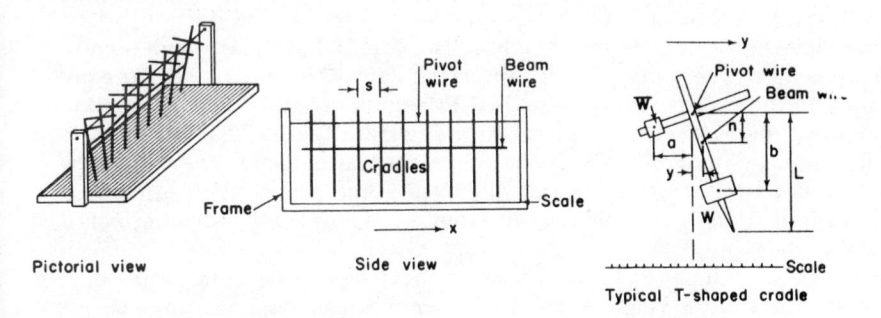

Fig. 422B.1. Analog apparatus for beam on elastic foundation.

See also section 422D.

400. MECHANICS OF MATERIALS (STEADY-STATE)

421C *Pendulum Analogy*

Glenn Murphy (C) has suggested the use of measurements of the angular displacement of a simple pendulum oscillating through small angles in the analysis of the lateral deflection of a uniform rod (pinned at each end) subjected to an axial load resulting in small displacements.

Characteristic equations:

ROD

$$\frac{d^2 y}{dx^2} + \left(\frac{P}{EI}\right) y = 0 \qquad (1)$$

PENDULUM

$$\frac{d^2 \theta}{dt^2} + \left(\frac{g}{L}\right) \theta = 0 \qquad (2)$$

where: y = lateral deflection
x = distance along the rod
P = axial load
E = modulus of elasticity
I = moment of inertia of the cross section

θ = angular displacement
t = time
g = acceleration of gravity
L = length of the pendulum

Thus, if the length of the pendulum is made proportional to EI/P, the variation in angular displacement of the pendulum with time during one oscillation (starting at the bottom of a swing) will correspond to the variation in lateral deflection of the rod with distance along the rod.

422 Beams

422A *The "Column" Analogy*

Cross developed an analytical analogy for determining the moment diagrams of statically indeterminate beams.

The beam to be analyzed is represented as a composite of a statically determinate beam and a correction beam. At the ends of the correction beam shears and moments are applied which, when combined with the reactions and loads of the statically determinate beam, will produce the same deformation as in the original statically indeterminate beam. The moment diagram of the correction beam will be equivalent to the stress distribution in a thin column loaded according to the moment diagram of the statically determinate beam. The two moment diagrams are combined to form the composite moment diagram of the statically indeterminate beam. Included in the report are examples of the application of this analogy to simple cases involving beams, bents, and arches.

Fok and Au have discussed the applications of the column analogy to the analysis of single span rigid frames whose supports may be fixed or hinged.

See also Lyle; W. Wright.

400. MECHANICS OF MATERIALS (STEADY-STATE)

422B *Mechanical Analogy*

A mechanical device to predict transverse deflections of beams resting on elastic foundations has been developed by Wells.
Characteristic equations:

BEAM

$$\frac{d^4y}{dx^4} - \left[f(x)\right]y + g(x) = 0 \quad (1)$$

where:
- x = axial coordinate
- y = transverse deflection of the beam
- f(x) = function of x proportional to the stiffness of the foundation
- g(x) = function of x proportional to the load distribution on the beam

MECHANICAL

$$\left(\frac{\pi E r^4}{4}\right)\frac{d^4y}{dx^4} - \left(\frac{W_x b_x}{sn^2}\right)y + \frac{w_x a_x}{sn} = 0 \quad (2)$$

- x = axial coordinate
- y = transverse deflection of the beam wire
- W_x = adjustable weights on the cradle legs as indicated
- b_x = distances of these weights from the pivot wire as indicated
- w_x = adjustable weights on the cradle arms as indicated
- a_x = distances of these weights from the pivot wire as indicated
- E = modulus of elasticity
- r = radius of the beam wire
- s = distance between cradles
- n = distance from the pivot wire to the beam wire

A series of the T-shaped cradles (hereafter designated as *tees*) is suspended from the pivot wire, stretched tightly between the ends of the frame, so that the tees are free to rotate about the pivot wire. Below the point of suspension of each wire is a hole through which passes the wire (designated as the beam wire) representing the beam. When the weights on the tees are adjusted so as to simulate the distributions of *f(x)* and *g(x)* in the actual beam, the horizontal displacement of the beam wire from the neutral position (directly below the pivot wire) at any point along the wire represents (approximately) the transverse deflection of the actual beam. The displacement of the wire is magnified and measured by measuring the horizontal displacement of the pointed ends of the legs of the tees, as indicated on the scale inscribed on the top surface of the frame base.

400. MECHANICS OF MATERIALS (STEADY-STATE)

The four types of boundary conditions are applied as follows:

1. For y the end tee is locked in the position giving the required deflection.

2. For $\frac{dy}{dx}$, two tees are locked together at a fixed distance apart to give the required slope to the wire. The tees are counterbalanced by weights above the pivot wire, so that no shear is applied to the end of the beam wire.

3. For $\frac{d^2y}{dx^2}$, a pure bending moment is applied by means of two adjacent tees loaded on opposite horizontal arms, and counterbalanced as for condition 2.

4. For $\frac{d^3y}{dx^3}$, a given shear force may be applied by means of a weight on the horizontal arm of any one tee, again counterbalanced as for condition 2.

Well's report includes experimental and comparable analytical results for a long beam of uniform stiffness to the center of which a concentrated force is applied normally.

422C *The "Moment Area" or "Conjugate Beam" Analogy*

The moment area principle of determining lateral deflections of beams was devised independently by Mohr (B) and Greene. Mohr's development of the principle was limited to simple beams, and Greene's development was applicable primarily to cantilever beams. "Later Mueller-Breslau (A, B) extended Mohr's original moment area principle in such a way that it became directly applicable to beams with any type of supports, and also to trusses. The extended method, applied to beams, includes as special cases both Mohr's and Greene's original principles. One of its main features is the use of an additional beam in which the bending moments are equal to or proportional to the deflections of the given beam. This beam, introduced by Mueller-Breslau, will here be called the 'conjugate beam', and, accordingly, we shall call the extended moment-area method the 'conjugate-beam method'." Westergaard (A).

Characteristic equations:

GIVEN BEAM	CONJUGATE BEAM
$\frac{d^2y}{dx^2} = \frac{M}{EI}$ (1)	$\frac{d^2M}{dx^2} = w$ (2)
where: y = lateral deflection	M = bending moment
x = longitudinal coordinate	x = longitudinal coordinate

400. MECHANICS OF MATERIALS (STEADY-STATE)

M = bending moment (a function of x)
E = modulus of elasticity
I = moment of inertia of the cross section (a function of x)

w = load distribution (a function of x)

Boundary conditions:

"At points where the deflection of the given beam is zero, the moment in the conjugate beam must be made zero, and where the slope of the given beam is zero, the shear in the conjugate beam must be made zero." Westergaard (A).

Thus, if the imaginary conjugate beam is considered as being loaded according to the $\frac{M}{EI}$ diagram of the given beam, then the deflections of the given beam correspond to the calculated bending moments of the conjugate beam. In addition, the slopes in the given beam correspond to the shears in the conjugate beam.

For Glenn Murphy's (A) extension of this analogy to plane frames, see section 472BA.

See also Mohr (A); Boyd.

422D Electrical Analogies

Criner and McCann have developed an electric circuit analog for the analysis of beams on elastic foundations that are subject to traveling loads. This analog was used to investigate the behavior of rails carrying relatively high carriage loads at very high speeds.

For further descriptions of electrical analogies for the analysis of beams, see Amenzade; Bluhm and Flanagan; Leicester; Malavard and Boscher; McCann and MacNeal; McCann (II, General References); Riesz and Swain; Roots, Scanlan (A, B); Shields and MacNeal; Trudsø; Uhrig.

See also sections 421A and 421B.

430 TRANSVERSE LOADS

431 Membrane Analogy A

Timoshenko has given two membrane analogies for the determination of the shearing stress distribution on a cross section of a uniform cantilever beam with a concentrated transverse load acting through the centroid of the cross section at the free end of the beam. The governing equation is the following:

$$\frac{\partial^2 \phi}{\partial x^2} + \frac{\partial^2 \phi}{\partial y^2} = \frac{\nu}{1+\nu} \frac{Py}{I} - \frac{df(y)}{dy} \tag{1}$$

400. MECHANICS OF MATERIALS (STEADY-STATE)

where: P = magnitude of the concentrated load

x = coordinate parallel to the load

y = transverse coordinate perpendicular to the load

I = moment of inertia of the cross section

ν = Poisson's ratio

f(y) = arbitrary function of y

ϕ = stress function defined by the following equations:

$$S_{xz} = \frac{\partial \phi}{\partial y} - \frac{Px^2}{2I} + f(y) \quad (2) \qquad S_{yz} = -\frac{\partial \phi}{\partial x} \quad (3)$$

where: z = longitudinal coordinate of the beam

S_{xz} = shearing stress acting on the cross section in the x-direction

S_{yz} = shearing stress acting on the cross section in the y-direction

At the boundary of a cross section, the following equation must be satisfied:

$$\frac{\partial \phi}{\partial s} = \left[\frac{Px^2}{2I} - f(y) \right] \frac{dy}{ds} \quad (4)$$

where: s = distance along the boundary.

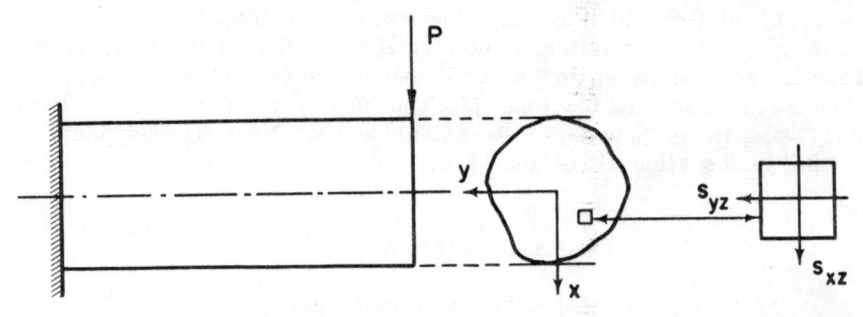

Fig. 431.1. Cantilever beam with applied load and resultant stress components.

The first of the two analogies involves a differential pressure membrane, ϕ being taken as constant along the boundary. Therefore, at the boundary $\frac{\partial \phi}{\partial s} = \left[\frac{Px^2}{2I} - f(y) \right] \frac{dy}{ds} = 0$. This equation can be satisfied if $\frac{Px^2}{2I} - f(y) = 0$. From this, the value of $f(y)$ can be determined by expressing the equation representing the cross-section boundary in

400. MECHANICS OF MATERIALS (STEADY-STATE)

terms of x^2. The corresponding value of $\dfrac{df(y)}{dy}$ is substituted into equation (1), whose right side may be designated as $g(y)$. Timoshenko has determined $f(y)$ and $g(y)$ for several shapes of cross sections.

Characteristic equations:

STRESS   MEMBRANE

$$\frac{\partial^2 \phi}{\partial x^2} + \frac{\partial^2 \phi}{\partial y^2} = g(y) \quad (5) \qquad \frac{\partial^2 z}{\partial x^2} + \frac{\partial^2 z}{\partial y^2} = p(y) \quad (6)$$

where: z = transverse displacement of the membrane

$p(y)$ = pressure differential (a function of y) between membrane surfaces

Boundary conditions:

1. The boundary of the membrane has the same shape as the boundary of the cross section.
2. At the boundary, ϕ = constant, therefore z = constant at the boundary.

Thus, the distribution of the stress function within the cross section may be determined by measurements of membrane displacement. Values of S_{xz} and S_{yz} can then be found by plotting ϕ along lines of constant x and y and measuring slopes to these curves.

According to Hetenyi (A), Cushman has made application of this analogy.

See also Cadambe and Tewari.

432 *Membrane Analogy B*

The second analogy for determining shearing stresses in a cantilever beam involves a pressureless membrane and was developed by Vening-Meinesz, according to Hetenyi (A).

In this analogy, $f(y)$ is taken equal to $\dfrac{\nu}{2(1+\nu)} \dfrac{Py^2}{I}$.

Characteristic equations:

STRESS   MEMBRANE

$$\frac{\partial^2 \phi}{\partial x^2} + \frac{\partial^2 \phi}{\partial y^2} = 0 \quad (1) \qquad \frac{\partial^2 z}{\partial x^2} + \frac{\partial^2 z}{\partial y^2} = 0 \quad (2)$$

where symbols are defined as in the previous analogy.

From equation (4) of the previous section,

$$\frac{\partial \phi}{\partial s} = \left[\frac{Px^2}{2I} - \frac{\nu}{2(1+\nu)} \frac{Py^2}{I} \right] \frac{dy}{ds}$$

400. MECHANICS OF MATERIALS (STEADY-STATE)

along the boundary of the cross section. Integrating this expression along the boundary produces $\phi = \frac{P}{I} \int \frac{x^2 dy}{2} - \frac{\nu}{2(1+\nu)} \frac{Py^3}{3I} + \text{constant}$.

Boundary conditions:

1. The boundary of the membrane has the same shape when projected onto the x, y plane as the shape of the boundary of the beam cross section.

2. The distribution of z along the boundary must be proportional to $\frac{P}{I} \int \frac{x^2 dy}{2} - \frac{\nu}{2(1+\nu)} \frac{Py^3}{3I} + \text{constant}$.

Timoshenko and Goodier have described a method, developed by Griffith and Taylor (A, C), of constructing such membrane boundaries. A hole is cut in a plate of celluloid, of such a shape that after the plate is bent the projection of the edge of the hole on the x, y plane has the same shape as the boundary of the cross section of the beam. The plate is fixed on vertical studs and adjusted by means of nuts and washers until the ordinates along the edge of the hole represent to a certain scale the values of ϕ along the boundary (Fig. 432.1). Another method developed by Cushman uses thin sheets of annealed brass.

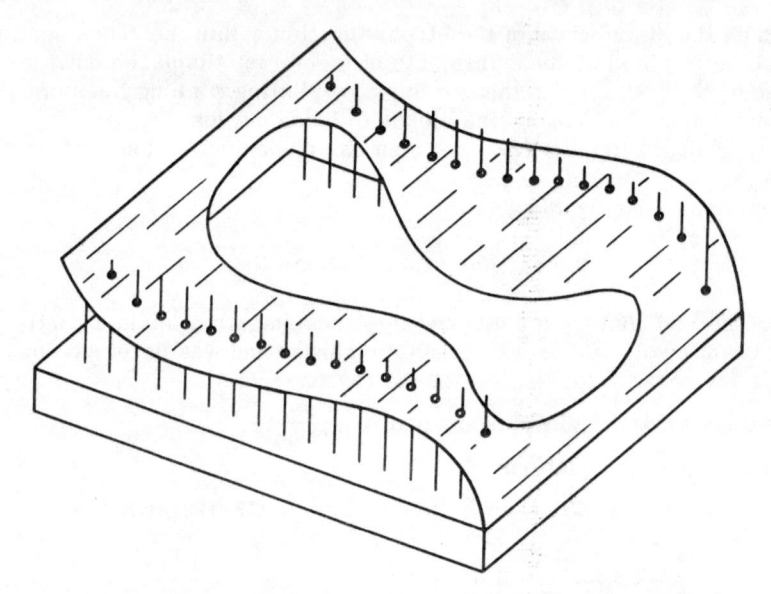

Fig. 432.1. A method of constructing membrane boundary ordinates.

Application of this analogy to circular and square sections was made by Cushman according to Hetenyi (A).

400. MECHANICS OF MATERIALS (STEADY-STATE)

433 *Electrical Analogy*

Hetenyi (A) suggests the use of a thin, constant thickness, electrically resistive plate (whose boundary is subject to an electric potential distribution) for determining the shearing stress distribution in a uniform cantilever beam such as in section 431. The effect could also be produced by using a shallow tank of electrolytic liquid instead of a plate. The stress equations of this analogy are the same as those in section 432.

Characteristic equations:

STRESS ELECTRICAL

$$\frac{\partial^2 \phi}{\partial x^2} + \frac{\partial^2 \phi}{\partial y^2} = 0 \qquad (1) \qquad \frac{\partial^2 V}{\partial x^2} + \frac{\partial^2 V}{\partial y^2} = 0 \qquad (2)$$

where: V = electric potential.

Boundary conditions:

1. The boundary of the plate or tank is the same shape as the boundary of the beam cross section.

2. Along the boundary, $\phi = \frac{P}{I} \int \frac{x^2 dy}{2} - \frac{\nu}{2(1+\nu)} \frac{Py^3}{3I} +$ constant. The distribution of V along the boundary must be proportional to this. For methods of effecting a variation of electric potential along a boundary, see sections 221AAA (Reinius and also Childs), 221AAB (Babbitt and Caldwell), 221B, and 441AD.

According to Hetenyi (A), Negoro (C) has applied an electrolytic tank in determining such shearing stresses in an equilateral triangular section, a circular ring, and an airfoil section. Negoro (A, B) has also applied the analogy to straight members with axial holes.

For an extension of the analogy, see Russell and MacNeal.

440 TORSIONAL LOADS

441 *Elastic*

441A *Uniform Bars*

441AA *Membrane Analogy A*

The "membrane analogy for torsion" was originated by Prandtl (C, D, E) who suggested using soap film membranes to analyze the shearing stress distributions on cross sections of uniform bars subject to elastic torsion.

400. MECHANICS OF MATERIALS (STEADY-STATE)

Characteristic equations:

STRESS

$$\frac{\partial^2 \phi}{\partial x^2} + \frac{\partial^2 \phi}{\partial y^2} = -\frac{2G\theta}{L} \quad (1)$$

where: G = modulus of rigidity of the bar material

$\frac{\theta}{L}$ = angle of twist per unit length of bar

x, y = coordinates of the bar cross section

ϕ = stress-potential function defined by the following equations:

$$S_{xz} = \frac{\partial \phi}{\partial y} \quad (3)$$

$$S_{yz} = -\frac{\partial \phi}{\partial x} \quad (4)$$

MEMBRANE

$$\frac{\partial^2 z}{\partial x^2} + \frac{\partial^2 z}{\partial y^2} = -\frac{p}{T} \quad (2)$$

p = pressure differential between membrane surfaces

T = surface tension in the membrane

x, y = coordinates of the membrane base

z = transverse displacement of the membrane

where: S_{xz}, S_{yz} = x, y components of shearing stress on a cross section.

Boundary conditions:

1. The boundary of the membrane is the same shape as the boundary of the bar cross section.

2. Along the boundary $\frac{d\phi}{dn} = 0$ and thus ϕ = constant (where n = distance along the boundary). This means that z = constant along the boundary of the membrane.

Thus the distribution of the stress function within the cross section may be determined by measurements of membrane displacement. Values of S_{xz} and S_{yz} can then be found by plotting ϕ along lines of constant x and y and measuring slopes to these curves.

Additional similar equations:

$$M = 2 \int \phi \, dA \quad (5) \qquad V = \int z \, dA \quad (6)$$

where: M = torsional moment developed in the bar

A = area of the bar cross section

V = displaced volume under the membrane

A = area of the membrane base

400. MECHANICS OF MATERIALS (STEADY-STATE)

Thus the torsional moment developed in the bar may be determined by measuring the displaced volume under the membrane.

For further discussion of the theory of this analogy, see Coker; Gilkey and Bergman; Griffith; Griffith and Taylor (A, B, D, E); B. G. Johnson; Marin; Soroka (A); Swinton; Trayer and March.

Anthes was possibly the first to apply this analogy experimentally and used it to find the torsional stress distributions in a circular, an elliptical, a triangular, a square, and five rectangular sections. With apparatus such as Anthes used, "Light transmitted from a black and white rectangular screen is reflected from the film, passes through a pinhole in the center of the screen and falls on a photographic plate. From photographs obtained before and after dilation of the film, the inclination of the normal to the surface is calculable."

The first experimenters to gain wide recognition for their application of this analogy were Griffith and Taylor who "made use of the membrane analogy to investigate the causes of aircraft engine crankshaft failures produced by excessive torsional stress." Included in their reports are "photographs and description of experimental apparatus ... comparison of experimentally and analytically obtained values of M [torque] and R [stress] for an equilateral triangle, square, two ellipses, two rectangles, an infinite strip, and several aircraft spar sections ...."

The analogy was used by Quest in analyzing the stress distributions in four notched circular sections, a cannon boring tool and a twist drill, three eccentric rectangular annuli, an angle, and an I-beam.

Reichenbacher obtained the stress distributions in a circular, a rectangular, a channel, and the two tool sections studied by Quest.

Pletta and Maher determined the torsional properties of two types of steel bar stock for helical springs. They modified the usual equipment by using a machined ring instead of a thin plate as a base for the membrane making it possible to change the shape being investigated without dismantling the apparatus completely.

[The foregoing material on applications of the differential pressure membrane analogy for torsion has been taken from Higgins (A).]

Trayer and March investigated the stress distributions in torsional "members having sections common in aircraft construction." Their report includes diagrams of contour elevations (and thus lines of constant ϕ) of soap films for T-sections and L-sections.

According to Hetenyi (A), Cushman applied this analogy to the torsion of circular and square sections.

Further use of the analogy was made by Neubauer and Boston who analyzed the torsional stress distributions in twenty-one different shapes of one-inch drills. Included in their report are diagrams of film elevation contours as well as numerical values for elevations, areas bounded by each contour line, and volumes bounded by successive adjacent contour planes and the film. In addition, they have presented tables of average stress and torque factors, film of the distribution of stress in an I-beam; the effect on stress of varying the radius of fillets.... In addition to using an autocollimator for measuring the inclinations of normals to the surface, Griffith and Taylor utilized the less accurate technique of sketching the "black spot" boundary as it spreads over the film for the determination of contour lines.

Biezeno and Rademaker employed this analogy in determining the stress distributions in a circular, a triangular, and a rectangular section.

Pirard has investigated the torsional stress distributions in bars of certain irregular cross sections, such as the circular segment, the split annular ring, and a wide flanged beam. The stress isostatics for these cases are pictured. [APPLIED MECHANICS REVIEW.]

To avoid difficulties encountered with soap films, Piccard and Baes proposed the use of a "liquid surface" membrane consisting of the separation surface of two immiscible liquids. "A plate with section cutout divides a cylindrical vessel in two; the lower part is filled with an electrically conducting liquid; a second liquid (non-conducting, of the same density, and nonmixable with the first) is carefully poured in; hydrostatic pressure applied to the lower liquid causes it to bulge through the opening, producing an interface or 'liquid surface' identical with that of a dilated membrane. Contour lines can be mapped with a sharp pointed micrometer screw, connected through headphone and voltage source to the conducting lower liquid, the circuit being completed when contact is established with the interface from above." [Higgins (A).]

By using such a liquid surface membrane, Piccard and Johner have investigated the stress distributions in square, triangular, and angle sections, for which they have presented photographs of lines of constant inclination of the liquid surface (hence, lines of constant stress). [Higgins (A).]

The liquid surface membrane was used also by Sunatani, Matuyama, and Hatamura, who have presented plots of the stress lines for a circular, a triangular, and a circularly pierced, irregular quadrilateral section. [Higgins (A).]

According to Higgins, "Kopf and Weber suggest use of a rubber diaphragm, stretched over a cutout S (cross section) in a plate and bulged into a mass of plastic paraffin of unit specific gravity by water pressure. On hardening, the paraffin provides a permanent cast of the bulged diaphragm. Hence, inclination of normal, stress lines, and other data can be taken at leisure." "Kopf and Weber obtained the stress lines for a square, for two rectangular areas with rounded corners and rectangular notches in each side, and for a curtate sector." For criticism of this method, see Thiel.

Drucker and Frocht have discussed the use of photoelastic scattering patterns as substitutes for the membrane analogy for torsion. Experiment has shown that each fringe is a line of constant membrane elevation or of constant value of the torsional stress function, and for a given cross section, the difference in "elevation" from one fringe to the next is a constant depending upon the twisting moment and material

400. MECHANICS OF MATERIALS (STEADY-STATE)

constants only. Included in the report are a description of experimental procedures followed, a photograph of a fringe pattern with superimposed membrane contours, and plots of theoretically — and experimentally — determined values of fringe order vs. length.

For a clear detailed account of classical torsion theory and the membrane analogy, see Zachrisson.

Olsyak has described the extension of this analogy for isotropic bars to apply to anisotropic bars by a suitable transformation of coordinates.

See also Boiten and Biezeno; Cadambe and Kaul (A); Dobie and Gent; Ikeda.

441AB *Membrane Analogy B*

A pressureless membrane analogy for elastic torsion has been developed in the following manner. [L. Föppl.]

If a companion function, ϕ_1, to the stress function (see the foregoing analogy) is defined as

$$\phi = \phi_1 + F(x,y) \tag{1}$$

where $F(x,y)$ is a function of x and y and is usually taken as

$$-\frac{G\theta}{2L}(x^2 + y^2),$$

then $\nabla^2[\phi_1 + F(x,y)] = \nabla^2 \phi = -\frac{2G\theta}{L}$ and thus $\nabla^2 \phi_1 = 0$ (where $\nabla^2 = \frac{\partial^2}{\partial x^2} + \frac{\partial^2}{\partial y^2}$).

The stresses S_{xz} and S_{yz} then become:

$$S_{xz} = \frac{\partial}{\partial y}\left[\phi_1 - \frac{G\theta}{2L}(x^2 + y^2)\right] = \frac{\partial \phi_1}{\partial y} - \frac{G\theta y}{L} \tag{2}$$

$$S_{yz} = -\frac{\partial}{\partial x}\left[\phi_1 - \frac{G\theta}{2L}(x^2 + y^2)\right] = -\frac{\partial \phi_1}{\partial x} + \frac{G\theta x}{L} \tag{3}$$

Characteristic equations:

STRESS   MEMBRANE

$$\frac{\partial^2 \phi_1}{\partial x^2} + \frac{\partial^2 \phi_1}{\partial y^2} = 0 \quad (4) \qquad \frac{\partial^2 z}{\partial x^2} + \frac{\partial^2 z}{\partial y^2} = 0 \quad (5)$$

where terms are defined as in the foregoing analogy.

Boundary conditions:

1. The boundary of the membrane is the same shape as the boundary of the bar cross section.

2. Along the boundary, $\phi_1 - \frac{G\theta}{2L}(x^2 + y^2) = \phi =$ constant. Thus $\phi_1 = \left[\frac{G\theta}{2L}(x^2 + y^2) + \text{constant}\right]$ and the distribution of z along the boundary must be proportional to this.

Methods for building boundaries to satisfy these conditions were developed by Griffith and Taylor (E) and Cushman. (See section 432.)

"Deutler pointed out that it is easier to construct the form over which the membrane is to be stretched if a parabolic boundary is used," i.e., let $F = -\frac{G\theta x^2}{L}$ or $-\frac{G\theta y^2}{L}$.

Thiel used this analogy to investigate the torsional stresses in a full and in a hollow square section.

A study of the variation of torsional stress with eccentricity in three eccentric annuli was made by Englemen who applied this analogy.

As with the differential pressure membrane analogy, a liquid surface has been used for the pressureless membrane analogy. Sunatani, Matuyama, and Hatamura applied this method in determining lines of constant elevation of the liquid surface (and thus lines of constant ϕ_1) for an irregularly shaped area.

[The foregoing material on applications of the pressureless membrane analogy for torsion has been taken from Higgins (A).]

Another membrane analogy which is different from the usual ones has been proposed by Barta (A).

441AC Electrical Analogy A

J. H. Weiner has utilized a resistor network as a finite difference analog for problems of elastic torsion of cylindrical bars with longitudinal holes. In the network analog, appropriate source currents are introduced at the nodes and the internal boundaries. (See section 441AA for the differential equations of torsion which have been approximated by the method of finite differences.)

Weiner, Salvadori, and Paschkis have extended this analogy to apply to torsion problems in which the stresses in portions of the members under load exceed the elastic limit.

Redshaw (E) has described the application of resistance networks to the solution of a variety of torsion problems.

See also Palmer (A).

441AD Electrical Analogy B

If the boundary of a uniformly thick sheet of metal or of electrolytic liquid confined in a shallow tank is subjected to a predetermined electric potential distribution, the distribution of potential within the plate or the liquid is proportional to the torsional stress potential function distribution in a uniform bar.

Characteristic equations:

400. MECHANICS OF MATERIALS (STEADY-STATE)

STRESS

$$\frac{\partial^2 \phi_1}{\partial x^2} + \frac{\partial^2 \phi_1}{\partial y^2} = 0 \quad (1)$$

ELECTRICAL

$$\frac{\partial^2 V}{\partial x^2} + \frac{\partial^2 V}{\partial y^2} = 0 \quad (2)$$

where V = electric potential, and other quantities are as defined in sections 441AA and 441AB.

Boundary conditions:

1. The sheet or tank must have the same cross section as that of the bar.

2. Along the boundary, $\phi_1 = \left[\frac{G\theta}{2L}(x^2 + y^2) + \text{constant}\right]$, and the distribution of z along the boundary must be proportional to this.

Meyer and Tank have done some development work on the experimental techniques of this analogy. They suggest that the boundary condition can be approximated quite closely "if the boundary wall of the tank be constructed of a weakly conducting material and the conducting probes be imbedded therein, whence a linear potential drop is secured between adjacent pairs of probes." For other methods of effecting a variation of electric potential along a boundary, see sections 221AAA (Reinius and also Childs), 221AAB (Babbitt and Caldwell), and 221B.

Biezeno and Koch used a manganin plate to determine values of ϕ and ϕ_1 for a rectangular cross section. These values of ϕ were compared with analytically determined values.

Cranz (B) obtained ϕ_1 for an ellipse and for the "Zweibogendreick" bounded by the arcs of two circles of equal radius, each of which is tangent to a diameter of the other at its center. These values of ϕ_1 were compared with analytically determined values.

An electrolytic tank was applied by Negoro (C) to the determination of ϕ_1 for an equilateral triangular section, an annulus, and a streamline section. See also Negoro (A, B).

Values of the function ϕ_1 for a T-section were obtained by Peres and Malavard (E).

Hohenemser remarked on the possibility of combining electrolytic tank and conformal transformation techniques to determine values of ϕ_1. Bradfield, Hooker, and Southwell advanced theory and procedure for the combining of electrolytic tank and conformal transformation techniques to determine values of ϕ. Analytically determined values of ϕ for an equilateral triangular section were compared with values determined by analogy. See also Cranz (C).

[The foregoing material on applications of the electrical analogy for torsion has been taken from Higgins (A).]

Waner and Soroka have determined stress concentrations of angle sections under torsion by use of conductive paper analogs. A similar study has been made by Friedmann, Yamamoto, and Rosenthal. See also Nakazawa.

Edamoto (A, B, C) has utilized both d.c. and a.c. electrolytic tanks in solving torsion problems. See also Edamoto and Kanayama. See also Stone.

441AE *Hydrodynamical Analogy A*

Greenhill suggested that an ideal fluid circulating with a uniform longitudinal vorticity 2ω in a uniform tube could be used to analyze the stress distribution in a uniform bar subject to torsion.

Characteristic equations:

STRESS

$$\frac{\partial^2 \phi}{\partial x^2} + \frac{\partial^2 \phi}{\partial y^2} = -\frac{2G\theta}{L} \quad (1)$$

where: stress quantities are defined as in section 441AA

FLUID

$$\frac{\partial^2 \psi}{\partial x^2} + \frac{\partial^2 \psi}{\partial y^2} = -2\omega \quad (2)$$

ω = constant angular velocity (vorticity) of the fluid

ψ = stream function defined by the following equations:

$$u = \frac{\partial \psi}{\partial y} \quad (3)$$

$$v = -\frac{\partial \psi}{\partial x} \quad (4)$$

where: u = x component of fluid velocity
v = y component of fluid velocity

Thus, it is evident that the components of torsional stress are analogous to the corresponding components of fluid velocity.

Boundary conditions:

1. The tube must have the same cross section as that of the bar.
2. Along the boundary, the stress must act in the direction of the boundary. This means that along the boundary, the fluid must flow in the direction of the boundary, obviously a condition which is automatically satisfied.

Higgins (A) states that since constant longitudinal vorticity is difficult (if at all possible) to produce, this analogy does not lend itself to experimental use. However, Carter and Oliphint have proposed a method of determining the maximum torsional shearing stress in a shaft with small flats, involving the mathematical computation of the

400. MECHANICS OF MATERIALS (STEADY-STATE)

maximum flow velocity across the flat of a similar tube and the application of this analogy.

441AF Hydrodynamical Analogy B

Thomson and Tait proposed the determination of the torsional-stress distribution in a uniform bar by the analysis of the velocity distribution in a body of ideal fluid in irrotational motion within a uniform tube rotating with constant angular velocity.

Characteristic equations:

1. Relative to axes rotating with the tube:

STRESS	FLUID
$\dfrac{\partial^2 \phi}{\partial x^2} + \dfrac{\partial^2 \phi}{\partial y^2} = -\dfrac{2G\theta}{L}$ (1)	$\dfrac{\partial^2 \psi}{\partial x^2} + \dfrac{\partial^2 \psi}{\partial y^2} = -2\omega$ (2)

2. Relative to stationary axes of the tube:

$\dfrac{\partial^2 \phi_1}{\partial x^2} + \dfrac{\partial^2 \phi_1}{\partial y^2} = 0$ (3)	$\dfrac{\partial^2 \psi}{\partial x^2} + \dfrac{\partial^2 \psi}{\partial y^2} = 0$ (4)

where quantities are defined as in the foregoing analogy and in section 441AB.

Thus, the components of stress are analogous to the corresponding components of fluid velocity observed relative to rotating axes and could also be computed from the corresponding components of fluid velocity observed relative to stationary axes, if such components could be accurately measured. However, as Higgins (A) states, "Though the general pattern of stress distribution is to be discerned, this procedure does not lend itself to accurate determination of numerical values."

Boundary conditions are the same as for the foregoing analogy.
Additional similar equations (relative to stationary axes):

$\dfrac{\partial^2 w}{\partial x^2} + \dfrac{\partial^2 w}{\partial y^2} = 0$ (5)	$\dfrac{\partial^2 \psi_1}{\partial x^2} + \dfrac{\partial^2 \psi_1}{\partial y^2} = 0$ (6)

where: w = longitudinal displacement of the bar cross section, components of the slope of which are:

ψ_1 = velocity potential function defined by the following equations:

$M_x = \dfrac{\partial w}{\partial x}$ (7)	$u = \dfrac{\partial \psi_1}{\partial x}$ (8)
$M_y = \dfrac{\partial w}{\partial y}$ (9)	$u = \dfrac{\partial \psi_1}{\partial y}$ (10)

This indicates that the shape of the warped cross section of the bar may be determined, since the components of slope of the cross section

400. MECHANICS OF MATERIALS (STEADY-STATE)

are analogous to the components of fluid velocity relative to stationary axes.

A further relation is that the torsional moment acting on the bar corresponds to twice the kinetic energy of the fluid relative to rotating axes.

Analytical application of this analogy to torsional members containing flaws and cavities has been made by Larmor.

Den Hartog and McGivern have provided experimental confirmation of this analogy. Aluminum particles were sprinkled on the surface of water contained in a vertical cylindrical tank which rotated about a vertical axis. The fluid surface was photographed by a camera mounted above the tank both (a) while rotating at the same speed as the tank and (b) while standing still. The lines formed by the apparent motion of the particles in photographs (a) follow the directions of the shear stresses in the corresponding bar, the lengths of the individual streaks being proportional to these stresses, and in photographs (b) follow the directions of the maximum slopes of the warped cross section, being perpendicular to the lines of equal height of that cross section. Both (a) and (b) photographs are shown for a square section and a circular section with a keyway, and an (a) photograph is shown for a narrow rectangular section. The effects of eddy currents developed (because of friction) by the use of an actual, rather than an ideal fluid, are diminished by taking the photographs within a very short time after the beginning of rotation.

Binder has presented a brief qualitative study of this analogy and how it may be applied in machine design.

441AG *Hydrodynamical Analogy C*

According to Timoshenko and Goodier, Boussinesq showed that the torsional stress distribution in a uniform bar may be determined from the velocity distribution in a laminar motion of a viscous fluid along a uniform tube.

Characteristic equations:

STRESS

$$\frac{\partial^2 \phi}{\partial x^2} + \frac{\partial^2 \phi}{\partial y^2} = -\frac{2G\theta}{L} \quad (1)$$

where: stress quantities are defined as in section 441AA

FLUID

$$\frac{\partial^2 V}{\partial x^2} + \frac{\partial^2 V}{\partial y^2} = -\frac{k}{\mu} \quad (2)$$

V = fluid velocity

k = pressure gradient or slope of the hydraulic gradient

μ = fluid viscosity

Boundary conditions:

1. The tube must have the same cross section as that of the bar.

2. Along the boundary, ϕ (and thus V) must be constant. To be more specific, V will actually be zero.

In addition, "$\dfrac{M}{\left(\dfrac{2G\theta}{L}\right)}$ [where M = torsional moment] is to be identified with the rate of flow through a given cross section of the tube. Paschoud (A), deriving mathematically that the same is true of the flow of a nonviscous liquid through a long narrow bore tube of cross section S, has suggested that M for the corresponding prism [bar] can be obtained by measuring the discharge from the tube. Conversely, from known solutions of the torsion problem Tumura has calculated the corresponding hydraulic discharge." [Higgins (A).]

See also Paschoud (B); Sander.

441AH Hydrodynamical Analogy D

Pestel (A, B) has proposed a hydrodynamical analog for torsion, consisting of the flow of fluid in a narrow slit of everywhere the same width over a region similar to the shape of the cross section to be studied. By changing the width of the slit, the effect of sources is simulated. The stress function appears as excess pressure in the slit and the unit source strength as proportional to the time derivative of the width, divided by its third power.

G. Grossman has described the use of this analogy.

441B Bodies of Revolution

441BA Electrical Analogy A

Jacobsen (A, B) analyzed the torsional shearing stress distributions in axially symmetric bars by measuring the electric potential distributions in metal plates (though electrolytic tanks may be used as well).

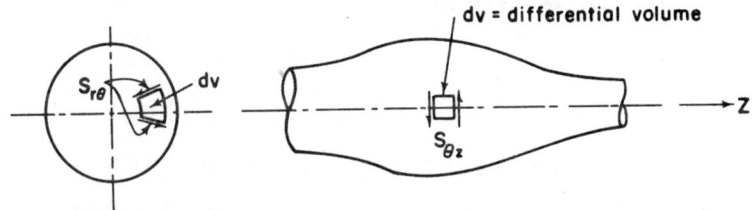

Fig. 441BA.1. Torsional shearing stresses in an axially symmetric bar.

Such an electric potential distribution may be expressed by the following equation:

$$\frac{\partial}{\partial x}\left(t\,\frac{\partial V}{\partial x}\right) + \frac{\partial}{\partial y}\left(t\,\frac{\partial V}{\partial y}\right) = 0 \qquad (1)$$

where: t = plate thickness
 V = electric potential
 x, y = coordinates

However, t may be made proportional to x^3.

400. MECHANICS OF MATERIALS (STEADY-STATE)

Characteristic equations:

<table>
<tr><th>STRESS</th><th>ELECTRICAL</th></tr>
</table>

$$\frac{\partial}{\partial r}\left(r^3 \frac{\partial \psi}{\partial r}\right) + \frac{\partial}{\partial z}\left(r^3 \frac{\partial \psi}{\partial z}\right) = 0 \quad (2) \qquad \frac{\partial}{\partial x}\left(x^3 \frac{\partial V}{\partial x}\right) + \frac{\partial}{\partial y}\left(x^3 \frac{\partial V}{\partial y}\right) = 0 \quad (3)$$

where: r = radial coordinate of the bar
z = axial coordinate of the bar
ψ = twist function defined by the following equations:

$$S_{r\theta} = Gr \frac{\partial \psi}{\partial r} \quad (4)$$

$$S_{\theta z} = Gr \frac{\partial \psi}{\partial z} \quad (5)$$

where: G = modulus of rigidity of the bar material
$S_{r\theta}$, $S_{\theta z}$ = shearing stresses as indicated in figure 441BA.1.

Boundary conditions:

1. The plate or tank must have the same profile as one equal half of a meridianal section (plane passing through the axis) in the r, z plane of the bar.

2. Since resultant shearing stresses are perpendicular to lines of constant ψ, the end boundaries of the meridianal section (assumed to be of infinite length) are lines of constant ψ and the side boundaries are lines of maximum change of ψ. In the plate or tank, this corresponds to equipotential end boundaries (with a constant potential drop between them) and insulated side boundaries (Fig. 441BA.2).

Thus the stress at a point is proportional to the product of the electric potential gradient and the distance from the axis.

The twist function, ψ, is physically equivalent to the angular rotation, $\frac{v}{r}$, of a differential ring of radius r, where v represents the tangential displacement of such a ring. Thus, this analogy may be used also to determine the angular rotation of any point in the bar.

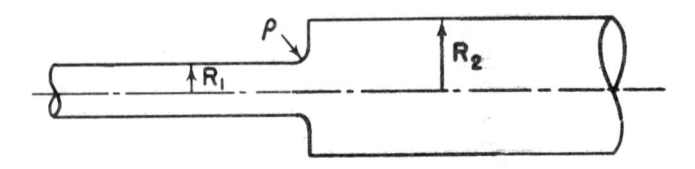

Fig. 441BA.2. Stepped shaft of two constant diameters.

400. MECHANICS OF MATERIALS (STEADY-STATE) 119

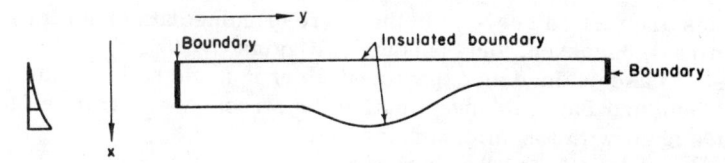

Fig. 441BA.3. Metal plate analog of axially symmetric bar of figure 441BA.1.

Jacobsen (A, B) employed this analogy in finding the stress distributions at fillets of stepped shafts (adjacent sections each of constant diameter). His reports include plots of stress concentration factor plotted against $\frac{\rho}{R_1}$ and $\frac{R_2}{R_1}$, where ρ, R_1, and R_2 are as indicated in figure 441BA.2.

441BB *Electrical Analogy B*

Recognizing certain limitations of the analogy used by Jacobsen (A, B) (see previous section), Thum and Bautz (A, B, C, D) used a slightly different analogy, employing an electrolytic tank (though a metal plate could be used as well). See also A. Föppl.

Rather than making t proportional to x^3 (see the foregoing analogy), Thum and Bautz made it proportional to x^{-3}.

Characteristic equations:

STRESS   ELECTRICAL

$$\frac{\partial}{\partial r}\left(r^{-3}\frac{\partial \phi}{\partial r}\right) + \frac{\partial}{\partial z}\left(r^{-3}\frac{\partial \phi}{\partial z}\right) = 0 \quad (1) \qquad \frac{\partial}{\partial x}\left(x^{-3}\frac{\partial V}{\partial x}\right) + \frac{\partial}{\partial y}\left(x^{-3}\frac{\partial V}{\partial y}\right) = 0 \quad (2)$$

where: ϕ = stress function defined by the following equations:

$$S_{r\theta} = -r^{-2}\frac{\partial \phi}{\partial z} \quad (3)$$

$$S_{\theta z} = r^{-2}\frac{\partial \phi}{\partial r} \quad (4)$$

Other quantities are as defined in the foregoing analogy.

Boundary conditions:

1. Same as for the foregoing analogy.
2. Resultant shearing stresses coincide with lines of constant ϕ. Thus, in the plate or tank, side boundaries are equipotential lines and end boundaries are insulated.

"... Thum and Bautz (D) studied the stress distribution in a series of grooved, slit, filleted, stepped, right stepped, and notched shafts. The results for the first four shapes are embodied in a set of curves of maximum stress as a function of geometrical parameters. On the same axes are plotted curves for the corresponding data of Willers, Wyszomirski, Jacobsen, and Sonntag." [Higgins (B).]

Salet (A) ascertained an experimental error made both by Jacobsen and by Thum and Bautz by investigating the stress-concentration factors for a shaft with a semicircular notch.

See also Salet (B) and Walther.

441C *Others*

Miles (A) developed an analogy among torsional rigidity, rotating fluid inertia and self-inductance. Results were derived for an ellipse, an equilateral triangle, a right isosceles triangle, and a rectangle.

442 *Plastic*

442A *"Sand-Heap" Analogy*

Nadai (A) was apparently the first to propose the use of a pile of sand resting on a flat horizontal plate to determine the shearing stress distribution in a bar subject to torsion entirely beyond the range of elastic stress into the range of plastic stress (assuming a horizontal stress strain curve in the plastic range).

Characteristic equations:

STRESS

$$\left(\frac{\partial \phi}{\partial x}\right)^2 + \left(\frac{\partial \phi}{\partial y}\right)^2 = S^2 \quad (1)$$

SAND HEAP

$$\left(\frac{\partial z}{\partial x}\right)^2 + \left(\frac{\partial z}{\partial y}\right)^2 = m^2 \quad (2)$$

where: x, y = coordinates of the bar cross section

S = constant magnitude of stress throughout the bar

ϕ = stress function defined by the following equations:

$$S_{xz} = \frac{\partial \phi}{\partial y} \quad (3)$$

$$S_{yz} = -\frac{\partial \phi}{\partial x} \quad (4)$$

x, y = coordinates of the base of the sand heap

m = constant slope of the sand heap lateral surface

z = vertical distance from the base to the lateral surface of the sand heap

where: S_{xz}, S_{yz} = x, y components of stress.

The condition that $S^2_{xz} + S^2_{yz} = S^2$ is satisfied in the analogy by the condition that the sand heap have the same slope everywhere.

Thus, it is seen that the direction of stress throughout the cross

400. MECHANICS OF MATERIALS (STEADY-STATE)

section coincides with the contours of the sand heap surface.

Boundary conditions:

1. The size and the shape of the base plate of the sand heap are the same as those of the bar cross section.
2. Along the boundary, $S_{yz} dx - S_{xz} dy = 0$ since $\frac{S_{yz}}{S_{xz}} = \frac{dy}{dx}$, which means that $\phi = 0$. This corresponds to the condition of zero elevation of the sand heap ($dz = \frac{\partial z}{\partial y} dy + \frac{\partial z}{\partial x} dx = 0$).

The torsional moment transmitted by the bar may be determined by the following equation:

$$T = 2 \left(\frac{S}{m}\right) V \qquad (5)$$

where: T = torsional moment
V = total volume of sand in the heap (measured by weighing the sand)
m = slope of the sand heap

Nadai (C) has extended this analogy to the determination of the extent of the elastic stress region in a bar, only part of which is stressed in the plastic range. First, the inside of a glass frame of the same shape as the lateral surface of the sand heap is coated with oil. A rubber membrane whose top surface is coated with a layer of powder is then fastened to the base edges of the glass frame. Finally, a pressure differential expands the membrane into the enclosed volume. As the pressure increases, the progression of the separation line between elastic and plastic stress in the bar is indicated by the progression of the inside boundary of the white surface where the membrane touches the glass frame. The white surface corresponds to the portion of the bar in plastic stress. Nadai's report includes photographs of such membrane frames and also sand heaps upon plates of various shapes.

Sadowsky has explained how the analogy is modified to apply to a bar containing axial holes. The base plate used contains holes corresponding to those in the bar. Into these holes are inserted tubes (with countersunk ends) that are movable vertically. While the sand heap is being built up, these tubes are in as high a position as possible. When the heap is completed, the tubes are slowly lowered until their entire lateral surfaces are covered by the sand, the excess sand falling through the tubes (Fig. 442A.1).

The torsional moment transmitted by the bar may then be expressed by the following equation:

$$T = 2 \left(\frac{S}{m}\right) (V_0 + h_1 A_1 + h_2 A_2 + \cdots + h_n A_n)$$

122   400. MECHANICS OF MATERIALS (STEADY-STATE)

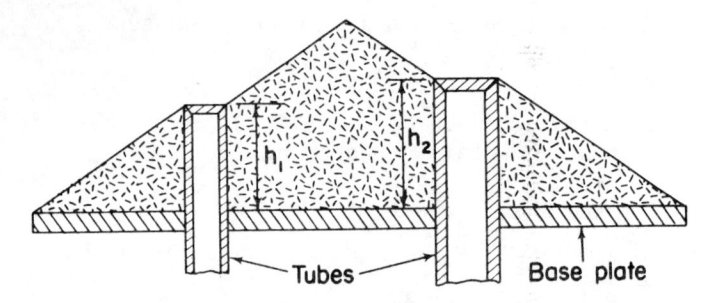

Fig. 442A.1. Sand heap to analyze bar with holes.

where: T = torsional moment
S = constant magnitude of stress
m = constant slope of the sand heap lateral surface
V_o = volume of the modified sand heap
h_i = height of the top of a tube above the base of the sand heap as indicated in the figure
A_i = area of the corresponding hole

See also Nadai (B); Swinton; Coker.

442B *Electrical Analogy*

Redshaw (E) has described electrical networks for the solution of problems of plastic torsion.

Weiner, Salvadori, and Paschkis have extended this analogy to apply to torsion problems in which the stresses in portions of the members under load exceed the elastic limit.

450 PLANE ELASTICITY

451 *The "Slab" Analogy*

"The analogy between the two dimensional field of stress and the transverse flexure of a thin plate slab was first applied by K. Wieghardt to the solution of a problem involving boundary loading of a simply-connected body." [Mindlin.] According to Westergaard (B), Wieghardt's investigation involved the bending of a brass plate to obtain information concerning the concentration of stresses at reentrant corners of a slice (of a cylindrical body bounded by parallel planes) loaded by two equal and opposite forces. Mindlin states that "Westergaard (B) introduced the useful terminology of slab and slice, free slice and constrained slice, and gave the boundary conditions for the slab when the slice is multipli-connected and is stressed by boundary loads having no resultant force on an internal boundary."

Characteristic equations:

400. MECHANICS OF MATERIALS (STEADY-STATE)

TWO-DIMENSIONAL SLICE

$$\frac{\partial^4 \phi}{\partial x^4} + 2\frac{\partial^4 \phi}{\partial x^2 \partial y^2} + \frac{\partial^4 \phi}{\partial y^4} = f(V, T) \quad (1)$$

where: x, y = coordinates

ϕ = Airy's stress function defined by the following equations:

$$S_x = \frac{\partial^2 \phi}{\partial y^2} + V \quad (3)$$

$$S_y = \frac{\partial^2 \phi}{\partial x^2} + V \quad (4)$$

$$S_{xy} = -\frac{\partial^2 \phi}{\partial x \partial y} \quad (5)$$

S_x, S_y = x, y components of the normal stress

S_{xy} = shearing stress

$f(V, T)$ = function of V and T

V = body-force potential

T = temperature

SLAB

$$\frac{\partial^4 w}{\partial x^4} + 2\frac{\partial^4 w}{\partial x^2 \partial y^2} + \frac{\partial^4 w}{\partial y^4} = \frac{q}{D} \quad (2)$$

x, y = coordinates

w = transverse deflection of the slab

q = transverse loading distribution on the slab

D = flexural rigidity of the slab = $\dfrac{Eh^3}{12(1-\mu^2)}$

E = modulus of elasticity of the slab material

h = slab thickness

μ = Poisson's ratio for the slab material

Boundary conditions:
1. The slab must have the same shape as the slice.
2. Other conditions depend on the type of loading on the slice.

Westergaard proposed the use of this analogy in the investigation of the stresses in the Boulder Canyon Dam. A test was made under his direction by V. P. Jensen, who studied the nonlinear distribution of stresses in a vertical slice of the dam. An extension of this test, made with the same apparatus, was made by the engineering staff of the Bureau of Reclamation [see "Slab Analogy Experiments" and "Stress Studies for Boulder Dam," also G. Murphy (B)]. The slab used in these tests was of 3/4 inch thick, medium grade rubber. Boundary

124   400. MECHANICS OF MATERIALS (STEADY-STATE)

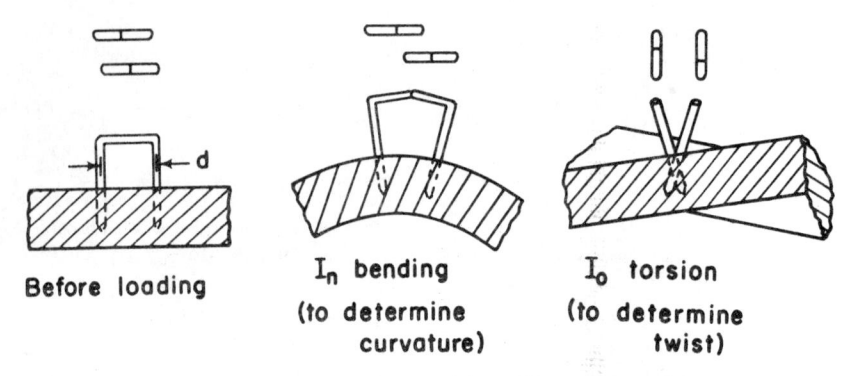

Fig. 451.1. Pins mounted in the rubber-slab analog.

deflection and slopes were produced by specially designed clamps. A cross-bar linkage with a carriage holding a microscope and a filar micrometer was used to measure curvature, twist, and deflection. Curvature and twist were measured between pins imbedded in pairs, normal to the surface throughout the slab (Fig. 451.1). Curvature was determined from measurements of the change in distance between the scratches on each pair of pins after loads were applied:

$$\frac{\partial^2 \phi}{\partial x^2} = (\Delta)(1/2)\left(\frac{1}{L}\right)\left(\frac{1}{d}\right) = \frac{\Delta}{2Ld}$$

where: $\frac{\partial^2 \phi}{\partial x^2}$ = curvature

Δ = change in distance between the scratches
L = moment arm measured from the center of the slab
d = gage distance between the pins

Twist was measured in a similar manner.

"Cranz (A) made improvements in technique in an application to the measurement of the stress concentration in a long notched plate under tension. He observed that, a free boundary such as abcde (Fig. 451.2) has a constant slope in the y direction, while the displacement varies linearly with y. Along a uniformly loaded boundary, such as a-a or e-e, the displacement follows a parabolic law while the slope,

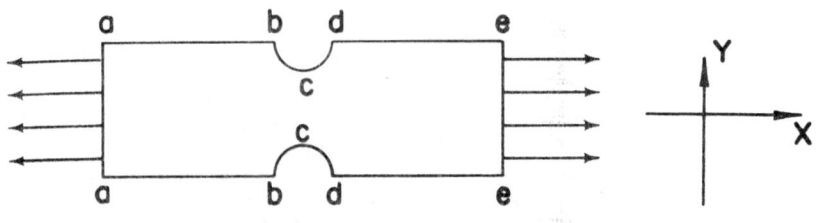

Fig. 451.2. Notched plate under tension.

400. MECHANICS OF MATERIALS (STEADY-STATE)

normal to the edge, is zero.... The slab material was a polished plexiglas sheet 1 or 1.5 mm. thick. Curvatures were measured by a direct optical method" described by Einsporn. A more direct and accurate optical instrument for measuring slab curvatures has been developed by A. Martinelli; [Hetenyi (A)].

Mindlin has presented the general boundary conditions for the slab when the slice is multipli-connected and is stressed by any combination of boundary loading, body forces, dislocations, and thermal dilatations.

Ryan has experimented with the same situation that Cranz (A) investigated.

Dantu has applied a new method of determining slab curvatures to the problem of a square slice compressed by two equal and opposite loads applied at the centers of opposite edges. Included in his report are comparable results obtained by photoelastic methods. [from APPLIED MECHANICS REVIEW]

Richards has determined slab geometry by means of a novel experimental technique developed by Ligtenberg. This optical method gave moiré fringe photographs, from which slopes of the plate were determined directly.

See also A. and L. Föppl; Love; Southwell; Vainberg; Westergaard (C); Yu.

452 *Membrane Analogy A*

Den Hartog pointed out the analogy between the deflection of a pressureless membrane and the sum of the magnitudes of the plane principal stresses at a point in a two-dimensional member.

Characteristic equations:

TWO-DIMENSIONAL MEMBER

$$\frac{\partial^2}{\partial x^2}(S_u + S_v) + \frac{\partial^2}{\partial y^2}(S_u + S_v) = 0 \qquad (1)$$

MEMBRANE

$$\frac{\partial^2 z}{\partial x^2} + \frac{\partial^2 z}{\partial y^2} = 0 \qquad (2)$$

where: x, y = coordinates

S_u, S_v = magnitudes of the principal stresses

x, y = coordinates

z = transverse deflection of the membrane

Boundary conditions:

1. The membrane boundary must be the same shape as the boundary of the two-dimensional member.

2. The distribution of z along the boundary of the membrane must be the same as the distribution of $(S_u + S_v)$ along the boundary of the two-dimensional member. (For the construction of such boundaries see section 432.)

The distribution of $(S_u - S_v)$ throughout the entire two-dimensional member may be determined by photoelastic analysis. Since $S_v = 0$ along the boundary, $(S_u + S_v) = (S_u - S_v)$ along the boundary. Thus the distribution of $(S_u + S_v)$ along the boundary is determined.

Mindlin states that Weibel (A) has applied this analogy using a soap film and that Biot and Smits have applied the analogy using a rubber membrane.

See also Biot; Pirard; Smits; Supper; Weibel (B).

453 Membrane Analogy B

The method of section 463 could also be applied to problems of plane elasticity.

454 Hydrodynamical Analogies

A brief review of analogies between plane stress and non-viscous fluid flow has been made by Hetenyi (B):

(1) The equipotential and stream lines of a flow pattern can always be considered as lines of equal lateral contractions (isopachics) and lines of equal rotations in a corresponding plane stress system. For discussion of possibilities of using the analogy in solving stress problems, see Doerfel.

(2) The velocities of a potential flow can be considered as displacements in a plane stress system. Barta (B) discussed this second analogy with reference to aeronautical problems.

(3) The orthogonal system of equipotential and stream lines are equivalent to system of stress trajectories. For further references of this third analogy, see Nemenyi (A, B); Neuber (A, B); Wegner.

Inoue (A, B) has described a hydrodynamical analogy for the state of stress in a plastic solid. See also Hill.

For other accounts of hydrodynamical analogies, see Baud; Smith, J.; Szebehely and Pletta.

455 Electrical Analogy A

Biezeno and Koch have described the use of an electrically conducting plate in determining the sums of the plane principal stresses at a point in a two-dimensional member.

Characteristic equations:

TWO-DIMENSIONAL MEMBER   ELECTRICAL

$$\frac{\partial^2}{\partial x^2}(S_u + S_v) + \frac{\partial^2}{\partial y^2}(S_u + S_v) = 0 \quad (1) \qquad \frac{\partial^2 V}{\partial x^2} + \frac{\partial^2 V}{\partial y^2} = 0 \quad (2)$$

where V = electric potential and other symbols are defined as in the previous analogy.

Thus the distribution of the electric potential corresponds to the distribution of the sum of the magnitudes of the principal stresses.

400. MECHANICS OF MATERIALS (STEADY-STATE)

Boundary conditions:

1. The electrical plate must be the same shape as the two-dimensional member.

2. The distribution of V along the boundary of the plate must be the same as the distribution of $(S_u + S_v)$ along the boundary of the two-dimensional member (see the previous analogy). For methods of effecting a variation of electric potential along a boundary, see sections 221AAA (Reinius and also Childs), 221AAB (Babbitt and Caldwell), 221B, and 441AD.

See also Biot; Malavard (E); Meyer and Tank; Paschkis (D).

Calvert has used an electrolytic tank in applying a modification of this analogy to the determination of stress concentrations in lamina of complicated shape.

456 *Electrical Analogy B*

Kron (B, G, L) has developed equivalent circuits representing the partial differential equations of the theory of elasticity for bodies of arbitrary shapes. The analysis may be used to study static, steady state or transient elastic field phenomena.

The equivalent circuit must satisfy three sets of elasticity equations, which are as follows:

1. Hooke's law — representing six relations between stresses and strains.
2. The equations of equilibrium — representing three relations among the stresses.
3. The conditions of compatibility — representing six relations among the strains.

These relations in elasticity correspond respectively to:

1. Ohm's law — determining the values of admittances of the circuit elements.
2. Kirchhoff's second law — stating that the sum of the currents entering each junction is zero.
3. Kirchhoff's first law — stating that the sum of the voltages around each mesh is zero.

Gross and Soroka (B) extended the analogy to three dimensions. For some applications of this analogy, see G. K. Carter.

457 *Electrical Analogy C*

Liebmann (F, G, K) has described the use of two resistance networks in cascade to solve the biharmonic equation $\nabla^4 \phi = 0$, which is

400. MECHANICS OF MATERIALS (STEADY-STATE)

satisfied by the Airy stress function for plane elasticity. Voltages appearing at the nodes of one network are applied to high value resistances connected to the corresponding nodes of the other network. This introduces "source" currents into the second network which are proportional to the applied voltages.

See also Boscher (B, C); Boscher and Malavard; Palmer and Redshaw (B); Redshaw (A); Redshaw and Rushton.

458 Other Electrical Analogies

For other electrical analogies to solve problems in plane elasticity, see Gagarina; Stokey and Hughes; Theocaris (A, B, C).

459 Water Diffusion Analogy

S. K. Clark and Hess have proposed a water diffusion analogy for the determination of transient thermal stresses in bodies.

The adsorption of water vapor by certain organic photoelastic materials causes swelling to occur. It is assumed that small water vapor concentration differences will "cause swelling which, inside a restricted range, is linearly proportional to the quantity of water vapor adsorbed, just as thermal expansion is proportional to the temperature."

Since heat diffusion and vapor diffusion (in organic solids) follow equations of identical form, and "since the equations of elasticity apply in both cases, then stresses should be set up due to vapor diffusion... which are exact analogs of those stresses arising from unsteady heat conduction."

Thus the stress (as determined photoelastically) in the transparent, organic-solid body exposed to water vapor should predict, at a given point and time, the stress in a similarly-shaped body through which heat is flowing, if similar boundary conditions are applied.

460 PLATES AND SHELLS

461 Electrical Analogy A

An electric circuit analogy to analyze the transverse deflection of constant-thickness plates under constant load, in transient vibration, or in normal mode vibration has been proposed by MacNeal (D).

The characteristic equation for the transverse deflection of a plate,

$$\frac{\partial^4 w}{\partial x^4} + 2 \frac{\partial^4 w}{\partial x^2 \partial y^2} + \frac{\partial^4 w}{\partial y^4} = \frac{q}{D} \tag{1}$$

where w is the lateral deflection of the plate, may be written as

400. MECHANICS OF MATERIALS (STEADY-STATE)

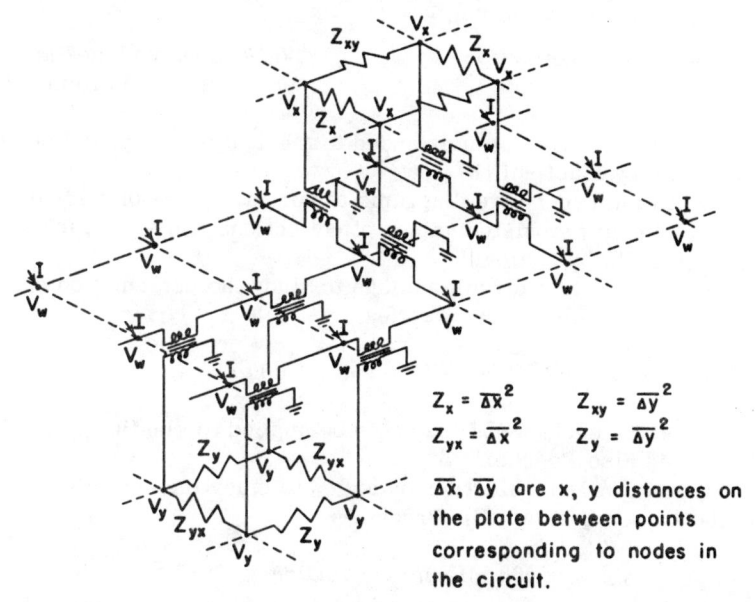

$Z_x = \overline{\Delta x}^2 \qquad Z_{xy} = \overline{\Delta y}^2$
$Z_{yx} = \overline{\Delta x}^2 \qquad Z_y = \overline{\Delta y}^2$

$\overline{\Delta x}, \overline{\Delta y}$ are x, y distances on the plate between points corresponding to nodes in the circuit.

Fig. 461.1. Portion of analog circuit for determining deflections in constant-thickness plates.

$$\frac{\partial}{\partial x}\left[\left(\frac{\partial^2}{\partial x^2} + \frac{\partial^2}{\partial y^2}\right)\frac{\partial w}{\partial x}\right] + \frac{\partial}{\partial y}\left[\left(\frac{\partial^2}{\partial x^2} + \frac{\partial^2}{\partial y^2}\right)\frac{\partial w}{\partial y}\right] = \frac{q}{D} \qquad (2)$$

This expression may be further modified by making the substitutions $\theta_x = \frac{\partial w}{\partial x}$ and $\theta_y = \frac{\partial w}{\partial y}$.

Characteristic equations:

PLATE

$$\frac{\partial}{\partial x}\left[\left(\frac{\partial^2}{\partial x^2} + \frac{\partial^2}{\partial y^2}\right)\theta_x\right] + \frac{\partial}{\partial y}\left[\left(\frac{\partial^2}{\partial x^2} + \frac{\partial^2}{\partial y^2}\right)\theta_y\right] = \frac{q}{D} \qquad (3)$$

ELECTRICAL

$$\frac{\partial}{\partial x}\left[\left(\frac{\partial^2}{\partial x^2} + \frac{\partial^2}{\partial y^2}\right)V_x\right] + \frac{\partial}{\partial y}\left[\left(\frac{\partial^2}{\partial x^2} + \frac{\partial^2}{\partial y^2}\right)V_y\right] = I \qquad (4)$$

where: x, y = displacement coordinates

θ_x, θ_y = x, y components of the slope of the plate

q = load density

x, y = displacement coordinates

V_x, V_y = node voltages as indicated in figure 461.1

I = junction currents as indicated in figure 461.1

400. MECHANICS OF MATERIALS (STEADY-STATE)

D = stiffness constant
w = plate deflection
V_w = node voltages as indicated in figure 461.1

This analogy has also been extended to the analysis of plates of variable thickness by MacNeal (F).

Included in the report are diagrams of normal modes of a square cantilever plate with tabulated values of deflections and slopes for various points (results by analogy).

For similar electric circuit analogs to study the deformation of elastic plates, see Shields and MacNeal.

462 Other Electrical Analogies

The method of section 457 may also be applied to elastic plate problems. See also Boscher (A).

For other reports on electric analogies to study the deflection of elastic plates, see Scanlan (B); Trudsø.

463 Membrane Analogy

The fourth order differential equation for the deflection of an elastic plate

$$D\nabla^4 w - q = 0 \tag{1}$$

may be replaced by the following two second order equations:

$$\frac{\partial^2 M}{\partial x^2} + \frac{\partial^2 M}{\partial y^2} + q = 0 \tag{2}$$

$$\frac{\partial^2 w}{\partial x^2} + \frac{\partial^2 w}{\partial y^2} + \frac{M}{D} = 0 \tag{3}$$

where M is a variable related to the bending moment and other quantities are as defined in section 461.

These last two equations are both analogous to the equation for the deflection of a membrane. Cadambe and Kaul (B) have thus described the solution of equation (1) by the application of the relaxation method to equations (2) and (3) in terms of the deflection of a membrane.

464 Plane Elasticity Analogy

Fridman has applied an analogy between plane elasticity and the deflection of a plate. His investigation was of a thin isotropic elastic plate supported by joints along the outer edge and coupled also by joints to rigid disks along the inner edges. The load consisted of a normal

400. MECHANICS OF MATERIALS (STEADY-STATE) 131

nonuniform pressure and of linear forces acting along the internal boundaries. This was found to be analogous to plane elasticity where the normal component of displacements and the tangential components of the external forces are given along the boundary, a method of solution being available for this case. [APPLIED MECHANICS REVIEW]

465 *Mechanical Analogy*

Wells has suggested the use of a mechanical analogy (see section 422B) to determine lateral displacements in the full heads of pressure vessels.

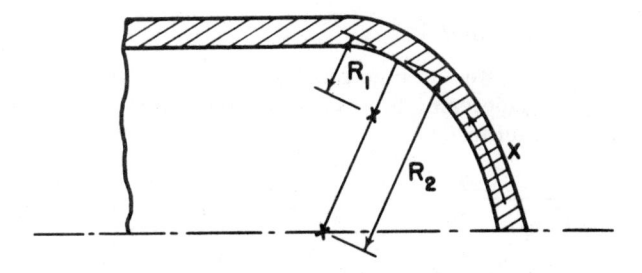

R_1 is measured in the plane of the paper.

R_2 is measured $\perp$ to the plane of the paper.

Fig. 465.1. Cross section of one symmetrical half of the full head of a pressure vessel.

Characteristic equations:

DISPLACEMENT

$$\frac{Et^3}{12(1-\sigma^2)} \frac{d^4y}{dx^4} - \left(\frac{Et}{R_2^2}\right) y + p \left(1 - \frac{R_2}{2R_1} - \frac{\sigma}{2}\right) = 0 \qquad (1)$$

MECHANICAL

$$\left(\frac{\pi E r^4}{4}\right) \frac{d^4y}{dx^4} - \frac{w_x b_x}{sn^2} y + \frac{w_x a_x}{sn^2} = 0 \qquad (2)$$

where: y = lateral displacement
x = distance along a meridian (Fig. 523.1)
t = head thickness
p = internal pressure
E = modulus of elasticity
σ = Poisson's ratio
R_1, R_2 = radii of curvature of the neutral surface of the head, in the plane of the meridian, and normal to the plane of the meridian, resp. (Fig. 463.1)
and other symbols are as defined in section 422B.

132   400. MECHANICS OF MATERIALS (STEADY-STATE)

The application of the apparatus is similar to that described in section 422B. In general, both R_1 and R_2 vary with x. Suitable boundary conditions are $\frac{dy}{dx} = \frac{d^3y}{dy^3} = 0$, at both ends.

Results obtained by analogy and by direct measurement are given for the stress distribution in a spherical drumhead.

470 FRAMES AND OTHER STRUCTURES

471 *Pin-connected*

471A *Statically Determinate*

Bush (B) has presented an electrical analogy for determining the stresses in statically determinate two- and three-dimensional, pin-connected frame structures.

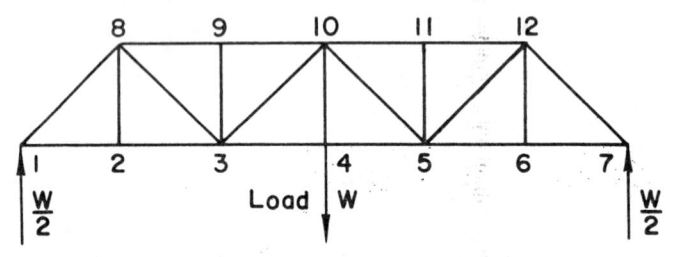

Fig. 471A.1. Pin-connected truss.

In this analogy, members lying in one of the coordinate directions are represented by perfect conductors, with a separate circuit for each

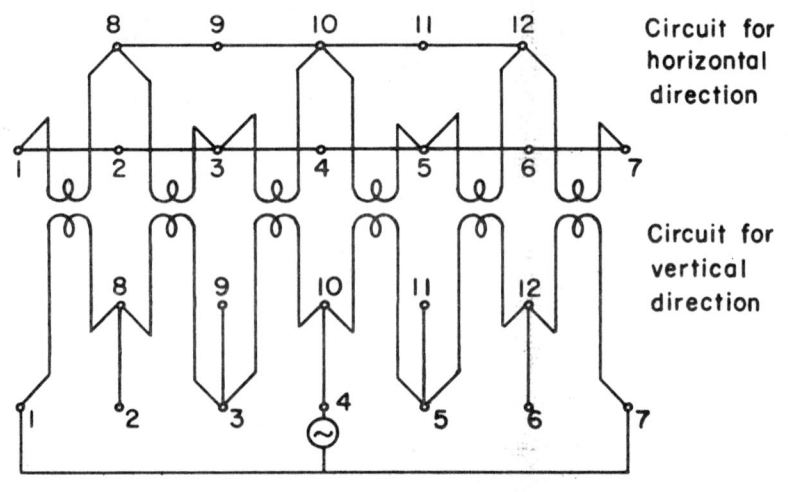

Fig. 471A.2. Analog circuit for pin-connected truss of figure 471A.1.

400. MECHANICS OF MATERIALS (STEADY-STATE)

coordinate direction. Oblique members are represented by perfect transformers, one coil of each connected to each circuit representing a coordinate direction in which the oblique member has a component. The ratio of each transformer is such that the current flowing through each coil of the transformer is proportional to the component of stress in the direction represented by the circuit in which the coil lies.

The current flowing through a conductor represents the stress in the corresponding member. Applied loads and reactions are represented by applied voltages (Figs. 471A.1, 471A.2).

For a similar analogy, see Svensson.

See also section 472.

471B *Statically Indeterminate*

Bush (B) has presented an analogy similar to the preceding for determining the stresses and deflections in statically indeterminate, pin-connected frames.

Characteristic equations (for each member):

FRAME		ELECTRICAL
$\dfrac{e}{S} = \alpha$ (1)		$\dfrac{V}{I} = R$ (2)

where: e = total strain
$\quad\quad\quad S$ = unit stress
$\quad\quad\quad \alpha$ = compliance = $\dfrac{L}{E}$
$\quad\quad\quad L$ = length
$\quad\quad\quad E$ = modulus of elasticity

V = voltage
I = current
R = resistance

In this analogy, members are represented by resistances proportional to the compliances of the corresponding members, rather than by perfect conductors (Figs. 471B.1, 471B.2). Thus the total strain and the unit stress of each member are proportional to the voltage across and the current through the corresponding resistance, respectively.

For other electrical analogies, see Kron (N); Ryder (D).

See also sections 472AA, 472AB, 472BC, 472BD.

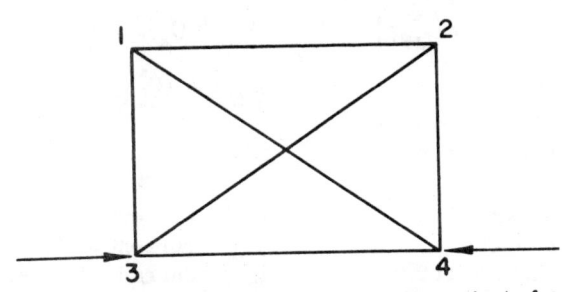

Fig. 471B.1. Pin-connected statically indeterminate frame.

400. MECHANICS OF MATERIALS (STEADY-STATE)

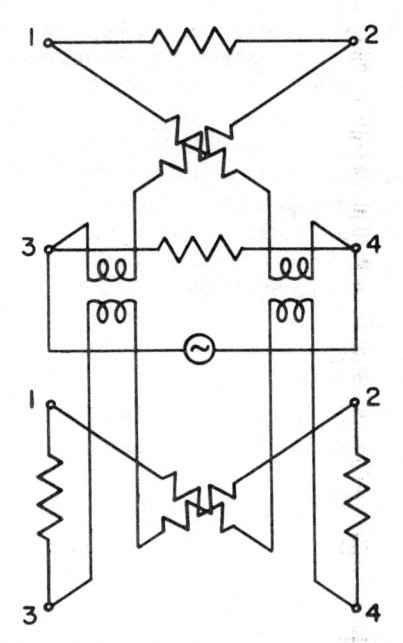

Fig. 471B.2. Analog circuit for pin-connected statically indeterminate frame of figure 471B.1.

472 *Rigidly Connected*

472A *Stress*

472AA *Electrical Analogy*

Bush (B) has presented an electrical analogy for analyzing two- and three-dimensional rigid frame structures.

Characteristic equations (for each member):

STRUCTURES   ELECTRICAL

$$M_A + M_B + SL = 0 \quad (1) \qquad (I_M)_A + (I_M)_B + I_s R_1 = 0 \quad (2)$$

$$\theta_A = \theta_B + M_A\left(\frac{L}{EI}\right) + S\left(\frac{L^2}{2EI}\right) \quad (3) \qquad (I_\theta)_A = (I_\theta)_B + (I_M)_A R_2 + I_s R_3 \quad (4)$$

$$\Delta_A - \Delta_B = \theta_B L \qquad\qquad (I_\Delta)_A - (I_\Delta)_B = (I_\theta)_B R_1$$

$$+ M_A\left(\frac{L^2}{6EI}\right) + S\left(\frac{L^3}{6EI}\right) \quad (5) \qquad\qquad + (I_M)_A R_3 + I_s R_4 \quad (6)$$

where: S = shear
M = moment
θ = angle of rotation
Δ = relative deflection

I_s = current at S terminals
I_M = current at M terminals
I_θ = current at θ terminals
I_Δ = current at Δ terminals

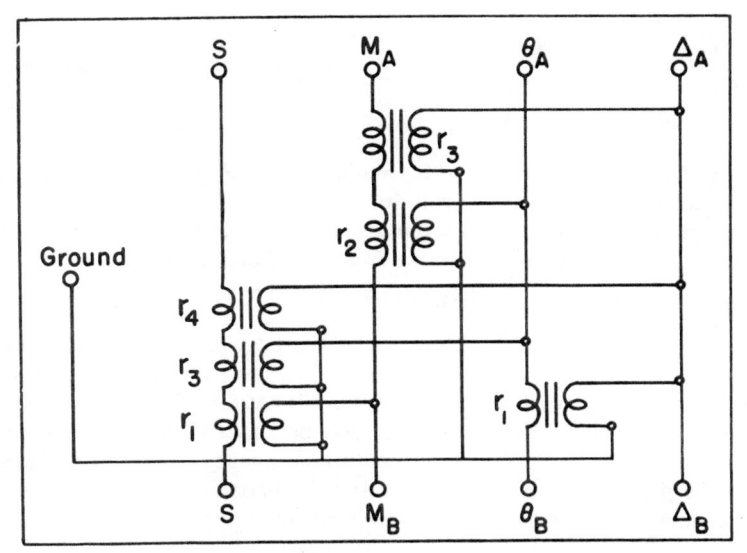

Fig. 472AA.1. Analog circuit for a single member.

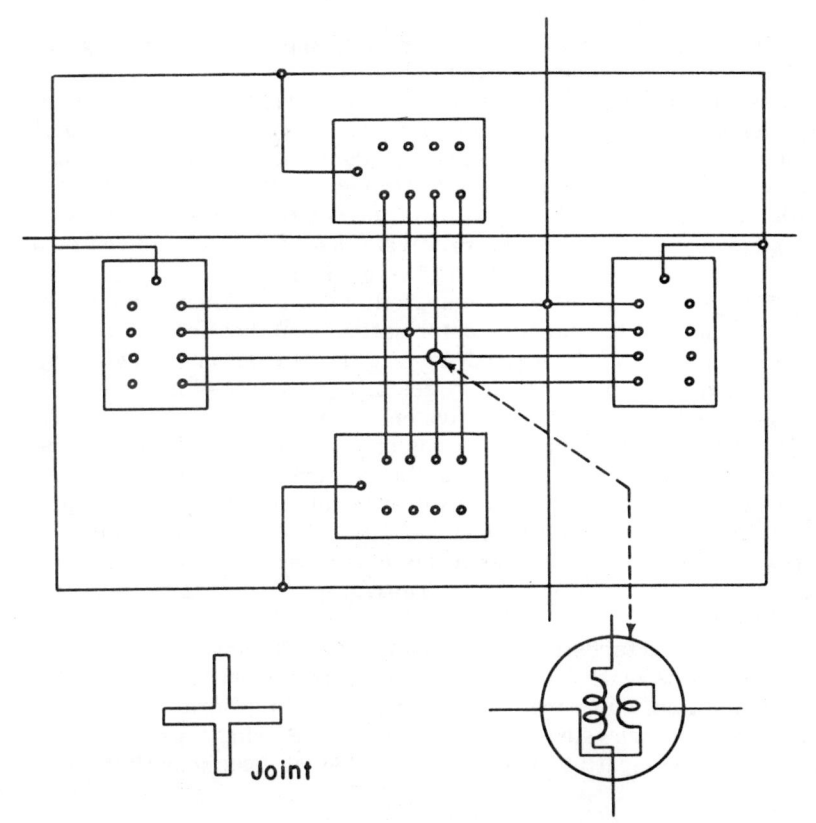

Fig. 472AA.2. Analog circuit for a joint.

136   400. MECHANICS OF MATERIALS (STEADY-STATE)

L = length
E = modulus of elasticity
I = moment of inertia of the member cross section
R = transformer ratio as indicated in figure 472AA.1

and subscripts A and B refer to the ends of the member, and subscripts 1, 2, 3, 4 differentiate among transformer ratios.

Fenn (A, B) has reported on the application of equivalent voltages and currents to the slope-deflection method of analysis.

472AB *Other Electrical Analogies*

Kron (N) had developed an electric circuit analogy for one-, two-, and three-dimensional elastic structures (frames) under steady small deformations and under forced or natural vibration. The elastic structures are assumed to consist of long, thin beams and rigid bodies having different elastic coefficients (tension and shear, bending and twist) along three perpendicular directions and interconnected into an arbitrary network either rigidly or by various constraints in any combination.

Carter and Kron (A) have presented numerical and analogic solutions for two frames.

Ryder (D) has utilized an unconventional approach in devising electrical analogs for structures. In his own words:

"Most practical electrical simulators of structures are based on the simulation of the relationships of force and moment equilibrium by current-continuity relationships, and on the simulation of compatibility of deformations by compatibility of voltages. Although equilibrium is readily simulated, the identification of the voltages with analogous deformations ordinarily requires relatively ingenious and involved theoretical analysis for each new type of structure.

"This type of simulator is compared in several illustrative cases with a second type in which the deformations and their compatibility are not considered explicitly, but instead the simulation of the energy in the structure is provided for by resistive power loss in the static case and by inductive and capacitive energy in the dynamic case. The validity of this type of analog depends on the direct use of certain electrical energy theorems that are similar to structural energy theorems, with the result that the simulator will by itself automatically compute and satisfy the compatibility requirements. This use is not to be confused with the use of energy theorems in the structure alone for the purpose of obtaining deformations preparatory to applying compatibility considerations.

"In most structures thus far considered the second type of approach has led to a drastically simpler theoretical treatment than the first type. In addition, since the analog circuitry is in general not unique, the energy approach often leads to a circuit requiring fewer components than the compatibility approach and having less sensitivity to

400. MECHANICS OF MATERIALS (STEADY-STATE)

imperfections of the components." For application of this procedure, see Ryder (A, B, C, E); Zaid and Ryder.

For other discussions of electrical analogies, see Bray; Brouwer and Van DerMeer; Il'enko (A, B); K. Kohler; Kuros; McCann; Riesz and Swain; Shields and MacNeal; Sved; Svensson; Yamauchi.

See also section 472BD.

472AC *The "Shear and Torsion" Analogy*

Baron and Michalos have developed an analytical procedure, called the "Shear and Torsion" analogy, for determining moments and shears in plane structures, continuous between two supports, whose loads are applied normal to the plane of structure and whose moments are about axes in this plane. This analogy might be called the counterpart of the "Column" analogy (Sec. 422A), used in analyzing members and frames loaded in their own planes. "The expressions [of the analogy] are somewhat similar in form to those of the column analogy, the results in both cases being interpreted in terms of a pressure line concept." These two analogies have also been combined and extended in this article for the analysis of structures curved or segmental in space and subjected to loads in any direction and to moments about any axis.

Baron has described a procedure called a "circuit analysis," an extension of the "Shear and Torsion" analogy. This procedure may be used to "determine the effects of loads normal to the plane of such structures as suspension bridge towers, delta wings of aircraft, grillages of beams, and substitute networks for slabs. In addition a circuit analysis is developed for determining the effects of loads lying in the plane of such structures as building frames, Vierendeel girders, and continuous arches or gabled structures on elastic piers."

See also Huang.

472AD *Frame Analogy A*

Prager and Hay have suggested the use of plane loaded frames in analyzing plane rigid frames loaded perpendicularly to their planes. Their first analogy relates the statics of plane loaded frames to the kinematics of space loaded frames.

Characteristic equations (for a member):

SPACE LOADED | | PLANE LOADED |

$\theta'_x - h_x = 0$ (1) $\qquad R'_x + f_x = 0$ (2)

$\theta'_y - h_y = 0$ (3) $\qquad R'_y + f_y = 0$ (4)

$u'_z - g_z + \theta_y = 0$ (5) $\qquad M'_z + c_z + R_y = 0$ (6)

$\theta_x(s+\epsilon) - \theta_x(s-\epsilon) - H_x = 0$ (7) $\qquad R_x(s+\epsilon) - R_x(s-\epsilon) + F_x = 0$ (8)

$\theta_y(s+\epsilon) - \theta_y(s-\epsilon) - H_y = 0$ (9) $\qquad R_y(s+\epsilon) - R_y(s-\epsilon) + F_y = 0$ (10)

$u_z(s+\epsilon) - u_z(s-\epsilon) - L_z = 0$ (11) $\qquad M_z(s+\epsilon) - M_z(s-\epsilon) + C_z = 0$ (12)

138   400. MECHANICS OF MATERIALS (STEADY-STATE)

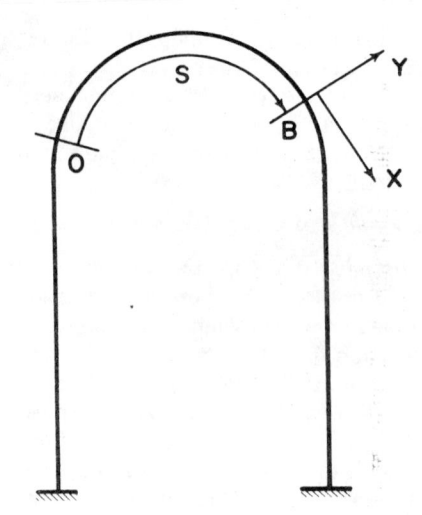

Fig. 472AD.1. Distorted member, showing coordinates.

where: θ = rotation at B (Fig. 472AD.1)
u = displacement at B
h_x = twist
h_y = bend
g_z = slip
H = concentrated distortion
L = concentrated distortion

R = equivalent force at the centroid
M = equivalent couple
f = distributed force at B
c = distributed couple at B
F = concentrated force at B
C = concentrated couple at B

and subscripts x, y, z indicate components in those respective directions. The direction of x is along the member; the direction of y is perpendicular to that of x but is in the plane of the frame. The symbol s stands for length, O is the origin, and primes indicate derivatives with respect to s (Fig. 472AD.1). The symbol ϵ stands for an arbitrary small length.

Some of the above quantities are explained further by the following relationships:

$$R_x = EAg_x \qquad R_y = \frac{GA}{k_y} g_y \qquad R_z = \frac{GA}{k_z} g_z \qquad (13)$$

$$M_x = \beta_x h_x \qquad M_y = EI_z h_y \qquad M_z = EI_y h_z \qquad (14)$$

where: E = modulus of elasticity
A = area of the member cross section
I = moment of inertia of the cross section
G = modulus of rigidity
β_x = torsional rigidity
k_y, k_z = constants depending on the shape of the cross section

g_x = stretch
g_y = slip
h_z = bend

and other symbols are as previously defined in this section.

472AE *Frame Analogy B*

The second analogy of Prager and Hay relates the kinematics of plane loaded frames to the statics of space loaded frames.

Characteristic equations:

PLANE LOADED		SPACE LOADED	
$u'_x - g_x = 0$	(1)	$M'_x + c_x = 0$	(2)
$u'_y - g_y - \theta_z = 0$	(3)	$M'_y + c_y - R_z = 0$	(4)
$\theta'_z - h_z = 0$	(5)	$R'_z + f_z = 0$	(6)
$U_x(s+\epsilon) - U_x(s-\epsilon) - L_x = 0$	(7)	$M_x(s+\epsilon) - M_x(s-\epsilon) + C_x = 0$	(8)
$U_y(s+\epsilon) - U_y(s-\epsilon) - L_y = 0$	(9)	$M_y(s+\epsilon) - M_y(s-\epsilon) + C_y = 0$	(10)
$\theta_z(s+\epsilon) - \theta_z(s-\epsilon) - H_z = 0$	(11)	$R_z(s+\epsilon) - R_z(s-\epsilon) + F_z = 0$	(12)

where symbols are defined as in the preceding analogy.

472AF *Other Frame Analogies*

For an analogy similar to those in sections 472AD and 472AE, see Guerra.

See also Vaughan.

472B *Deflection*

472BA *Conjugate Frame Analogy*

Glenn Murphy (A) has extended the "Moment area" analogy of section 422C to the analysis of the deflections of plane loaded plane frames by utilizing a conjugate frame..

The analogy is developed from the expression for the deflection of point E (Figs. 472BA.1, 472BA.2) in any arbitrary direction y:

$$y = \Delta + \phi_A x_A + \Sigma \phi_i x_i \tag{1}$$

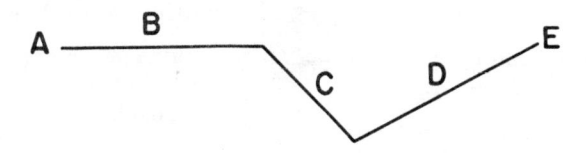

Fig. 472BA.1. Undeformed plane frame.

140   400. MECHANICS OF MATERIALS (STEADY-STATE)

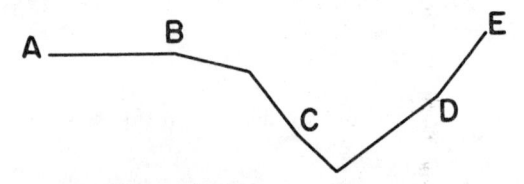

Fig. 472BA.2. Plane frame of figure 472BA.1 deformed by bending.

Since bending is usually continuous rather than concentrated, the relative angle $d\phi_i$ of rotation between two points a distance ds apart along the axis of the frame is:

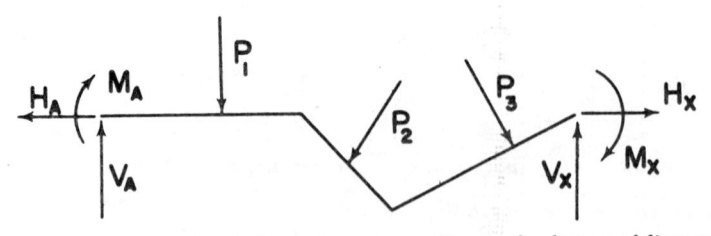

$$d\phi_i = \frac{Mds}{EI} \qquad (2)$$

Fig. 472BA.3. Conjugate frame corresponding to the frame of figures 472BA.1 and 472BA.2.

Characteristic equations:

PLANE LOADED FRAME   CONJUGATE FRAME

$$y = \Delta + \phi_A x_A + \int_A^E \frac{Mds}{EI} x_i \quad (3) \qquad M_x = M_A + V_A x'_A + \Sigma P_i x'_i \quad (4)$$

where: y = y component of the displacement of E

M_x = bending moment at E (see Fig. 472BA.3)

Δ = y component of the displacement at A

M_A = bending moment at A

ϕ_A = change in the angle of the frame at A

V_A = shear at A (the resultant force in the direction AE)

x_A = y component of the distance AE

x'_A = moment arm of V_A with respect to an axis through E

M = bending moment

E = modulus of elasticity of the frame material

P_i = magnitude of a concentrated load

400. MECHANICS OF MATERIALS (STEADY-STATE)

I = moment of inertia of the frame cross section

x_i = y component of the distance from any point of deformation to E

x'_i = moment arm of P_i with respect to an axis through E

The characteristic equations suggest the following conditions:

CONDITION	ACTUAL FRAME	CONJUGATE FRAME
1	y	M_x
2	Δ	M_A
3	ϕ_A	V_A
4	$\int_A^E \frac{M ds}{EI}$	P_i
5	x_A	x'_A
6	x_i	x'_i

Conditions 1 and 2 indicate that a component moment in the original structure corresponds to a component of deflection in the conjugate frame. Condition 3 indicates that the shear in the conjugate frame is analogous to change in slope of the original structure. Condition 4 is satisfied by loading the conjugate frame with the $\frac{M}{EI}$ diagram of the original structure, the loading being perpendicular to the plane in which the conjugate frame lies.

Conditions 5 and 6 are satisfied by making the conjugate frame geometrically similar to the original structure.

End conditions of the conjugate frames may be derived from the design conditions 2 to 6.

472BB *Other Frame Analogies*

See sections 472AD and 472AE.

472BC *Electrical Analogy A*

By using choke coils, transformers, and condensers, Schoenfeld (B) has developed "an electrical model of a straight bar in which the terminals represent the ends of the bar, the currents are proportional to the forces and moments impressed on the end sections, and the potentials of the terminals (which are the measured quantities) are proportional to the linear and angular displacements of the end sections. A frame is represented by an assembly of a number of such models with their terminals connected according to the position and boundary conditions of the bars in the frame. Two illustrated examples are given." [APPLIED MECHANICS REVIEW]

472BD *Other Electrical Analogies*

See sections 472AA and 472AB, especially Ryder (D).

400. MECHANICS OF MATERIALS (STEADY-STATE)

480 REINFORCED SHEETS, PANELS, AND SHELLS

481 *Electrical Analogy A*

Goran has developed a d.c. electrical circuit analogy for the stress distributions in stiffened shells.

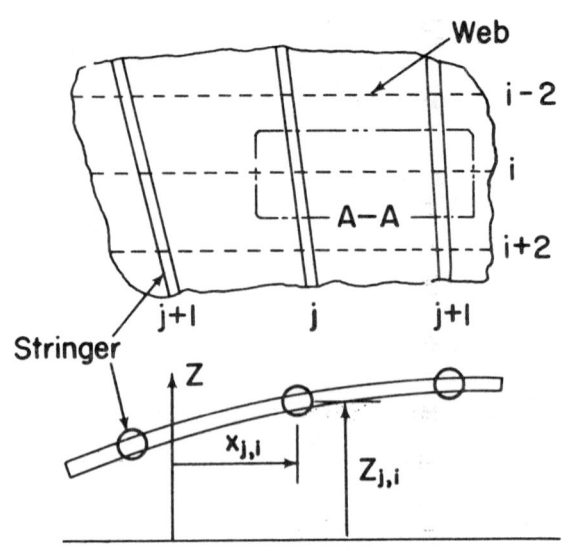

Fig. 481.1. Section of a stiffened shell.

Characteristic equations:

STRESS

$$P_{j,i} \frac{L_i}{A_{j,i} E} + x_{j,i} X_i + z_{j,i} Z_i + q_{j,i-1} \left(\frac{w}{t}\right)_{j,i-1} \left(\frac{1}{G}\right) + 2\overline{A}_{j,i-1} \frac{\theta_{i-1}}{L_{i-1}}$$

$$- P_{j-1,i} \frac{L_i}{A_{j-1,i} E} - x_{j-1,i} X_i - z_{j-1,i} Z_i - q_{j,i+1} \left(\frac{w}{t}\right)_{j,i+1} \left(\frac{1}{G}\right)$$

$$- 2\overline{A}_{j,i+1} \frac{\theta_{i+1}}{L_{i+1}} = 0 \qquad (1)$$

ELECTRICAL

$$I_{j,i} R_{j,i} + x_{j,i} X_i + z_i Z_i + i_{j,i-1} r_{j,i-1} + v_{j,i-1} \theta_{i-1} - I_{j-1,i} R_{j-1,i}$$

$$- x_{j-1,i} X_i - z_{j-1,i} Z_i - i_{j,i+1} r_{j,i+1} - v_{j,i+1} \theta_{i+1} = 0 \qquad (2)$$

400. MECHANICS OF MATERIALS (STEADY-STATE)

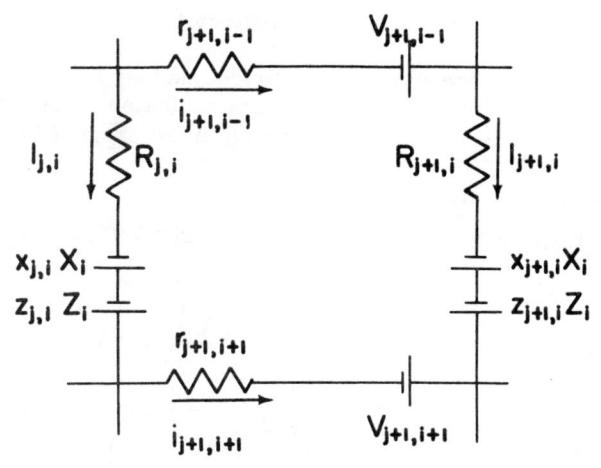

Fig. 481.2. Analog circuit for region A-A of stiffened shell section (figure 481.1).

where: P = stringer axial load   I = longitudinal current
        L = length of stringer or
            web element
        A = cross-sectional   R = longitudinal resistance
            area of stringer
        E = modulus of elasticity
        q = web shear flow   i = transverse current
        w = developed width of
            shear web element
        t = sheet thickness   r = transverse resistance
        G = modulus of rigidity
X, Z, θ = Lagrangian   X, Z, θ = voltage factors
            multipliers
     x, z = stringer coordinates   x, z = longitudinal voltages
     $\overline{A}$ = area enclosed by web   v = transverse voltage
            median line and
            lines joining the ends
            of the web to the tor-
            sional reference axis

An a.c. circuit has been used in a modification of this analogy.

144   400. MECHANICS OF MATERIALS (STEADY-STATE)

482 *Electrical Analogy B*

Newton (A, B) has presented an electrical analogy for the shear lag problem — the determination of the stress distribution in sheets longitudinally reinforced by stringers and loaded in the plane of the sheet by longitudinal forces.

Characteristic equations:

SHEAR LAG   ELECTRICAL

$$P_{m,n} = P_{m,n-1} + (q_{m-1,n}\Delta y) - (q_{m,n}\Delta y) \quad (1)$$

$$I_{m,n} = I_{m,n-1} + i_{m-1,n} - i_{m,n} \quad (2)$$

$$u_{m,n-1} - u_{m,n} = P_{m,n}\frac{\Delta y_n}{EA_{m,n}} \quad (3)$$

$$V_{m,n-1} - V_m = I_{m,n} R_{m,n} \quad (4)$$

$$u_{m+1,n} - u_{m,n} = q_{m,n}\Delta y \frac{w_m}{G\Delta y t_{m,n}} \quad (5)$$

$$V_{m+1,n} - V_{m,n} = (i_{m,n})(r_{m,n}) \quad (6)$$

where: $P_{m,n}$ = axial load in stringer m at station n

$I_{m,n}$ = current carried by longitudinal conductor m over length n

$q_{m,n}$ = shear flow in panel m, bay n

$i_{m,n}$ = current carried by transverse conductor n over width m

Δy = length of bay

$u_{m,n}$ = y displacement of the point on stringer m at the middle of bay n

$V_{m,n}$ = potential at m,n

Δy_n = distance from the middle of bay n to the middle of bay n+1

$R_{m,n}$ = resistance of longitudinal conductor m over length n

E = modulus of elasticity

$A_{m,n}$ = area of stringer m at station n

w_m = width of panel m

G = modulus of rigidity

$t_{m,n}$ = thickness of the sheet in panel m, bay n

$r_{m,n}$ = resistance of transverse conductor n over width m

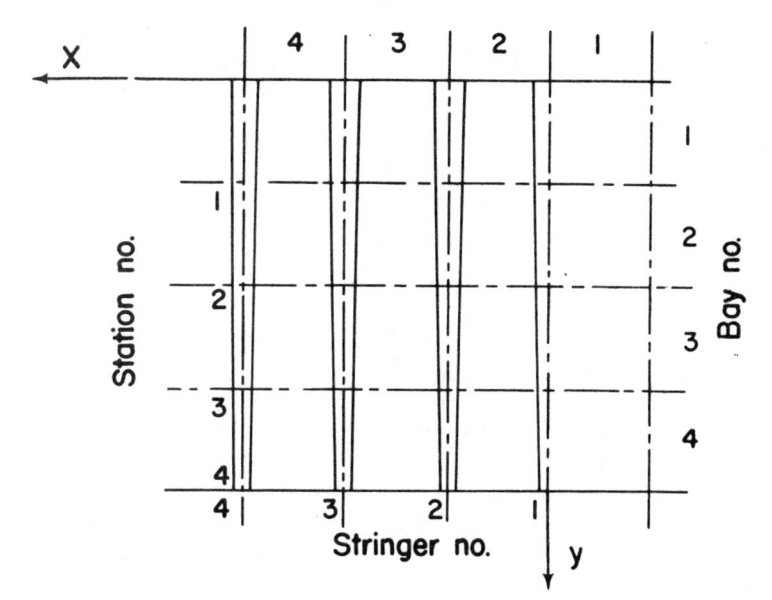

Fig. 482.1. Half panel in shear lag.

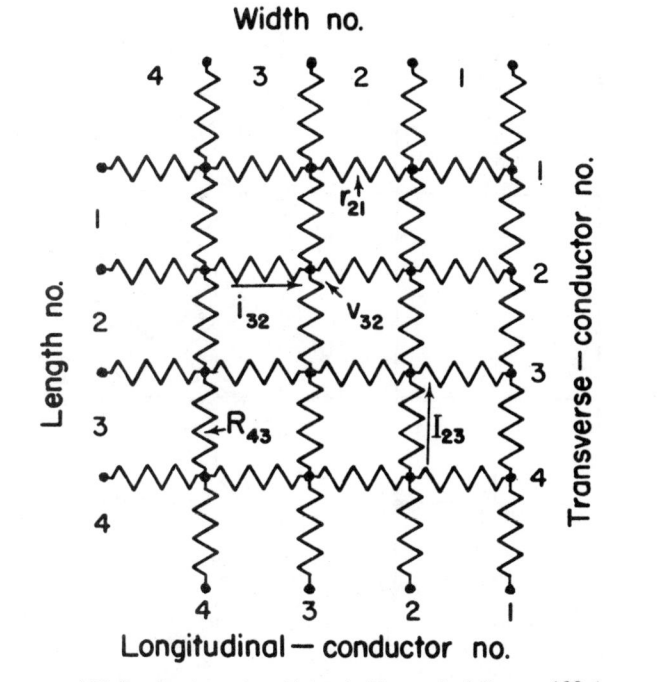

Fig. 482.2. Analog circuit for half panel of figure 482.1.

146   400. MECHANICS OF MATERIALS (STEADY-STATE)

and m is the general subscript for stringer, longitudinal conductors, panels, and widths, n is the general subscript for stations, lengths, transverse conductors, and bays.

The current through a resistor is found by measuring the voltage across the resistor and dividing by the resistance. This is then converted to force and finally to stress.

Included in the report are plots of axial stress vs. longitudinal distance from analytical, direct-experimental, and analogic solutions for panels tested in tension and as beams.

See also Engle and Newton.

483 *Electrical Analogy C*

An electric network analogy for analyzing displacements and forces in reinforced panels has been suggested by Hoff and Libby. They have assumed that the transverse stiffeners are infinitely rigid so that the vertical, or longitudinal, displacements u alone need be determined. The portion of the sheet covering considered effective in tension or compression is added to the cross-sectional area of each stringer, and the panels of each sheet are assumed to carry shear stresses only. A consequence of these assumptions is that the shearing stress must be constant in each panel.

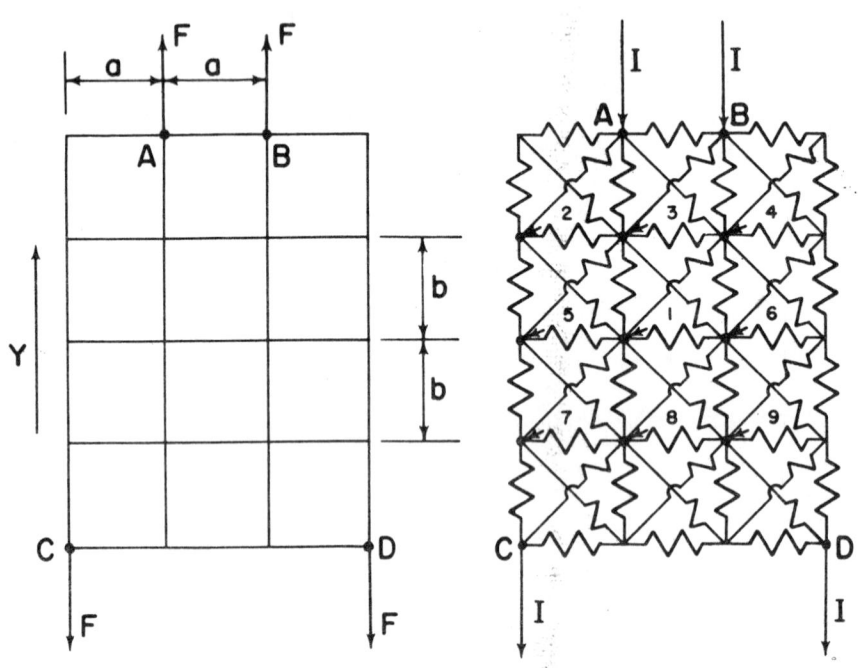

Fig. 483.1. Section of a reinforced panel.

Fig. 483.2. Analog circuit for reinforced panel of figure 483.1.

400. MECHANICS OF MATERIALS (STEADY-STATE)

Characteristic equations:

PANELS		ELECTRICAL	
$\Sigma F = 0$ (At each junction)	(1)	$\Sigma I = 0$	(2)
$F_{81} = u\dfrac{EA}{b} - \dfrac{Gbt}{2q}$	(3)	$I_{81} = VC_{81}$	(4)
$F_{91} = u\dfrac{Gbt}{4a}$ Typical Relationships	(5)	$I_{91} = VC_{91}$	(6)
$F_{61} = u\dfrac{Gbt}{2a}$	(7)	$I_{61} = VC_{61}$	(8)

where: F = force through a stringer

u = displacement between two points

I = current

V = voltage

Influence coefficients:

E = modulus of elasticity

A = cross-sectional area of a stringer

b = distance between adjacent transverse stiffeners

G = modulus of rigidity

a = distance between adjacent longitudinal stringers

C = conductance

484 *Other Electrical Analogies*

For other electrical analogies for the study of reinforced sheets, panels, and shells, see MacNeal (B, C); Ryder (D); [Sec. 472AC].

490 OTHER

491 *Visco-Elasticity*

By using electrical instead of mechanical components, simple linear or non-linear models can be assembled rapidly so that creep curves, stress-strain characteristics or dynamic performance of elastic materials can be obtained; thus some unusual characteristics may be reproduced and special force-time relationships may be observed.

400. MECHANICS OF MATERIALS (STEADY-STATE)

Stambaugh has described the use of electrical models to simulate mechanical systems which are approximations to specimens of rubber or other long chain polymers. In the article, both creep and vibration tests were made; and photographs, diagram of apparatus, diagrams of models and oscillograms of results were given.

Gross and Pelyer showed how an electric network analog provided an illustration of the relation between the relaxation spectrum and the creep spectrum of a visco-elastic material.

For further references, see Alfrey; Burgers; Fromm; Gast; Horton; A. J. Kennedy (A, B); Olshak and Litvinishin.

492 *Miscellaneous*

Kvitka, Agarev, and Umanski have developed an electric circuit model for the analysis of the stress condition in an elastic body of rotation under the action of an axially symmetrical, superficial and three-dimensional load and unsymmetrical heating setting up an axially symmetrical temperature field.

For other analogies pertaining to mechanics of materials, see Chatterjee and Dutt; Colin; Fenn (A, B); Jasper (B); Kihara and Masabuchi; Kron (D, N); Rawlings.

Section 500. Mechanical Vibrations and Transients

510 AXIAL SYSTEMS

511 *Concentrated Masses*

511A *Electrical Analogy A*

There are basically two different electrical analogies for the axial vibration of concentrated masses.

The "traditional" analogy relates mass velocity with electric current. A complete list of analogous quantities is given below.

MECHANICAL	ELECTRICAL
x = displacement of the mass	q = charge stored in the capacitor
v = velocity of the mass	i = current through an element ($= \frac{dq}{dt}$)
t = time	t = time
m = concentrated mass	L = inductance
r = viscous damping coefficient	R = resistance
k = spring constant	$\frac{1}{C} = \frac{1}{\text{capacitance}}$
f = force acting on the mass	e = electric potential
p = momentum of the mass ($= \int f \, dt$)	$\int e \, dt$

A typical application of this analogy is in the analysis of a "parallel" vibrating mass system of one degree of freedom (as illustrated in figure 511A.1), upon which a force, $f(t)$, varying with time is acting.

150   500. MECHANICAL VIBRATIONS AND TRANSIENTS

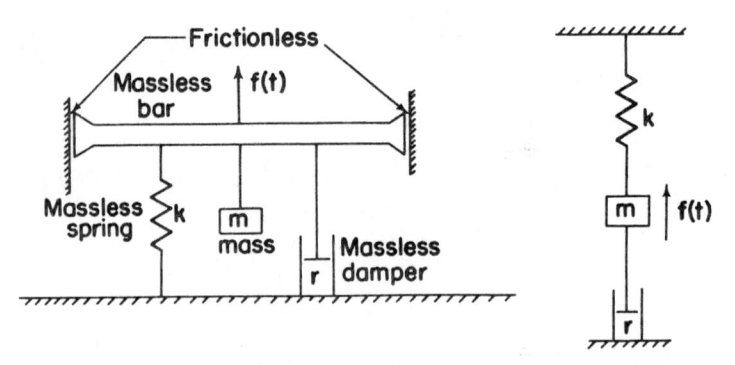

Fig. 511A.1. Two idealized representations of a "parallel" vibrating mass system of one degree of freedom.

The electrical analog for this system is shown in figure 511A.2. The characteristic equations for this system are given below.

$$m \frac{d^2x}{dt^2} + r \frac{dx}{dt} + kx = f(t) \quad (1) \qquad L \frac{d^2q}{dt^2} + R \frac{dq}{dt} + \frac{1}{C} q = e(t) \quad (2)$$

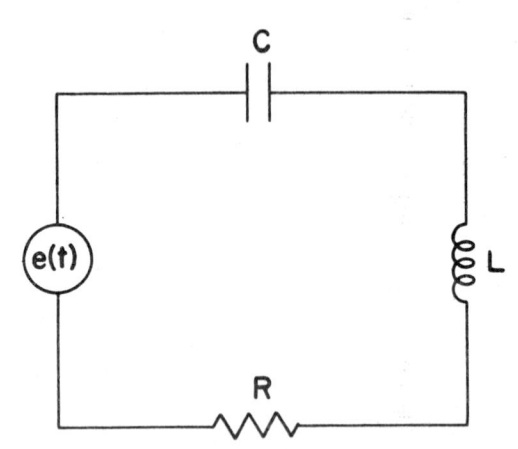

Fig. 511A.2. Series analog circuit for vibrating mass system of one degree of freedom.

The electrical analog for a "series" vibrating mass system of one degree of freedom (Fig. 511A.3) is shown in figure 511A.4. The characteristic equations for this system are given below.

$$\frac{1}{k} \frac{df}{dt} + \frac{1}{r} f + \frac{1}{m} \int f dt = v(t) \quad (3) \qquad C \frac{de}{dt} + \frac{1}{R} e + \frac{1}{L} \int e dt = i(t) \quad (4)$$

The analogy may be extended to systems which are combinations of the above mentioned in several degrees of freedom.

500. MECHANICAL VIBRATIONS AND TRANSIENTS 151

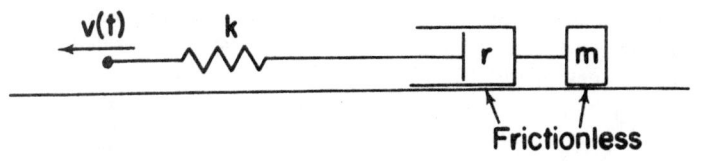

Fig. 511A.3. Idealized representation of a "series" vibrating mass system of one degree of freedom.

For further discussion of the theory of this analogy, see Bloch (A, B); Bordoni; Bush (A, B); Cherry; Criner, McCann, and Warren; Freberg; Gehlshøj; Graham; Jefferson; Kemler (B); Kemler and Freberg; Klotter; Le Corbeiller and Yeung; Manley; Miles (B); Nickle; Olson; Pawley; Sanial; Say; Warshavsky and Fedorovich.

Housner and McCann have applied the analogy (with multidegrees of freedom) to the analysis of strong motion earthquakes. Included are oscillograms of typical oscillations.

Mechanical elements of recording apparatus have been studied by Maxfield and Harrison by this analogy.

McCann and Criner (A, B) have presented examples of typical applications of the analogy, including sketches, circuit diagrams, and oscillograms.

Use of the analogy to study vibrating machines has been made by Soroka (B).

Welch has described how the analogy was used to investigate the basic principle of a new type of shock recorder used to measure mechanical shock in field tests on ships, airplanes, and other vehicles. The report includes curves of amplitude vs. time. The results verified that damping should be approximately 0.5 critical and showed that errors to be expected from the required computations should not exceed 5%.

The results of a general study (by use of the analogy) of linear systems with two degrees of freedom have been presented by McCann and Kopper. See also Gustafson, von Karman and Biot.

M. J. Murphy, Bycroft, and Harrison have modified this analogy for

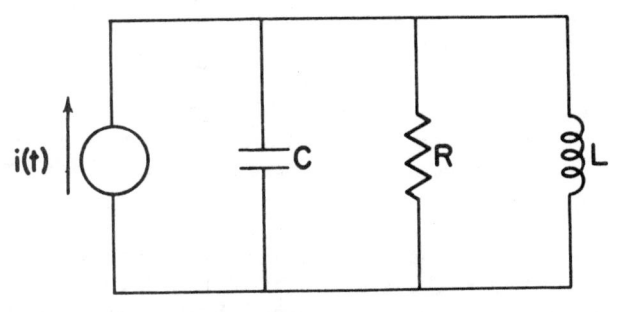

Fig. 511A.4. Parallel analog circuit for vebrating mass system of one degree of freedom.

152   500. MECHANICAL VIBRATIONS AND TRANSIENTS

the investigation of stresses developed in buildings due to earthquakes. Bycroft, Murphy, and Brown have extended this type of investigation in a study of the effects of the 1940 El Centro earthquake.

511B *Electrical Analogy B*

A "new" electrical analogy for the axial vibration of concentrated masses has been proposed by several authors, notably among them Firestone (A, B, C). This analogy relates mass velocity with electric potential. A complete list of analogous quantities is given below.

MECHANICAL	ELECTRICAL
x	$\int e\,dt$
v	e
t	t
m	C
r	$\dfrac{1}{R}$
k	$\dfrac{1}{L}$
f	i
p	q

where symbols are defined as in section 511A.

In this analogy, the electrical analog for the "parallel" mass system of figure 511A.1 is the parallel electric circuit of figure 511A.4. The characteristic equations for this system are given below.

$$m\frac{dv}{dt} + rv + k\int v\,dt = f(t) \quad (1) \qquad C\frac{de}{dt} + \frac{1}{R}e + \frac{1}{L}\int e\,dt = i(t) \quad (2)$$

The electrical analog for the "series" mass system of figure 511A.3 is now the series circuit of figure 511A.2. Characteristic equations for this system are given below.

$$\frac{1}{k}\frac{df}{dt} + \frac{1}{r}f + \frac{1}{m}\int f\,dt = v(t) \quad (3) \qquad L\frac{di}{dt} + Ri + \frac{1}{C}\int i\,dt = e(t) \quad (4)$$

This analogy also may be extended to systems which are combinations of the above mentioned in several degrees of freedom.

For further discussion of the theory of this analogy, see Criner, McCann, and Warren; Hahnle; Le Corbeiller and Yeung; McCann and Criner (A, B); Saxton; Toupin; Trent (A).

For other references dealing with this analogy or the analogy of section 511A, see Bauer (C); Belgaumkar (B, C); Fagin; Kettenacker; Koffmann; the article, "Mechanical Circuits"; Reichhardt; Schönfeld (A); Trent (C); Whitehead.

500. MECHANICAL VIBRATIONS AND TRANSIENTS

511C *Other Analogies*

Canning has proposed a rolling ball analogy for pitching and yawing motions.

Kron (E) developed equivalent circuits for some special oscillating systems.

Berry has used a resistor network to study the oscillation of a mass connected to a nonlinear spring.

See also Caughey and Hudson.

512 *Distributed Masses*

512A *Frictionless*

The analogy between an electrical transmission line without resistance and a longitudinally vibrating rod has been presented by Pipes (C).

Characteristic equations:

MECHANICAL

$$\frac{1}{AE} \frac{\partial F}{\partial t} = - \frac{\partial v}{\partial x} \quad (1)$$

$$\rho A \frac{\partial v}{\partial t} = - \frac{\partial F}{\partial x} \quad (3)$$

$$\frac{\rho}{E} \frac{\partial^2 F}{\partial t^2} = \frac{\partial^2 F}{\partial x^2} \quad (5)$$

ELECTRICAL

$$C \frac{\partial e}{\partial t} = - \frac{\partial i}{\partial x} \quad (2)$$

$$L \frac{\partial i}{\partial t} = - \frac{\partial e}{\partial x} \quad (4)$$

$$LC \frac{\partial^2 e}{\partial t^2} = \frac{\partial^2 e}{\partial x^2} \quad (6)$$

where: A = cross sectional area of the rod
E = modulus of elasticity
ρ = density of the rod material
F = longitudinal force
v = longitudinal velocity of any point of the rod
t = time
x = longitudinal coordinate

C = capacitance per unit length
L = inductance per unit length
e = voltage
i = current
t = time
x = longitudinal coordinate

Pipes indicates the development of the operational equations of transmission lines and their application to the analysis of special cases of rod vibration.

For this same analogy, Simonyi shows that various end conditions can be represented in terms of impedance so that mechanical problems are reduced to electrical problems. The problems considered include

154   500. MECHANICAL VIBRATIONS AND TRANSIENTS

reflection and transmission at the interface between rods of different density and elastic modulus, tapered rods and rods with discontinuous diameter, and natural frequencies of such compound rods. An approximation method is proposed for estimating the sizes of cavities or cracks in rods by measuring reflection and transmission properties. The striking of a hammer on an anvil with or without a plastically deforming object interposed is considered by idealizing the problem considerably.

512B *With Friction*

An electric network analogy has been applied by Kemler (A) in determining the loads and velocities of sucker rods of oil well pumping units.

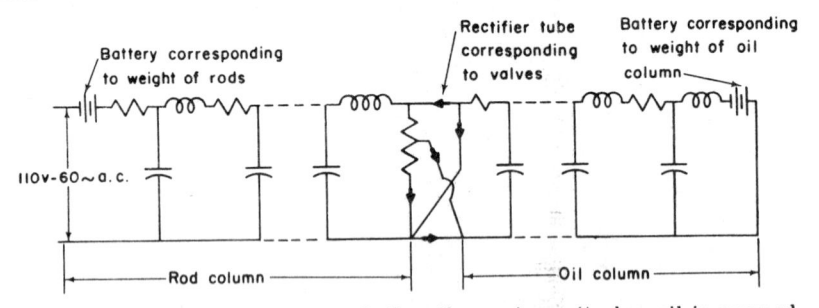

Fig. 512B.1. Electrical analog of oil well pumping unit when oil is pumped on both the upstroke and the downstroke.

For a sucker rod:

$$\frac{W}{g}\frac{\partial^2 y}{\partial t^2} + r\frac{\partial y}{\partial t} = AE\frac{\partial^2 y}{\partial x^2} + W \qquad (1)$$

where y = unit elongation of the rod.
However, the following substitution is made:

$$y = \bar{y} - \frac{Wx^2}{2AE} + \frac{Whx}{AE} - \frac{Wh^2}{2AE}$$

where h = length of rod.
Characteristic equations:

MECHANICAL

$$\frac{W}{g}\frac{\partial^2 \bar{y}}{\partial t^2} + r\frac{\partial \bar{y}}{\partial t} = AE\frac{\partial^2 \bar{y}}{\partial x^2} \quad (2)$$

ELECTRICAL

$$L\frac{\partial^2 Q}{\partial t^2} + R\frac{\partial Q}{\partial t} = \frac{1}{C}\frac{\partial^2 Q}{\partial x^2} \quad (3)$$

where: W = rod weight per unit length

L = inductance per unit length

500. MECHANICAL VIBRATIONS AND TRANSIENTS

g = acceleration of gravity
r = resistance constant
A = rod cross sectional area
E = modulus of elasticity
$\bar{y}$ = (defined above)
t = time
x = longitudinal coordinate

R = resistance per unit length
C = capacitance per unit length
Q = charge
t = time
x = longitudinal coordinate

Included in the report are typical oscillographs of load and velocity vs. time for the top and the bottom of the well.

See also Kemler (C).

For further discussion of electrical analogies for longitudinal vibrations, see Pipes (A), who has analyzed the longitudinal motions of trains by use of an electrical analogy.

See also Lazaryan.

520 BEAMS AND PLATES (FLEXURE)

521 *Electrical Analogy A*

McCann and MacNeal have presented an electrical analogy for the analysis of the steady state and transient vibration of members in uncoupled flexure, with particular application to aircraft wings.

For the uncoupled flexural vibration of members:

$$\frac{\partial^2}{\partial x^2} \left(EI \frac{\partial^2 y}{\partial x^2} \right) + m \frac{\partial^2 y}{\partial t^2} = 0 \qquad (1)$$

where: y = lateral deflection
x = axial coordinate
E = modulus of elasticity
I = moment of inertia of the cross section
t = time
m = mass per unit length

By making the substitutions,

$$\theta = \frac{\partial y}{\partial x} \qquad (2)$$

$$V = \frac{\partial}{\partial x} \left(EI \frac{\partial \theta}{\partial x} \right) \qquad (3)$$

where θ = slope and V = shear, equation (1) becomes

500. MECHANICAL VIBRATIONS AND TRANSIENTS

$$\frac{\partial V}{\partial x} + m \frac{\partial^2 y}{\partial t^2} = 0 \qquad (4)$$

These last three equations may be replaced by corresponding finite difference equations.

Characteristic equations:

FLEXURAL

$$\theta_{n+\frac{1}{2}} = \frac{y_{n+1} - y_n}{\overline{\Delta x}} \qquad (5)$$

$$\overline{\Delta x}\,(V)_{n+\frac{1}{2}} = (\theta_{n+\frac{3}{2}} - \theta_{n+\frac{1}{2}})\left(\frac{EI_{n+1}}{\overline{\Delta x}}\right) + (\theta_{n-\frac{1}{2}} - \theta_{n+\frac{1}{2}})\left(\frac{EI_n}{\overline{\Delta x}}\right) \qquad (7)$$

$$(V_{n+\frac{1}{2}} - V_{n-\frac{1}{2}}) + (\overline{\Delta x}\,m_n)\frac{\partial^2 y_n}{\partial t^2} = 0 \qquad (9)$$

ELECTRICAL

$$(V_\alpha)_{n+\frac{1}{2}} = \frac{(V_\beta)_{n+1} - (V_\beta)_n}{\overline{\Delta x}} \qquad (6)$$

$$\overline{\Delta x}\left(\frac{di_\beta}{dt}\right)_{n+\frac{1}{2}} = \left[(V_\alpha)_{n+\frac{3}{2}} - (V_\alpha)_{n+\frac{1}{2}}\right]\left(\frac{1}{L_{n+1}}\right) + \left[(V_\alpha)_{n-\frac{1}{2}} - (V_\alpha)_{n+\frac{1}{2}}\right]\left(\frac{1}{L_n}\right) \qquad (8)$$

$$\left[\left(\frac{di_\beta}{dt}\right)_{n+\frac{1}{2}} - \left(\frac{di_\beta}{dt}\right)_{n-\frac{1}{2}}\right] + (C_n)\frac{\partial^2 (V_\beta)_n}{\partial t^2} = 0 \qquad (10)$$

where subscripts represent nodes, $\overline{\Delta x}$ equals distance between nodes, and electrical symbols are as indicated in figure 521.1.

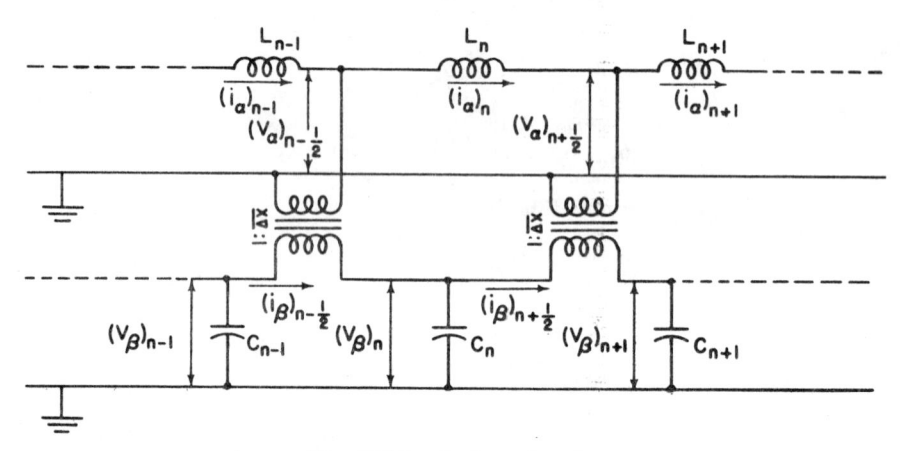

Fig. 521.1. Analog circuit.

500. MECHANICAL VIBRATIONS AND TRANSIENTS

Thus, if inductance is made inversely proportional to the product EI and capacitance made directly proportional to mass per unit length, the following flexural quantities will be proportional to the corresponding electrical quantities.

FLEXURAL	ELECTRICAL
θ	V_α
y	V_β
V	i_β
M	$\dfrac{di_\alpha}{dt}$

Oscillograms of deflections and slopes vs. time for various stations (nodes) are included in the report. In addition, there are plots of symmetric and unsymmetric node shapes and frequencies.

For a similar analogy, see MacNeal, McCann, and Wilts. Their report includes oscillographic solutions (by analogy) of an airplane wing subjected to vertical gust loading.

522 *Electrical Analogy B*

Another analogy quite similar to the previous one has been indicated by McCann and MacNeal.

Characteristic equations:

FLEXURAL

$$\theta_{n+\frac{1}{2}} = \frac{y_{n+1} - y_n}{\overline{\Delta x}} \tag{1}$$

$$\overline{\Delta x}\,(V)_{n+\frac{1}{2}} = (\theta_{n+\frac{3}{2}} - \theta_{n+\frac{1}{2}})\left(\frac{EI_{n+1}}{\overline{\Delta x}}\right) + (\theta_{n-\frac{1}{2}} - \theta_{n+\frac{1}{2}})\left(\frac{EI_n}{\overline{\Delta x}}\right) \tag{3}$$

$$(V_{n+\frac{1}{2}} - V_{n-\frac{1}{2}}) + (\overline{\Delta x m}_n)\frac{\partial^2 y_n}{\partial t^2} = 0 \tag{5}$$

ELECTRICAL

$$(q_\beta)_{n+\frac{1}{2}} = \frac{(q_\alpha)_{n+1} - (q_\alpha)_n}{\overline{\Delta x}} \tag{2}$$

$$\overline{\Delta x}\,(V_\alpha)_{n+\frac{1}{2}} = \left[(q_\beta)_{n+\frac{3}{2}} - (q_\beta)_{n+\frac{1}{2}}\right]\left(\frac{1}{C_{n+1}}\right) + \left[(q_\beta)_{n-\frac{1}{2}} - (q_\beta)_{n+\frac{1}{2}}\right]\left(\frac{1}{C_n}\right) \tag{4}$$

$$\left[(V_\alpha)_{n+\frac{1}{2}} - (V_\alpha)_{n-\frac{1}{2}}\right] + (L_n)\frac{\partial^2 (q_\alpha)_n}{\partial t^2} = 0 \tag{6}$$

158   500. MECHANICAL VIBRATIONS AND TRANSIENTS

where $q = \int i\,dt$ and other symbols are as indicated in the analogy of section 521. The same analog circuit is used.

523 *Electrical Analogy C*

Trent (B) has indicated an electrical analogy for vibrating beams, taking into account the effect of shear.

Characteristic equations:

FLEXURAL

$$\left(\frac{dy}{dt}\right)_2 - \left(\frac{dy}{dt}\right)_1 = \left[\left(\frac{dy}{dt}\right)_{2s} - \left(\frac{dy}{dt}\right)_{1s}\right] + \left[\left(\frac{dy}{dt}\right)_{2b} - \left(\frac{dy}{dt}\right)_{1b}\right] \quad (1)$$

$$V_1 - V_2 = \left(\frac{\mu \overline{\Delta x}}{2}\right)\left[\left(\frac{d^2y}{dt^2}\right)_1 + \left(\frac{d^2y}{dt^2}\right)_2\right] \quad (3)$$

$$\left(\frac{d\gamma}{dt}\right) = \frac{1}{\overline{\Delta x}}\left[\left(\frac{dy}{dt}\right)_{2b} - \left(\frac{dy}{dt}\right)_{1b}\right] \quad (5)$$

$$M_2 - M_1 = \left[V_1 - \frac{\mu \overline{\Delta x}}{2}\left(\frac{d^2y}{dt^2}\right)_1\right]\overline{\Delta x} + (\mu \rho^2 \overline{\Delta x})\frac{d^2\gamma}{dt^2} \quad (7)$$

$$\left(\frac{dy}{dt}\right)_2 - \left(\frac{dy}{dt}\right)_1 = \left(\frac{\overline{\Delta x}}{2EI}\right)\left[\left(\frac{dM}{dt}\right)_1 + \left(\frac{dM}{dt}\right)_2\right] \quad (9)$$

$$\frac{d}{dt}\left[V_1 - \left(\frac{\mu \overline{\Delta x}}{2}\right)\left(\frac{d^2y}{dt^2}\right)_1\right] = \left(\frac{dy}{dt}\right)_{1s} - \left(\frac{dy}{dt}\right)_{2s} \quad (11)$$

ELECTRICAL

$$(V_y)_2 - (V_y)_1 = \left[(V_y)_{2s} - (V_y)_{1s}\right] + \left[(V_y)_{2b} - (V_y)_{1b}\right] \quad (2)$$

$$(i_y)_1 - (i_y)_2 = C_y\left[\left(\frac{dV_y}{dt}\right)_1 + \left(\frac{dV_y}{dt}\right)_2\right] \quad (4)$$

$$V_\gamma = \frac{1}{\overline{\Delta x}}\left[(V_y)_{2b} - (V_y)_{1b}\right] \quad (6)$$

$$(i_\gamma)_2 - (i_\gamma)_1 = \left[(i_y)_1 - C_y\left(\frac{dV_y}{dt}\right)_1\right]\overline{\Delta x} + C_\gamma \frac{dV_\gamma}{dt} \quad (8)$$

$$(V_\gamma)_2 - (V_\gamma)_1 = L_\gamma\left[\left(\frac{di_\gamma}{dt}\right)_1 + \left(\frac{di_\gamma}{dt}\right)_2\right] \quad (10)$$

$$\frac{d}{dt}\left[(i_y)_1 - C_y\left(\frac{dV_y}{dt}\right)_1\right]L_y = (V_y)_{1s} - (V_y)_{2s} \quad (12)$$

500. MECHANICAL VIBRATIONS AND TRANSIENTS

where:
- y = lateral deflection
- x = axial coordinate
- t = time
- V = shear force
- μ = average mass per unit length
- γ = slope of the neutral axis if bending alone exists
- M = bending moment
- ρ = radius of gyration about the mass center
- E = modulus of elasticity
- I = moment of inertia of the cross section
- A = area of the cross section
- G = modulus of rigidity
- $K = \dfrac{\text{average shear strain over the cross section}}{\text{shear strain at the neutral axis}}$

Subscripts 1 and 2 denote stations, s denotes shear motion alone, b denotes bending motion alone, and electrical symbols are as indicated in figure 523.1.

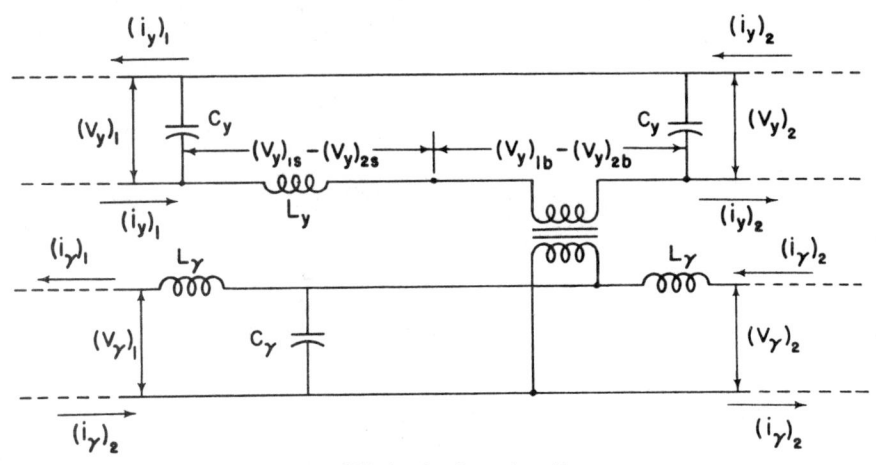

Fig. 523.1. Analog circuit.

524 *Other Electrical Analogies*

An electrical apparatus to determine wing flutter velocities is described in the article, "Electrical Computer Solves Wing Flutter Problems." The wing is considered as a cantilever beam fastened to a fuselage of infinite mass.

Decomposing a beam into a number of finite masses joined together by an elastic massless cantilever, Cahen has described a generalization of known electromechanical analogies by means of which it is possible to measure the eigenfrequencies of this system. To each mass corresponds an electrical network composed of two circuits, one

500. MECHANICAL VIBRATIONS AND TRANSIENTS

analogous to shear force and displacement (due to bending and shearing), the other to bending moment and rotation. Frequency of the electrical system has the inverted value of that of the mechanical one in order to realize the coupling of the two circuits. Extension is made to torsional vibration. [APPLIED MECHANICS REVIEW.]

Baruch and Lang have described electric circuits for the simulation of loudspeakers.

See also Benscoter and MacNeal (A, B, C); Bisplinghoff, Pian, and Levy; Biot and Wianko; Bondar (A, B); Chegolin (B); Hagg and Warner; Jennings; Kharlamov; Kron (D); MacNeal (A); MacNeal and Benscoter (A, B); E. F. Murphy (A); Polya; Scanlan (B); Wilts.

See also sections 461, 472AB, and 532C.

530 TORSIONAL MEMBERS

531 *Concentrated Masses*

531A *Electrical Analogy A*

Many applications have been made of the "inertia inductance" analogy for investigating the torsional vibration of concentrated masses connected by massless shafts.

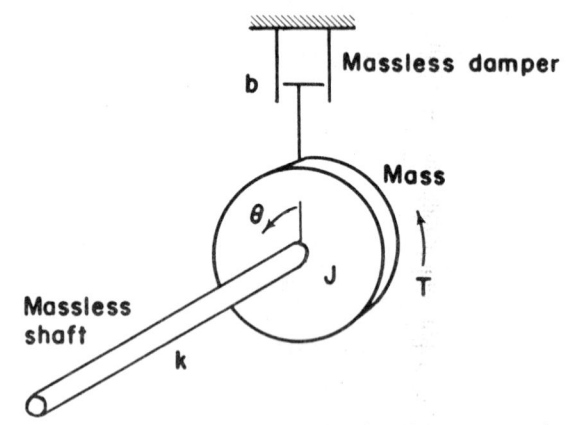

Fig. 531A.1. Typical torsionally vibrating mass system.

Characteristic equations for a typical system (Fig. 531A.1) (single degree of freedom):

$$J \frac{d^2 \theta}{dt^2} + b \frac{d\theta}{dt} + k\theta = T(t) \quad (1) \qquad L \frac{d^2 q}{dt^2} + R \frac{dq}{dt} + \frac{1}{C} q = e(t) \quad (2)$$

where: θ = angular displacement of the mass q = charge stored in the capacitor

500. MECHANICAL VIBRATIONS AND TRANSIENTS 161

t = time	t = time
J = polar moment of inertia of the mass	L = inductance
b = viscous damping coefficient	R = resistance
k = torsional flexibility of the shaft	C = capacitance
T(t) = applied torque (a function of time)	e(t) = applied electric potential (a function of time)

The analog circuit for this system is shown in figure 531A.2.

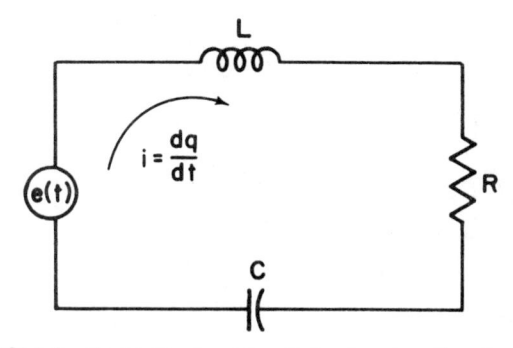

Fig. 531A.2. Series analog circuit for torsionally vibrating mass system of figure 531A.1.

Most of the applications of this analogy involve two or more distributed masses. Such applications involve slightly modified equations.

Pipes (B) has used this analogy to study the vibrational characteristics of torsional systems involving several disks on the same shaft. He has discussed the application of the analogy in determining the natural frequencies of multi-cylinder engines, analyzing geared systems, and studying forced oscillation.

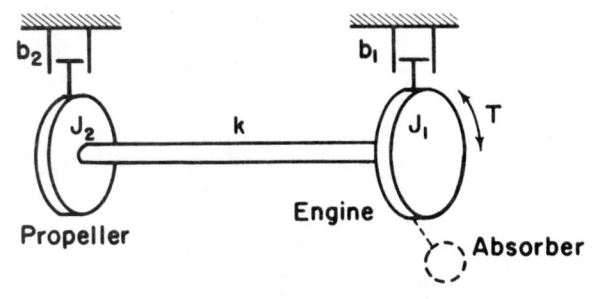

Fig. 531A.3. Schematic diagram of torsional system with dynamic absorbers.

162   500. MECHANICAL VIBRATIONS AND TRANSIENTS

Anderson and Soroka have utilized this analogy in solving vibration problems involving centrifugally tuned dynamic absorbers. For a representative dynamical system of this nature, consisting of a propeller, crankshaft, and engine:

TORSIONAL   ELECTRICAL

$$J_1 \frac{d^2\theta_1}{dt^2} + b_1 \frac{d\theta_1}{dt} + k(\theta_1 - \theta_2) = T \quad (3) \qquad L_1 \frac{d^2 q_1}{dt^2} + R_1 \frac{dq_1}{dt} + \frac{1}{C}(q_1 - q_2) = E \quad (4)$$

$$J_2 \frac{d^2\theta_2}{dt^2} + b_2 \frac{d\theta_2}{dt} + k(\theta_2 - \theta_1) = 0 \quad (5) \qquad L_2 \frac{d^2 q_2}{dt^2} + R_2 \frac{dq_2}{dt} + \frac{1}{C}(q_2 - q_1) = 0 \quad (6)$$

where quantities are as in the figures.

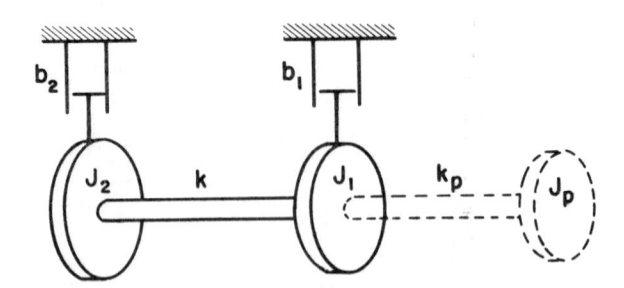

Fig. 531A.4. Schematic diagram of a system equivalent to that of figure 531A.3.

Application of the analogy to problems in torsional oscillations of ship drives has been made by Concordia (A).

Gross and Soroka (A) have utilized this analogy in simulating a two degree of freedom oscillatory torsional system containing a pre-set spring. The pre-set spring introduces a sharp break, or knee, in the load deflection curve of the shaft containing this spring. A considerable nonlinearity is present, the ratio of the slopes of the two straight lines meeting at the break being 1:50. An electronic circuit consisting

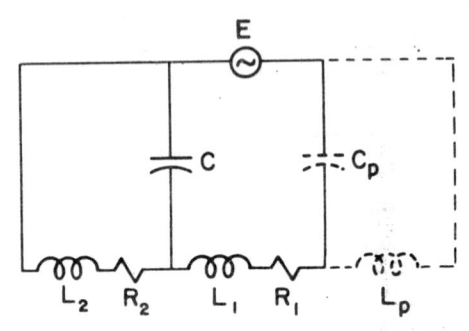

Fig. 531A.5. Analog circuit for system of figure 531A.4.

500. MECHANICAL VIBRATIONS AND TRANSIENTS 163

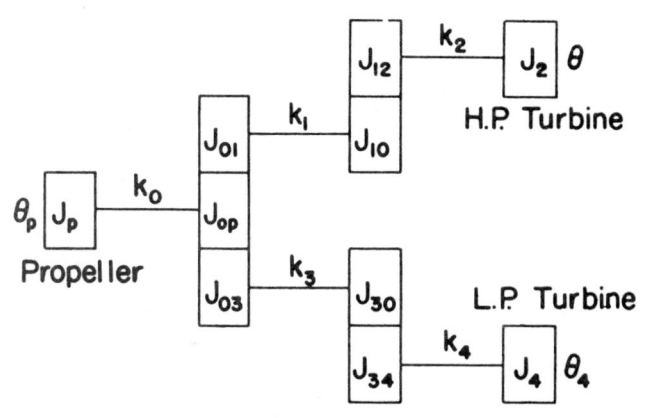

Fig. 531A.6. Schematic diagram of a ship drive.

of a cathode follower, Schmitt trigger, and pentode current generator is used to operate a high-speed mechanical relay for introducing the knee characteristic in the analog circuit. Response curves are shown for two representative values of exciting force.

Skilling has applied the analogy to the solution of problems involving torsional vibration dampers, i.e., flywheels which fit loosely on their shafts. Up to a certain magnitude of torque, this type of flywheel rotates at the same speeds as its shaft. When the torque exceeds this, the flywheel damper slides on the shaft, causing a constant resisting torque regardless of any further increase in shaft torque. This type of action can be simulated by a mercury arc and by glow discharge under special conditions. Raytheon gaseous conduction tubes were employed in Skilling's investigation. Two tubes were used, one to pass current in each direction. In the circuit, they are represented as a leak.

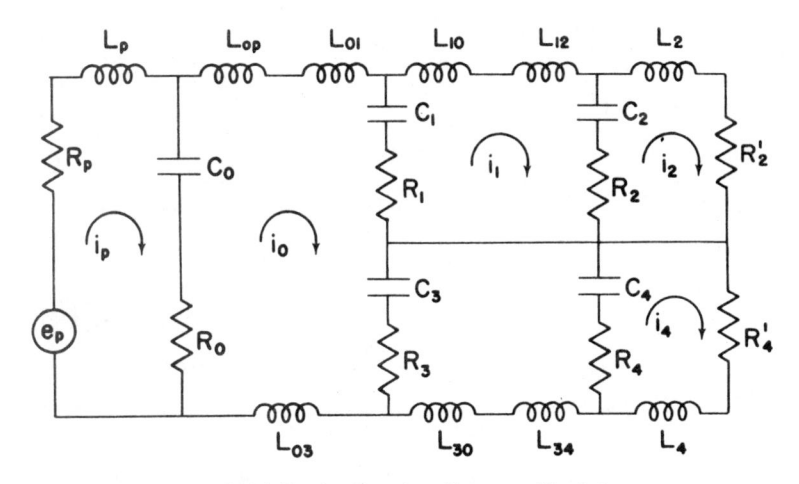

Fig. 531A.7. Analog circuit for a ship drive.

164   500. MECHANICAL VIBRATIONS AND TRANSIENTS

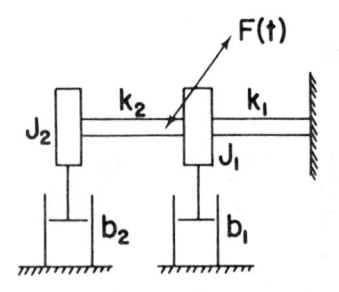

Fig. 531A.8. Torsional system with two degrees of freedom.

The equations representing the torsional system and its analogous circuit are as follows:

(1) when the flywheel is not slipping (no leakage):

TORSIONAL   ELECTRICAL

$$J\frac{d^2\theta}{dt^2} + J'\frac{d^2\theta}{dt^2} + k\theta = T_0 \sin\omega t \quad (7) \qquad L\frac{d^2q}{dt^2} + L'\frac{d^2q}{dt^2} + \frac{1}{C}q = E_0 \sin\omega t \quad (8)$$

(2) when the flywheel is slipping (leakage):

$$J\frac{d^2\theta}{dt^2} + k\theta \pm \overline{T} = T_0 \sin\omega t \quad (9) \qquad L\frac{d^2q}{dt^2} + \frac{1}{C}q \pm \overline{E} = E_0 \sin\omega t \quad (10)$$

where: T = magnitude of the constant resisting torque of the damper

E = magnitude of the constant resisting voltage of the tubes

For given relative sizes of each element of the torsional system, there will be a certain resonant frequency which will cause a maximum displacement (and thus torque) and charge (and thus voltage). By changing the relative sizes of the elements, different maximum values of torque

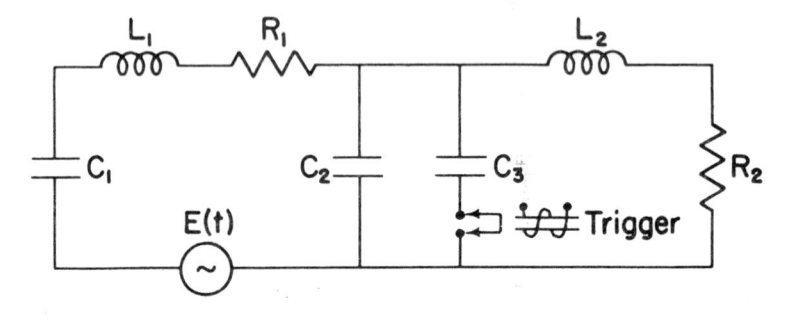

Fig. 531A.9. Analog circuit for torsional system with two degrees of freedom.

500. MECHANICAL VIBRATIONS AND TRANSIENTS

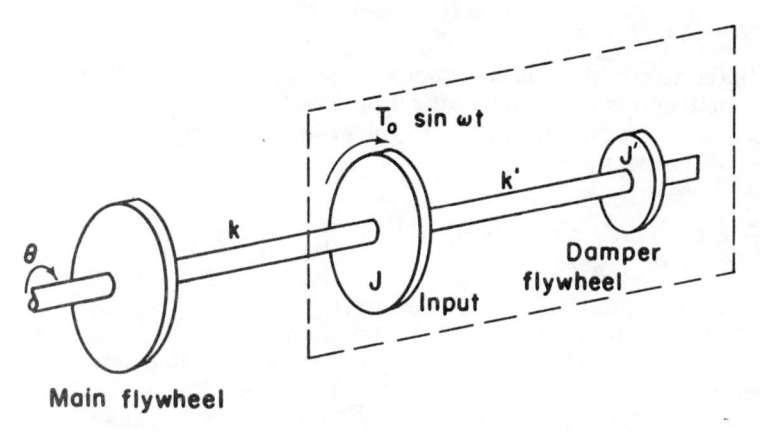

Fig. 531A.10. Torsional system with vibration damper.

and voltage will be obtained. By plotting $\frac{\bar{E}_{max}}{E_0}$ $\left(=\frac{T_{max}}{T_0}\right)$ against $\frac{\bar{E}}{E_0}$ $\left(=\frac{T}{T_0}\right)$ [where E_{max} = maximum voltage across condenser C and T_{max} maximum torque in the shaft], the optimum size of damper can be evaluated by noting the minimum value of $\frac{T_{max}}{T_0}$. Curves of typical results are included in the report.

Use of this analogy has been made by McCann, Warren, and Criner in analyzing shaft torques in turbine generators during transient disturbances such as electric short circuits and synchronizing out of phase. Their report includes representative curves and oscillograms.

Litman has investigated the damping effect of d.c. marine propulsion motors on vibrations produced in drive shafts by large propellers. He has discussed various cases involving one and two motors.

See also Concordia (B); Duncan; McCann and Bennett; Nickle; Shen and Packer.

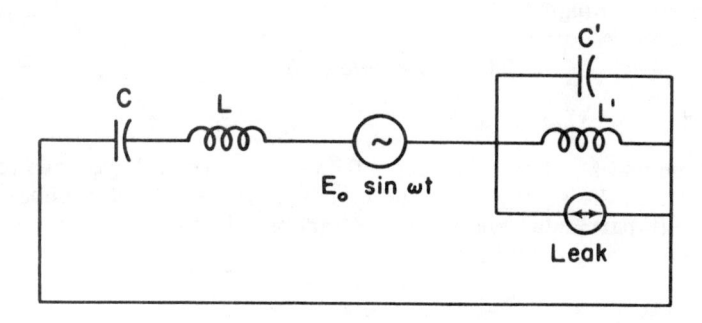

Fig. 531A.11. Analog circuit of torsional system with vibration damper.

531B *Electrical Analogy B*

Related to the "inertia inductance" analogy is the "inertia capacitance" analogy for torsionally vibrating masses.

Characteristic equations for a typical system (Fig. 531A.1) (single degree of freedom):

$$J \frac{d\omega}{dt} + b\omega + k \int \omega dt = T(t) \quad (1) \qquad C \frac{de}{dt} + \frac{1}{R} e + \frac{1}{L} \int e\, dt = i(t) \quad (2)$$

where: ω = angular velocity of the mass
$i(t)$ = applied current (a function of time)
e = potential difference existing across the elements

and other symbols are defined as in section 531A. The analog circuit for this system by the analogy of the present section is shown in figure 531B.1.

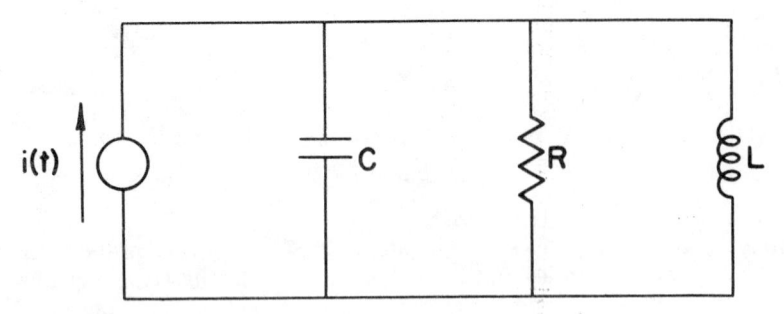

Fig. 531B.1. Parallel analog circuit for torsionally vibrating mass system of figure 531A.1.

Stewart has applied this analogy in analyzing the torsional vibration of aircraft-engine crankshafts. His report includes a resonance curve and plots of relative amplitudes vs. length.

See also Jennings.

532 *Distributed Masses*

532A *Electrical Analogy A*

McCann and MacNeal have described two electrical analogies for the analysis of the steady state and transient vibrations of members in torsion, with particular application to aircraft wings.

For the torsional vibration of members:

$$\frac{\partial}{\partial x}\left(k \frac{\partial \theta}{\partial x}\right) = J \frac{\partial^2 \theta}{\partial t^2} \qquad (1)$$

500. MECHANICAL VIBRATIONS AND TRANSIENTS

where: x = axial coordinate
k = torsional rigidity
θ = angular displacement
J = polar moment of inertia per unit length

This equation may be replaced by an equivalent finite difference equation.

Characteristic equations:

TORSIONAL

$$\frac{k_{n+\frac{1}{2}}}{\overline{\Delta x}}(\theta_{n+1} - \theta_n) + \frac{k_{n-\frac{1}{2}}}{\overline{\Delta x}}(\theta_{n-1} - \theta_n) = (J_n \overline{\Delta x})\frac{\partial^2 \theta_n}{\partial t^2} \quad (2)$$

ELECTRICAL

$$(\frac{1}{L_{n+\frac{1}{2}}})(V_{n+1} - V_n) + (\frac{1}{L_{n-\frac{1}{2}}})(V_{n-1} - V_n) = (C_n)\frac{\partial^2 V_n}{\partial t^2} \quad (3)$$

where subscripts represent nodes, $\overline{\Delta x}$ equals distance between nodes, and electrical symbols are as indicated in figure 532A.1.

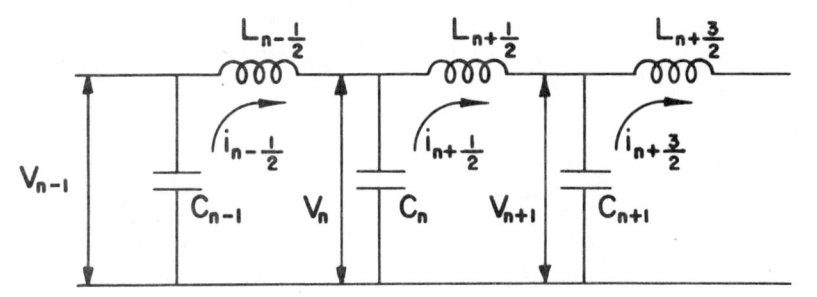

Fig. 532A.1. Analog circuit for torsionally vibrating members.

532B *Electrical Analogy B*

A companion analogy to the previous one has been presented by McCann and MacNeal.

Characteristic equations:

$$(\frac{k_{n+\frac{1}{2}}}{\overline{\Delta x}})(\theta_{n+1} - \theta_n) + \frac{k_{n-\frac{1}{2}}}{\overline{\Delta x}}(\theta_{n-1} - \theta_n) = (J_n \overline{\Delta x})\frac{\partial^2 \theta_n}{\partial t^2} \quad (1)$$

$$\frac{1}{C_{n+1}}(q_{n+\frac{3}{2}} - q_{n+\frac{1}{2}}) + (\frac{1}{C_n})(q_{n-\frac{1}{2}} - q_{n+\frac{1}{2}}) = L_{n+\frac{1}{2}}\frac{\partial^2 q_{n+\frac{1}{2}}}{\partial t^2} \quad (2)$$

where $q = \int i\, dt$ and other symbols are as defined in the previous analogy. The same analog circuit is used.

532C *Other Electrical Analogies*

For other electrical analogies for the torsional vibration of distributed mass members, see Ballet; Biot and Wianko; Cahen; "Electrical Computer Solves Wing Flutter Problems"; Kharlamov; Lazaryan; MacNeal, McCann, and Wilts; Wilts.

See also section 524.

540 STRUCTURES

Jasper (A) has utilized electric circuits in studying ship vibration problems.

Frost has used electric circuits to study the acoustic behavior of wall panels.

See also MacNeal (G).

See also sections 472AB, 524, and 532C.

Section 600.   Electricity and Magnetism

610 ELECTROSTATIC FIELDS

611 *Thermal Analogy*

Skinner described the use of heat flow to study electricity. He made observation of isothermal lines (as well as electric equipotential lines) by heating the end of a sheet of tinned iron and noting the distinct lines formed by the edge of melted tin. Arrangement of tin crystals upon solidification corresponds to such lines and can be brought out by washing the surface with a mixture of bleaching powder and hydrochloric acid. For further explanation of this analogy, see de Sénarmont (A, B); Voigt.

See also Gohar.

612 *Electrodynamic Analogy*

Adams probably was the first to study equipotential lines of electric fields by making use of the electrolytic tank method. Copper sulphate and zinc sulphate were tried as electrolytes. A polished metal plate in the electrolytic cell will discolor near electrodes. Guebhard (A, B) indicated that curves of constant color correspond to equipotential lines.

Fortescue and Farnsworth have used salt water as electrolyte and a deep, three-dimensional tank to plot equipotential surfaces in an electrostatic field between two terminals. Rice applied the same method to determine the electrostatic field around high-voltage bushings. In his system, water was used as electrolyte.

By measuring the resistance of electrolyte contained between cylindrical conductors, Teasdale and Higgins have determined the capacitance of conductive configurations of the same shape. See also M. K. Weston.

Cohn (B) describes the electrolytic tank determination of the dielectric constant of a medium containing obstacles. Measurements are made on a conducting model of the obstacle.

By means of an electrolytic tank, Pearce has investigated the permittivity of mixtures, particularly emulsions of water in oil.

Among the first to use a wedge-shaped electrolytic tank to study "circularly symmetric" systems were Bowman-Manifold and Nicoll.

P. E. Green has used a tank with a servo-controlled probe, which traversed equipotential lines automatically.

For further study of electrostatic fields by the electrolytic tank method, see D. H. Andrews; Atkinson (A, B); Balachowsky; Balachowsky and Tirroloy; Barkhausen and von Bruck; Dadda (A); Estorff; Förster; Hartill (A, C); Himpan (A, B); Kennelly and Whiting; D. McDonald (A, B); Mach (A); Ramberg and Morton; Sacchetto; Schmidl (A, B); Zschaage.

Miller and Kennedy have used conductive paper analogs to determine the field distortion in free air ionization chambers.

By the use of analogs, Schaefer and Stackowiack have investigated the electric field distribution in multilayer dielectrics in a condenser field excited by ultra-short electric waves.

See also Abramson; Carr; Cramp; Douglas and Kane; Emanueli; Hutter; Peplow; Simond and Kron.

613 *Hydrodynamical Analogy*

One of the earliest investigations of the electric field in three phase cables was made by Thornton and Williams. Their apparatus was of the type described in section 221AF, with glycerine flowing in at a point corresponding to one cable and flowing out at points corresponding to the other two cables.

See also A. D. Moore (A, B, C, E).

614 *Pendulum Analogy*

Emersleben has noticed that in terms of suitable parameters and coordinates, the period of a pendulum and the potential in a plane due to a charge ring have a formal analogy.

620 ELECTRIC CIRCUITS

621 *Basic Principles*

Blake designed a mechanical device analogous to an electric oscillator. This device can be used to show variations in the "Q" (quality factor) of oscillatory circuits (i.e. variations of energy loss) or to illustrate resonance between an oscillator and a second tuned circuit.

A mechanical apparatus to simulate and demonstrate sinusoidal amplitude modulation without harmonics of the modulating frequency has been described by Hardie. See Ladner for the description of a similar apparatus.

600. ELECTRICITY AND MAGNETISM

Rogers has given some simple hydraulic analogies to illustrate electrical phenomena.

A general method for designing networks with assigned gain or phase characteristics has been outlined by Darlington. This method is based on the analogy between gain and phase of linear networks and two-dimensional potential and stream functions.

Wagner has described a mechanical apparatus to simulate electric phenomena of traveling waves on power transmission lines. By means of this apparatus, various types of surges may be visualized and relations fundamental in transmission theory may be deduced. See also the article "Electrical Waves in Slow Motion; Apparatus Designed and Built by C. F. Wagner."

A mechanical device by Griscom simulates circuit vectors in order to analyze transmission lines.

Greenwood has developed a mechanical analog for solving certain power system stability problems.

For further references dealing with devices for demonstrating the principles of electric circuits, see Cremer and Klotter; the article "Electrical Quantities; a Mechanical Analogy"; Pawley; Masten; the article "Models and Analogies for Demonstrating Electrical Principles"; Rawcliffe; Roehmann; Wall.

622 *Coupled Circuits*

622A *Rolling Ball Analogy*

Magnusson (A, B) gives an analogy consisting of a freely rolling ball on a curved surface, wherein the path followed by the ball corresponds to the response of an inductively coupled two-mesh electrical network. The parallel nature of the analogous systems appears in their potential and kinetic energies.

Characteristic equations:

ELECTRICAL

$$W_L = \frac{1}{2}\left[\left(\sqrt{L_1}\frac{dq_1}{dt}\right)^2 + \left(\sqrt{L_2}\frac{dq_2}{dt}\right)^2 - 2\left(\sqrt{L_1}\frac{dq_1}{dt}\right)\left(\sqrt{L_2}\frac{dq_2}{dt}\right)\left(\frac{M}{\sqrt{L_1 L_2}}\right)\right] \quad (1)$$

ROLLING BALL

$$T = k_t\left[\left(\frac{dx_1}{dt}\right)^2 + \left(\frac{dx_2}{dt}\right)^2 - 2\left(\frac{dx_1}{dt}\right)\left(\frac{dx_2}{dt}\right)\cos(180° - \beta)\right] \quad (2)$$

$$W_C = \frac{1}{2}\left[\frac{(\sqrt{L_1}q_1)^2}{L_1 C_1} + \frac{(\sqrt{L_2}q_2)^2}{L_2 C_2}\right] \quad (3) \qquad U = k_U\left[\frac{(x_1)^2}{a} + \frac{(x_2)^2}{b}\right] \quad (4)$$

where: W_L = kinetic energy of the network

W_C = potential energy of the network

T = kinetic energy of the ball

U = potential energy of the ball

600. ELECTRICITY AND MAGNETISM

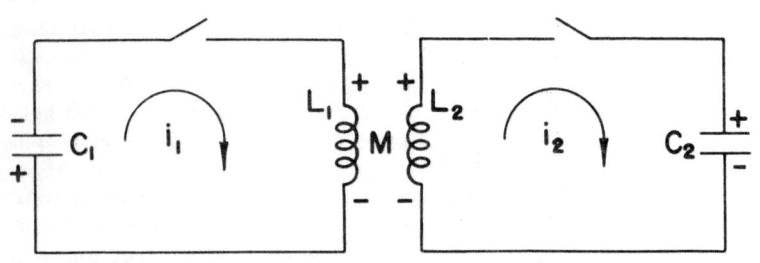

Fig. 622A.1. Inductively-coupled two mesh electrical network.

q_1, q_2 = charges across each capacitor

t = time

M = mutual inductance of L_1 and L_2

L_1, L_2 = magnitudes of the inductances

C_1, C_2 = magnitudes of the capacitances

x_1, x_2 = coordinates

t = time

β = angle between the directions of x_1 and x_2

a, b, k_T, k_U = constants

Thus, the charges across the capacitors are proportional to the coordinates, x_1 and x_2, of the displacement of the ball.

The above expression for the kinetic energy of the ball (neglecting vertical components of velocity and gyroscopic effects) is developed from the following:

$$T = \frac{1}{2} m \left(\frac{d\bar{x}}{dt} \right)^2 \qquad (5)$$

where $\bar{x}$ is the displacement of the ball. Equation (2) comes out of this expression when k_T is substituted for $\frac{1}{2}m$ and $\bar{x}$ is expanded into its components, x_1 and x_2, as in figure 622A.2.

The curved surface is so constructed as to follow the equation $h = k \left(\frac{x_1^2}{a} + \frac{x_2^2}{b} \right)$ where h is the elevation of the surface above an arbitrary datum level. Thus, the potential energy, U, of the ball is proportional to the right-hand side of the expression for h (since U is proportional to h), and equation (4) is formed. It is found that contours of the surface are ellipses.

The expressions for W_L and W_C are substituted into Lagrange's equation to obtain the equations of motion, which are then solved, the arbitrary constants and initial conditions being evaluated subsequently.

Diagrams are given of the confines of motion of the ball, and two typical paths of motion are presented. One simulates a simultaneous

600. ELECTRICITY AND MAGNETISM

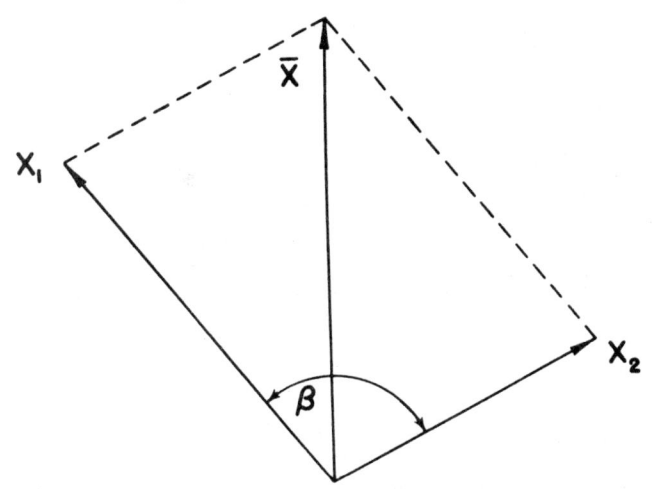

Fig. 622A.2. Vector diagram of the displacement, $\bar{X}$, and its components, X_1 and X_2.

closing of switches and the other a short time delay in the closing of one switch.

Various analogous effects and the extension of the analogy to other systems are discussed.

622B *Pendulum Analogy*

Chaffee and also Reich have detailed the use of coupled pendulum systems in the study of capacitively and inductively coupled electric circuits. See also Mandelstam.

The analogous equations are the same as those in section 531A.

630 MAGNETIC FIELDS

631 *Hydrodynamical Analogy*

The use of a thin sheet of viscous fluid in the investigation of magnetic fields has been described by Hele-Shaw and Hay.

Since the lines of induction correspond to the fluid streamlines, the alternate clear and colored bands of glycerine will clearly indicate the lines of induction. (For a description of the apparatus used, see section 221AF.)

Boundary conditions:
 The lateral boundaries of the flow region must be lines of induction and thus streamlines.

Since magnetic permeability is proportional to the cube of the thickness of the flow region, variations in permeability throughout the magnetic region to be investigated may be effected by coating one plate of glass with a thin layer of paraffin. Then portions of the paraffin corresponding to places of high permeability, such as the presence of a piece of iron, may be removed.

In this manner, Hele-Shaw and Hay have investigated the lines of magnetic induction around and within two-dimensional objects of various shapes and permeabilities due to the presence of a magnetic field.

Hele-Shaw, Hay, and Powell have used the same method in a study of the magnetic flux distributions in toothed core armatures.

Powell has also determined by this method the air gap correction coefficient for semi-closed slots. [From Hague.]

For further references, see Douglas; Hansen; A. D. Moore (A, B, C, E); Roeterink; Smoot; Thornton.

632 Electrical Analogy A. Sources (Scalar Potential)

Ahmed (A) has made use of an electrolytic tank to analyze the magnetic field of salient poles, especially to determine the leakage.

J. R. Barker has suggested the use of conductive paper to study magnetic fields.

Brouwer has used an electric circuit analog to study the eddy current distribution and domain boundary motion in ferromagnetic cores.

Parker has utilized electric circuits to study permanent magnets.

By the use of a conductive mixture of graphite and wax, Godsey determined equipotential lines and flux lines for magnetic fields. In the analog, regions of different conductivity simulated regions of different permeability in the prototype.

Cohn (B) has given the determination of the permeability of a medium containing obstacles by use of a nonconducting model of the obstacle in an electrolytic tank.

Beaver used an electrolytic tank of hyperbolic cross section to plot flux tubes directly in axially symmetric potential fields such as in magnetostatics.

Liebmann (A) determined the field distribution within magnetic electron lenses by means of electrical networks. See also Liebmann and Grad.

Douglas and Voith have used electrolytic tank and conducting paper analogs to determine the inductances of a.c. magnets.

See also Dadda (B); Douglas and Kane; Germain (A); Hahneman and Ehret; Levi; Striegel.

633 Electrical Analogy B. Vortices (Vector Potential)

"The determination of the magnetic field due to currents flowing in conductors requires the mapping of the magnetic vector potential. In the presence of iron a mathematical analysis becomes almost impossible, but it was shown by Peres and Malavard (E) and by Peierls that the problem can be solved experimentally with the help of the electrolytic tank analog in those cases where all currents flow in straight parallel conductors. The problem is then a two-dimensional one, and only one component (e.g. the z component) of the vector potential exists, which satisfied a Poisson equation in x and y. In the electrolytic tank analog the iron circuits are represented by wax models; the current conductors, supposed to be perpendicular to the surface of the electrolyte, are represented by a number of probes dipping into the electrolyte and feeding the appropriate currents into the tank.

"This model representation cannot be applied to rotationally symmetrical problems, where all currents flow in circles concentric with the axis of the system. The vector potential has then only an angular component A_θ which has to satisfy this equation:

$$\frac{\partial^2 A_\theta}{\partial z^2} + \frac{\partial^2 A_\theta}{\partial r^2} + \frac{1}{r}\frac{\partial A_\theta}{\partial r} - \frac{1}{r^2} A_\theta = -\frac{4\pi}{10} J \tag{1}$$

where J is the current density in the conductors, in amps/cm.^{2}

"Writing

$$rA_\theta = \Psi \tag{2}$$

this equation can be put into the form

$$\frac{\partial^2 \Psi}{\partial z^2} + \frac{\partial^2 \Psi}{\partial r^2} - \frac{1}{r}\frac{\partial \Psi}{\partial r} = -\frac{4\pi}{10} rJ. \tag{3}$$

"It was recently shown by Peierls and Skyrme that an experimental solution for this function Ψ can be obtained by using two tank measurements, first setting up the model in a tank with equal height of electrolyte, and then in a tilted or wedge tank. The difference between the two measurements can be used as a correction for the results obtained in the first tank, provided the correction is small, i.e. measurements are taken not too near the axis." [Liebmann (B).]

Liebmann (B) then goes on to discuss resistor networks which can be used for direct determination of either the vector potential A_θ or of the function (rA_θ).

See also Müllner.

640 ELECTROMAGNETIC FIELDS

641 *Electrical Network Analogies*

Kron (H) has developed equivalent electrical circuits for the field equations of Maxwell (for an electromagnetic field containing conductors and bound charges). Examples of transient and sinusoidal field phenomena which may now be studied by the network analyzer or by numerical and analytical circuit methods are the following: radiation from antennas, propagation through wave guides and cavity resonators

600. ELECTRICITY AND MAGNETISM

of arbitrary shapes, eddy currents in conductors, stresses in current-carrying conductors.

For additional references of equivalent circuits of the electromagnetic field, see Kron (B, J, L); McAllister.

Kron (H) also presented several networks for Maxwell's equations and proofs of certain equations and laws by use of the networks.

Spangenberg, Walters and Schott had a general discussion of design, construction, and operation of electrical network analyzers capable of obtaining solutions of the wave equation. The general network configuration is shown in figure 641.1.

For a two-dimensional, axially symmetrical case, in cylindrical coordinates, the network equations are:

$$\frac{\partial I_r}{\partial r} + \frac{\partial I_z}{\partial z} = - Y_\phi V_\phi$$

$$\frac{\partial V_\phi}{\partial r} = - I_r Z_r$$

$$\frac{\partial V_\phi}{\partial z} = I_z Z_z$$

where: I_r = current in radial element

I_z = current in longitudinal element

V_ϕ = voltage across a shunt element, i.e., between a lattice junction and the ground plane

Z_r = impedance per unit length of a radial element in the r-z lattice

Z_z = impedance per unit length of a longitudinal element in the r-z lattice

Y_ϕ = shunt admittance per unit area

Analog quantities:

(1) Cylindrical coordinates:

TE modes	Network	TM modes
E_r	I_z	$-H_r$
$-E_z$	I_r	H_z
rH_ϕ	V_ϕ	rE_ϕ
$j\omega\epsilon_r$	$Z_r = j\omega L$	$j\omega\mu_r$
$j\omega\epsilon_r$	$Z_z = j\omega L$	$j\omega\mu_r$
$\frac{j\omega\mu}{r}$	$Y_\phi = j\omega C$	$\frac{j\omega\epsilon}{r}$

600. ELECTRICITY AND MAGNETISM

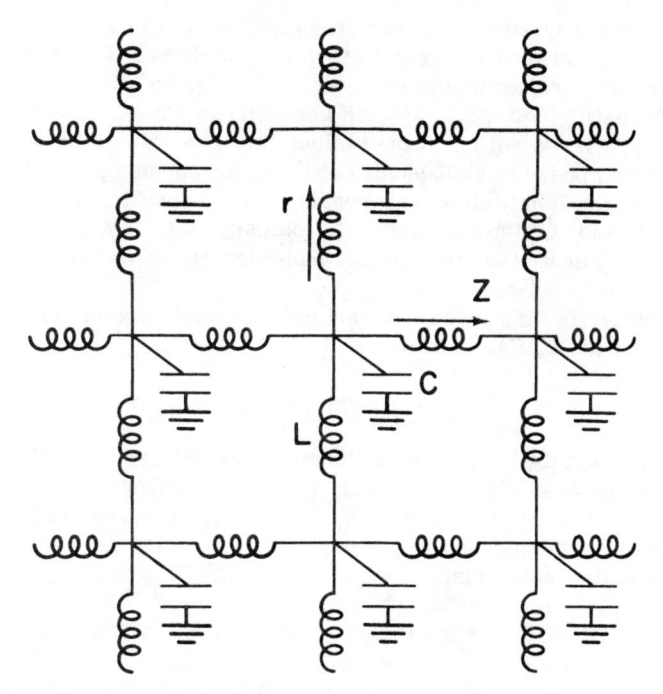

Fig. 641.1. General network configuration.

(2) Rectangular coordinates:

TE modes	Network	TM modes
E_y	I_x	$-H_y$
$-E_x$	I_y	H_x
H_z	V_z	E_z
$j\omega\epsilon$	$Z_x = j\omega L$	$j\omega\mu$
$j\omega\epsilon$	$Z_y = j\omega L$	$j\omega\mu$
	$Y_z = j\omega C$	$j\omega\epsilon$

Liebmann (N) has used resistance networks to solve waveguide and cavity resonator problems.

For further descriptions of electrical networks used as analogs to analyze electromagnetic field problems, see Whinnery and others; Whinnery and Ramo.

642 *Other Electrical Analogies*

Clement and Johnson have given an electrical analog for waveguides of arbitrary cross section. The analog has two electrically conductive

178 600. ELECTRICITY AND MAGNETISM

plates, geometrically similar to the cross section of the waveguide, separated by a sheet of dielectric material. The device is essentially a two-dimensional transmission line.

If an alternating voltage is applied between the plates, an electromagnetic wave will be set up in the dielectric. The voltage between the plates at a given position represents, at the corresponding position in the waveguide, the longitudinal component of the magnetic field intensity for a TE mode or the electric field intensity for a TM mode.

Boothroyd indicated an analogy between electric wave filters and electrolytic tanks.

For electrolytic tank investigations of microwave problems, see Bertram; Cohn (A, B, C).

643 *Membrane Analogy*

Makinson described a membrane analogy for electromagnetic waveguides. In transverse electromagnetic waves in a waveguide of rectangular section of the kind classed as H_{on} waves, the electric and magnetic vectors are independent of y.

Characteristic equations:

ELECTROMAGNETIC FIELD

$$\frac{\partial^2 E_y}{\partial x^2} + \frac{\partial^2 E_y}{\partial z^2} = \frac{\epsilon \mu}{c^2} \frac{\partial^2 E_y}{\partial t^2}$$

where: E_y = y component of electrical field strength

ϵ = dielectric constant

μ = permeability

c = velocity of light

t = time

VIBRATING MEMBRANE

$$\frac{\partial^2 h}{\partial x^2} + \frac{\partial^2 h}{\partial z^2} = \frac{\rho}{T} \frac{\partial^2 h}{\partial t^2}$$

h = transverse displacement of membrane (in y direction)

ρ = density of material

T = tension of material

t = time

Boundary conditions: $y = 0$ at $x = 0, b$. Thus, the membrane must be clamped along the edges. See figure 643.1.

According to the method of excitation of the membrane, waves corresponding to any of the H_{on} waves, or to any superposition of them, can be produced.

Since $\dfrac{\partial H_x}{\partial t} = \dfrac{c}{\mu} \dfrac{\partial E_y}{\partial z}$ and $\dfrac{\partial H_z}{\partial t} = \dfrac{-c}{\mu} \dfrac{\partial E_y}{\partial x}$

the magnetic field strength H is proportional to the slope of the membrane.

See also Allen, Fox, Motz, and Southwell; Knol and Diemer.

600. ELECTRICITY AND MAGNETISM

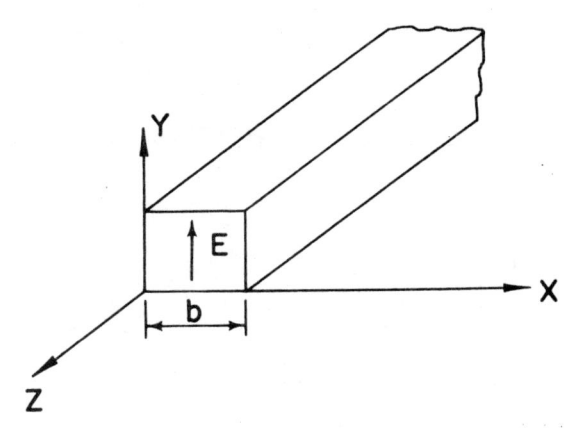

Fig. 643.1. Wave guide.

644 *Acoustical Analogy*

Canac has suggested the study of electromagnetic waves by the use of visible sonic waves.

650 MOTION OF CHARGED PARTICLES IN ELECTRIC AND MAGNETIC FIELDS

651 *General*

Perhaps the first to investigate the motion of a charged particle by analogy were Bruche and Recknagel, who constructed a mechanical model for electron motion in axially symmetric magnetic fields.

651A *Gyroscope Analogy*

The similarity between the force on an electron due to its motion transverse to a magnetic field, and the force on a gyroscope due to the angular velocity of its axis was suggested in 1933 by E. C. Kemble (according to Rose).

Rose gave a model for the two-dimensional motion of electrons acted upon by a two-dimensional electric field in a plane normal to a uniform axial magnetic field, the essential part of the model being a gyroscope mounted with its axis in the direction corresponding to the direction of the magnetic field.

Characteristic equations:

ELECTRON

GYROSCOPE

$$\frac{d^2r}{dt^2} - r\left(\frac{d\phi}{dt}\right)^2 = -r\frac{d\phi}{dt}\left(H\frac{e}{m}\right) - E_r\frac{e}{m} \quad (1) \qquad \frac{d^2r}{dt^2} - r\left(\frac{d\phi}{dt}\right)^2 = -r\frac{d\phi}{dt}\frac{d\Psi}{dt}\frac{C}{A} + F_r\frac{L^2}{A} \quad (2)$$

180   600. ELECTRICITY AND MAGNETISM

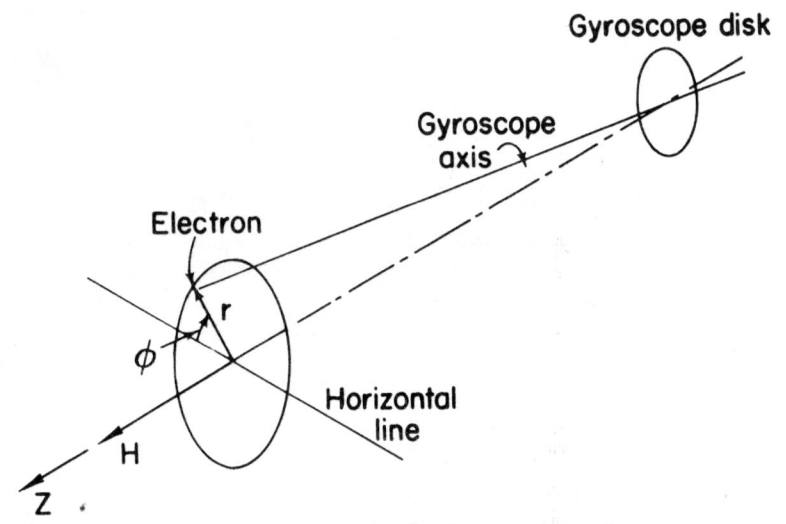

Fig. 651.1. Schematic sketch of gyroscope analogy.

$$r\frac{d^2\phi}{dt^2} + 2\frac{dr}{dt}\frac{d\phi}{dt} = \frac{dr}{dt}\left(H\frac{e}{m}\right) - E_\phi \frac{e}{m} \quad (3)$$

$$r\frac{d^2\phi}{dt^2} + 2\frac{dr}{dt}\frac{d\phi}{dt} = \frac{dr}{dt}\frac{d\Psi}{dt}\frac{C}{A} + F_\phi \frac{L^2}{A} \quad (4)$$

$$\frac{d^2 z}{dt^2} = 0 \quad (5) \qquad \frac{d^2 \Psi}{dt^2} = 0 \quad (6)$$

where: r, ϕ = position coordinates as shown in figure 651A.1

r, ϕ = position coordinates as shown in figure 651A.1

H = uniform magnetic field in the axial direction

$\frac{d\Psi}{dt}$ = initial spin velocity

e = charge on the electron

C = moment of inertia about the gyroscope axis

m = mass of the electron

A = moment of inertia about the other principal axis

E = electric field intensity

L = moment arm

F = force

Subscripts indicate components in the respective directions.

The force field analogous to the electrostatic field (which might be of several types) is effected by means of electromagnets with specially shaped pole faces. The electron is represented by one pole of a long, thin permanent magnet mounted so that one end is at the end of the gyroscope axis and the other end near the center of the gyroscope. A

600. ELECTRICITY AND MAGNETISM

small flashlight bulb and battery are mounted at the other end of the axis to make possible the photographing of the path. Included in the report are photographic patterns representing the paths of electrons in various types of fields.

651B *Rolling Ball Analogy A*

Vineyard has shown that the motion of a charged particle in a magnetic field can be simulated by a ball rolling on a rotating surface. The theory is given for the case of a warped surface undergoing arbitrary rotation about a fixed axis and translation perpendicular thereto, while the system from which the ball is observed partakes of similar but independent motion. With approximations based on not too large departure of the surface from flatness, the following cases can be simulated: (a) The magnetic field is homogeneous and constant. The electric field is perpendicular to the magnetic field, and irrotational, but otherwise arbitrarily spatially dependent, and arbitrarily time dependent with certain limits. (b) The magnetic field is homogeneous but arbitrarily time dependent. The electric field is perpendicular to the magnetic field and may have a variety of space and time dependences, including a part which encircles the axis and has just the right magnitude to produce acceleration in a circular orbit as in the betatron.

Results of rudimentary experiments are presented which indicate that the method is capable of good accuracy.

651C *Rolling Ball Analogy B*

Kleynen has given the analogy between the motion of a small particle sliding without friction on a rubber membrane and the motion of an electron in a two-dimensional electrostatic field.

Characteristic equations:

ELECTRON   PARTICLE

$$m_e \frac{d^2x}{dt^2} = -e \frac{\partial V}{\partial x} \quad (1) \qquad m_p \frac{d^2x}{dt^2} = m_p g \frac{\partial h}{\partial x} \quad (2)$$

$$m_e \frac{d^2y}{dt^2} = -e \frac{\partial V}{\partial y} \quad (3) \qquad m_p \frac{d^2y}{dt^2} = -m_p g \frac{\partial h}{\partial y} \quad (4)$$

where: x, y = components of displacement of the electron

t = time

V = electrostatic potential

m_e = mass of the electron

e = charge of the electron

x, y = horizontal components of the displacement of the particle

t = time

h = vertical displacement of the membrane

m_p = mass of the particle

g = acceleration of gravity

Boundary conditions:

1. The boundary of the membrane must have the same shape as the electrode boundaries of the field.

2. The distribution of elevation along the membrane boundary must be proportional to the distribution of potential along the electrode boundaries.

Thus, the path of an electron in the field is determined by the horizontal projection of the path of a small particle (approximated in practice by a ball rolling on the membrane).

In addition, contour lines on the membrane correspond to equipotential lines in the electrostatic field, and the maximum slope of the membrane at any point is proportional to the field strength at the corresponding point in the electrostatic field. The charge on an electrode is proportional to the force that the membrane exerts on the support at the corresponding point.

See also Moon and Oliphant; Rajchman; Ramberg and Morton.

C. L. Andrews has developed a rolling ball analog for the motion of electrons in a high-frequency longitudinal field. The balls, injected at regular time intervals, plotted photographically a family of graphs of distance against time, indicating the electron bunching, velocity modulation, and density modulation. See also the article "Tracing Electron Paths."

651D *Electrical Analogy*

Samuel used an electrolytic tank to study the potential distribution in the vicinity of the electron beam in an electron gun.

Perhaps the first to employ a mechanical device in conjunction with an electrolytic tank to study electron trajectories was Gabor. At about the same time, D. B. Langmuir (A, B) developed a similar device, automatic rather than manual, for the same purpose. Others who described the use of similar automatic plotting devices are Baker; Hollway (B, C); Marvaud (A, B); Sander, Oatley, and Yates; Sponsler (A, B).

In most of the above mentioned studies, the effects of space charge are neglected. In an attempt to overcome this limitation, Musson-Genon utilized a tank in which the depth of the liquid was varied according to the charge density. Later Alma, Diemer, and Groendijk solved space charge problems by adding distributed forces to models based on another analogy (membrane). Hollway (A) approached this problem by using discrete current sources to simulate clouds of distributed charge.

See also Ertaud; Pierce.

651E *Electromechanical Analogy*

Barbier studied the betatron oscillations in accelerators by the use of an electromechanical analog, which has two degrees of freedom susceptible to being excited as a function of place or time.

600. ELECTRICITY AND MAGNETISM

652 *In Lenses*

Prebus, Zlotowski, and Kron have shown that, by suitable correlation of the variables, the trajectory equation

$$\phi(z) \frac{d^2 r}{dz^2} + \frac{1}{2} \phi'(z) \frac{dr}{dz} + \frac{1}{4} \left[\phi''(z) + \frac{e}{2mc^2} H^2(z) \right] r = 0$$

of an electron lens may be identified with the differential equation describing the dependence of the voltage distribution upon the impedance characteristics of a simple type of ideal inhomogeneous transmission line. By choosing the specific impedances as functions of the independent variable z (distance along the axis of symmetry) and the parameters $H(z)$ and $\phi(z)$ (z component of the magnetic field and electrostatic potential respectively, on the axis), the values of the dependent variable $r(z)$ (radial displacement of the electron) correspond to voltages along the line. The equivalent network provides a rapid means of obtaining accurate numerical solutions of the trajectory equation. Application of this method is demonstrated for the case of a strong magnetic lens by comparing the results with the exact solutions given by Glaser. The usefulness of a two-dimensional network for determining the electrode configuration of an electrostatic lens from a given axial potential distribution is discussed.

In addition, Zlotowski and Prebus have applied network analysis to the specification of the cardinal points of electrostatic lenses. The advantages of this method of numerical integration of the lens trajectory equation are shown by comparing it with other methods currently in use. The requisite axial potential distributions of the lenses are obtained from accurate electrolytic trough measurements. Data of practical value are provided which show the dependence of optical properties on electrode parameters of some familiar types of electrostatic lenses. The inverse problem of determining the axial potential distributions corresponding to a given trajectory is investigated.

See also section 651D.

653 *In Gases*

Yarnold has shown that the steady motion of a stream of electrons through a gas in which a uniform electric field is maintained may be adequately represented for demonstration purposes by the two-dimensional motion of a steel sphere on an inclined plane into which has been driven at random a large number of nails.

The model used consists of a thick sheet of laminated wood 100 cm. square into which were driven 400 good quality wire nails each approximately 2 mm. in diameter; to obtain a random distribution of nails, the plane was marked out into large squares of side 20 cm., and small squares of side 1 cm.; 16 nails were allotted to each square, coordinates being chosen by cutting two packs of cards.

Eight hundred runs were made for each of three inclinations of the board. Included in the report are a table and curves of the distribution of velocity at the instant the ball left the board. These results are compared with the Maxwellian distribution corresponding to the same mean velocity.

Section 700.   Modern Physics

710 VIBRATION OF MOLECULES

Kron (A) developed electric circuit models of polyatomic molecules to be used in determining their vibration characteristics (normal frequencies and normal modes). Networks were set up without equations from physical considerations alone. Networks also allowed the establishment of the resultant secular equations of the molecule by simple inspection.

See also Carter and Kron (B).

720 QUANTUM THEORY

For waves associated with particles, see Kron (C), Kron and Carter.

730 NUCLEAR REACTORS

Nagao has proposed a continuous-medium electrical analog to investigate problems of diffusion within nuclear reactors.

Liebmann (L) applied electrical networks to the solution of such problems.

Mossop and McGhee have described the use of wax models in applying the analogy between the temperature distribution in a cooling solid and the calculated one-group neutron flux distribution in a reactor.

Melcher, in an M.S. thesis at Iowa State University, established an analogy between the diffusion of neutrons in a diffusing medium and the behavior of an electrical resonance cavity. The analogy is useful for determining the affect of control rods in systems having irregular boundaries.

Appendix I. Types of Potential Fields

A. Steady-State Harmonic

1. Without sources (Laplace equation: $\nabla^2 \phi = 0$)

    a. Two-dimensional
    See sections 221AAA, 221AB, 221AC, 221AD, 221AE, 221AF, 221B, 222B, 313A, 313B, 411, 412, 413, 432, 433, 441AB, 441AD, 441AF, 442A, 452, 455, 611, 612, 613, 631, 632.

    b. Axially-symmetric
    See sections 221AAB, 612, 631.

    c. General three-dimensional
    See sections 221AAC, 221AB.

2. With sources

    a. Uniformly-distributed ($\nabla^2 \phi = -k$)
    See sections 211AD, 312A, 312B, 312C, 312D, 441AA, 441AC, 441AE, 441AF, 441AG.

    b. Other $[\nabla^2 \phi = f(x,y,z)]$
    See sections 312B, 312D, 431, 453, 463.

B. Steady-State Pseudo Harmonic $\left[\frac{\partial}{\partial x}(\lambda \frac{\partial \phi}{\partial x}) + \frac{\partial}{\partial y}(\lambda \frac{\partial \phi}{\partial y}) + \frac{\partial}{\partial z}(\lambda \frac{\partial \phi}{\partial z}) = f(x,y,z) \right] [\lambda = \lambda(x,y,z)]$

1. Without sources
    See sections 211AAA, 211AAB, 211AAC, 211AAD, 441BA, 441BB, 631, 632.

2. With sources
    See sections 222A, 312D.

C. Biharmonic ($\nabla^4 \phi = K$)
See sections 222B, 451, 457, 461, 462, 464.

D. Wave

1. Instantaneous magnitude

    a. Ordinary ($\frac{\partial^2 \phi}{\partial t^2} = a^2 \nabla^2 \phi$)

    See sections 111, 113, 121AAA.

    b. Damped
    See sections 111, 121AAB.

2. Amplitude only ($\nabla^2\phi + k\phi = 0$)
   See sections 641, 642, 643, 730 (Neutron diffusion).

E. Transient potential ($\nabla^2\phi = \gamma \frac{\partial\phi}{\partial t}$)
   See sections 212A, 221AG, 311A, 311B, 311C, 459.

Appendix II. Analogs for Potential Fields

A. Membrane

   1. Stationary

      a. Pressureless (Laplace equation, except *)
         Sections *211AC, 221AC, 313B, 411, 432, 441AB, 452, *651C.

      b. With differential pressure (Poisson equation)
         Sections 211AD, 312A, 431, 441AA.
         For further references on techniques of this analog, see Fulop (A, B); Schneider and Cambel (A).

   2. Vibrating (wave equation — amplitude only)
      Section 643.

B. Hydrodynamic (fluid mappers)

   1. Without sources; constant thickness flow space (Laplace equation)
      Sections 221AF, 412.

   2. Uniformly-distributed source (Poisson equation)
      Section 312B.

   3. Variable thickness flow space (pseudo harmonic)
      Sections 613, 631.

C. Fluid network

   1. Steady state
      Section 322AB.

   2. Transient
      Sections 311B, 311C, 321B.

D. Electrical — geometric

   1. Electrolytic tank

      a. Two-dimensional

         1. Constant depth without sources (Laplace equation)
            Sections 211AAA, 211AAB, 221AAA, 221AB, 221B, 313A, 441AD, 455, 612, 632, 642, 651D.

APPENDIX II

   2. Variable depth without sources (pseudo harmonic)
      Sections 313A, 632, 651D.

   3. Constant depth with sources
      Sections 212A, 651D.

   4. Variable depth with sources
      Sections 222A, 312D.

   5. Wedge-shaped (for problems of axial symmetry)
      Sections 221AAB, 612, 632, 633.

  b. Three-dimensional
     Sections 221AAC, 221AB.

For further references on electrolytic tank techniques, see Ahmed (B); Basu; Boothroyd, Cherry, and Makar; Brower and DeRienzo; Coet; Diggle and Hartill (A, B); Dix; Egorov; Ehrenfried; Einstein; Farr and Wilson; Gaugin; Gelfand, Shinn, and Tuteur; Germain (D); Greenland and Holden; Hacques (A); Hansen and Lundstrom; Hartill (B, D); Hartill, McQueen, and Robson; Hepp; Huggins; Ip; Isobe and Nihei; D. A. Jones; Kennedy and Kent; Kunin; Makar, Boothroyd, and Cherry; Mickelson; Norbury and Platt; Platt and Norbury; Prudkovskii (A, B); Renard; Sander; Sander and Yates; Tanabe and Yamada.

2. Solid conductor

   a. Single layer, without sources

      1. Constant resistivity or thickness
         Sections 221AAA, 313A, 413, 433, 441AD, 455, 612, 632.

      2. Variable resistivity or thickness
         Sections 221AAB, 441BA, 441BB.

   b. Single layer, with sources
      Sections 222A, 312D.

   c. Capacitively coupled

      1. Sources
         Section 312D.

      2. Wave amplitude or neutron flux distribution
         Sections 642, 730.

      3. Transient
         Section 311A.

   d. Three-dimensional
      Section 221AAC.

For other references on techniques of solid conductive sheets, see Dahlin; Foster and Lodge; Kirchhoff; Murray and Hollway; Quincke; Schwedoff; W. R. Smith.

E. Electrical networks

Electrical networks have been used to solve the finite-difference form of all the equations of Appendix I. For further references on techniques of finite-difference networks, see "Analogue for Engineering Research"; "Analogue for Heat Flow Studies"; Culver (C); de Packh; Dove; Dusinberre; Francken; Friedmann (B); Fritzsche; Gair; Hazen, Schurig, and Gardner; Hutcheon (A, B); Johnson and Alley; Karplus; Kron (K, M); Kuehni and Lorraine; Landau (A, B); Liebmann (D, E, H, J); Liebmann and Bailey; McCann, Wilts, and Locanthi; MacNeal (A, E); Many and Meiboom; Palmer and Redshaw (A); Redshaw (B, C); Servranckx; Shiu-Jen Kuh; Stephenson and Starke; Su; Swenson (B); Swenson and Higgins.

General References

Analogies in technology. Symposium of the Adriatic Electrical Society Conference, Venice. (I Modelli Nella Tecnica), Rome, Italy. 1956.

Ashton, A. W. Analogy in engineering. J. Instn. Elec. Engrs. 90 (I):120-124. 1943.

Elberty, R. S. Fluid and electrical analogies work both ways. Mach. Des. 16:137-139. 1944.

Ferrell, E. B. Electrical and mechanical analogies. Bell Laboratories Rec. 24:372-373. 1946.

Fuchs, H. O. Model tests and laws for predicting performance. Prod. Engr. 13:556-559. 1942.

Ghaswala, S. K. Models and analogies in structural engineering. Civil Eng. (London). 47(547):54-56; 47(548):134-136; 47(549):225-227; 47(550):315-317; 47(551):404-406; 47(552):488-489. 1952.

Green, P. E. Methods of plotting electrostatic fields. Engrg. School Bul. North Carolina State College. 45. 1945.

Hague, B. Determining the distribution of electric and magnetic fields. Electrician 102:185-187, 315-317. 1929.

———. Méthodes analytiques, graphiques, et expérimentales utilisées pour l'étude des champs magnétiques et électriques dans les machines et appareils électriques. Comptes Rendus du Congrés International d'Électricité 4(3,1): 47-111. 1932.

Hansen, W. W., and Lundstrom, O. C. Experimental determination of impedance functions by the use of an electrolytic tank. Proc. IRE 33:528-534. 1945.

Hetényi, M. Handbook of experimental stress analysis. John Wiley and Sons, New York. 1950.

———. On similarities between stress and flow patterns. J. Appl. Phys. 12: 592-595. 1941.

Higgins, T. J. Analogic experimental methods in stress analysis as exemplified by Saint-Venant's torsion problem. Proc. Soc. Exp. Stress Anal. 2(2):17-27. 1945.

———. Electroanalogic methods. Appl. Mech. Rev. 9:1-4, 49-55. 1956. 10:49-54, 331-335, 443-448. 1957. 11:203-206. 1958.

———. Stress analysis of shafting exemplified by Saint-Venant's torsion problem. Proc. Soc. Exp. Stress Anal. 3(1):94-101. 1945.

Hillier, J. See Zworykin, V. K., Morton, G. A., Ramberg, E., and Vance, A. W.

Jakob, M. Heat transfer. John Wiley and Sons, New York. 1949.

Johnson, E. F. (ed.) Analogs as aids in process studies, (4 papers.) Ind. Eng. Chem. 47(3): 396-421. 1955.

Journeaux, D. Analogies and energies. Allis-Chalmers Elec. Rev. 17:30-34. 1952.

Karplus, W. J. Analog simulation; simulation of field problems. McGraw-Hill, New York. 1958.

———, and Soroka, W. W. Analog methods: computation and simulation. 2nd ed. McGraw-Hill, New York. 1959.

Kozlov, E. S., and Nikolaev, N. S. The approximate solution of partial differential equations via electrical models (In Russian.) Automation and Remote Control (USSR) 17:993-999. 1956. (Translation of Avtomatika i Telemekhanica 17:890-896)

Kuchemann, D., and Redshaw, S. C. Some problems in aerodynamics and their solution by electric analogy. J. Roy. Aero. Soc. 60:191-196. 1956.

Lakaye, L. Quelques applications de l'analyse dimensionelle et de la similitude en electricite. Assn. des Ingenieurs Electriciens sortis de l'Institut Electrotechnique Montefiore, Bul. 61 (5-6-7-8):219-257. 1948.

Layton, R. E. From Hardy Cross to electronics; Redding, California. Water Works Eng. 112:718-719. 1959.

Lefkowitz, I. Application of analogs to dynamic analysis. J. Inst. Soc. Amer. 2:301-302. 1955.

Liebmann, G. Electrical analogues. Brit. J. Appl. Phys. 4:193-200. 1953.

———. Field plotting and ray tracing in electron optics; a review of numerical methods. Advances in Electronics 2:101-149. 1950.

List, B. H. See McMaster, R. C., and Merrill, R. L.

Lundstrom, O. C. See Hansen, W. W.

McCann, G. D. Designing analogy circuits from test data. ISA J. 3:201-205. 1956.

———. The direct analogy electric analog computer. ISA J. 3:112-118. 1956.

———. Electric analogies for mechanical structures. ISA J. 3:161-165. 1956.

McMaster, R. C., Merrill, R. L., and List, B. H. Analogous systems in engineering design. Product Eng. 24:184-195. 1953.

Macagno, E. O. Mechanical analogy and its application to hydraulics (In Spanish). Cienc. y. Tecn. 116 (585):91-120. 1951.

Malavard, L. Electric analogies in elasticity (In French). Laboratories (8):19-21, 23, 25, 27, 29. 1953.

———. See Pérès, J.

———. Use of rheoelectrical analogies in aerodynamics. Advisory Group Aeronautical Research and Development (NATO) — AGARDograph 18. 1956.

Merrill, R. L. See McMaster, R. C., and List, B. H.

Meyer, M. L. Stress investigation by model tests and analogues. Trans., Liverpool Eng. Soc. 76:114-132. 1955.

Mindlin, Raymond D. A review of the photoelastic method of stress analysis. J. Appl. Phys. 10:273. 1939.

Morton, G. A. See Zworykin, V. K., Ramberg, E., Hillier, J., and Vance, A. W.

Murphy, Glenn. Similitude in Engineering. Ronald Press, New York. 1950.

Muskat, M. Flow of homogeneous fluids through porous media. 139-141, 145, McGraw-Hill, New York. 1937.

Nikolaev, N. S. See Kozlov, E. S.

Northrup, Edwin F. The use of analogies in viewing physical phenomena. J. Franklin Inst. 1:10. 1908.

Pérès, J. Les methodes d'analogie en mecanique appliquee. Proceedings of the Fifth International Congress for Applied Mechanics (Cambridge, 1938). 9-19.

———, and Malavard, L. La method d'analogies rheographiques et rheometriques. Bul. Soc. Fran. d. Elec. (5) 8:715-744. 1938.

Ramberg, E. See Zworykin, V. K., Morton, G. A., Hillier, J., and Vance, A. W.

Redshaw, S. C. See Kuchemann, D.

GENERAL REFERENCES

Schneider, P. J. Conduction heat transfer. Addison-Wesley, Cambridge. 1955.
Schoenfeld, J. C. Analogy of hydraulic, mechanical, acoustic, and electric systems. Appl. Sci. Research Sec. B, 3:417-450. 1954.
Soroka, W. W. Analog methods in computation and simulation. McGraw-Hill, New York. 1954.
──── . Analogs; tools of engineering. J. Eng. Educ. 40:514-520. 1950.
──── . Experimental aids in engineering design analysis. Mech. Eng. 71:831-837. 1949.
Spangenberg, K. R. Vacuum tubes. McGraw-Hill, New York. 72-82. 1948.
Strobel, C. Elektrische darstellung mathematischer funktionen. Archiv für Electrotechnik 34 (6):334-338. 1940.
Tribus, M. Use of analogs and analog computers in heat transfer. Oklahoma State University — Engineering Experiment Station — Publ. 100. 1958.
Vance, A. W. See Zworykin, V. K., Morton, G. A., Ramberg, E., and Hillier, J.
Weber, E. Mapping of fields. Elec. Engr. 53:1563-1570.
Willoughby, E. O. Some applications of field plotting. J. Instn. Elec. Engrs. 93 (III):275-293. 1946.
Yorke, R. Dynamical analogies. Engineering 176:763-764. 1953.
Zworykin, V. K., Morton, G. A., Ramberg, E., Hillier, J., and Vance, A. W. Electron optics and the electron microscope. John Wiley and Sons, New York. Ch. 11. 1945.

References

Abbott, B. L. Home-made network calculator takes guesswork out of planning. Am. Gas J. 184:18-21. 1957. (122BAC)

Abramov, F. A., and Podolsky, V. A. The investigation of mine shaft ventilation by means of models. (In Russian.) Izv. Dnepropetr. Gorn. In-ta 23:133-139. 1956. (122BB)

Abramson, H. N. Theoretical investigations of the four-electrode crevasse detector. Trans. Amer. Geophys. Un. 38:849-856. 1957. (612)

Abul-Fetough, Abdel-Hadi. See Rouse, H. (221AAB)

Ackert, J. See Gerber, H. (221AAA)

Ackroyd, R. T., Houstoun, J., Lynn, J. W., and Mann, E. Electrical analogue for heat waves in an exothermic medium. Proc. Inst. E. E. 108B:33-36. 1961. (311A)

Adams, E. Q. See Langmuir, I., and Mickle, G. S. (313A)

Adams, W. G. On the forms of equipotential curves and surfaces and lines of electric force. Proc. Roy. Soc. London 25:280-284. 1875. (612)

Agarev, V. A. See Kvitka, A. L., and Umanski, E. S. (492)

Ahmed, A. A. (A) Investigation by electrolytic means of the magnetic leakage of salient poles. World Power 2:334-340. 1924. 3:91-99, 157-161, 218-222. 1925. (632)

Aitken, M. J. An electrical analogy to a high vacuum system. Brit. J. Appl. Phys. 4:188. 1953. (211C)

Alfrey, T. Non-homogeneous stresses in viscoelastic media. Q. of Appl. Math. 2:113-119. 1944. (491)

Allder, J. R. See Karplus, W. J. (221D)

Allen, Fox, Motz, and Southwell. Free transverse vibrations of membranes with an application (by analogy) to two-dimensional oscillations in an electromagnetic system. Phil. Trans. Roy. Soc. London Series A. 239:488-500. 1945. (643)

Alley, R. E., Jr. See Johnson, W. C. (Appendix II E)

Anderson, J. E., and Soroka, W. W. Electrical analog solution for centrifugally-tuned pendulum absorber system. J. Aero. Sci. 17:349-355. 1950. (531A)

Andrews, D. H. Automatically plotting electrostatic field lines. Electronics 27:182-183. 1954. (612)

Andrews, R. V. Solving conductive heat transfer problems with electrical-analogue shape factors. Chem. Eng. Progress 51:67-73. 1955. (313A)

Anthes, H. Versuchsmethode zur Ermittlung der Spannungsverteilung bei Torsion Prismatischer Stabe. Singlers Poly. J., pp. 321, 342-345, 356-359, 388-392, 441-444, 455-459, 471-475. 1906. (441AA)

REFERENCES

Appleyard, V. A. Use of McIlroy fluid network analyzer in Philadelphia. J. New England Water Works Assn. 71:139-148. 1957. (122AB)

―――, and Linaweaver, F. P., Jr. The McIlroy fluid analyzer in water works practice. J. Am. Water Works Assn. 49:15-20. 1957. (122AB)

Aronofsky, J. S. Mobility ratio — Its influence on flood patterns during water encroachment. Trans. Am. Inst. Min. Metal. Engrs. 195:15. 1952. (221AAA)

Atkinson, R. W. (A) The dielectric field in an electric power cable. Trans. AIEE 38:971-1036. 1919. (612)

―――. (B) Dielectric field in an electric power cable — II. Trans. AIEE 43:966-981. 1924. (612)

Au, T. See Fok, T. D. Y. (422A)

Avrami, M. and Paschkis, V. Application of an electrical model to the study of two-dimensional heat flow. Trans. AIChE 38:631-652. 1942. (311A)

Awberry, J. H., and Schofield, F. H. The effect of shape on the heat loss through insulation. Fifth Int. Cong. on Refrig., Rome. 1928. (313A)

Babbitt, H. E., and Caldwell, D. H. The free surface around, and interference between, gravity wells. Univ. of Ill. Engr. Exp. Sta. Bul. 374, Urbana, Ill. 1948. (221AAB)

Babister, A. W., Marshall, W. S. D., Lilley, G. M., Sills, E. C., and Deards, S. R. The use of a potential flow tank for testing axisymmetric contraction shapes suitable for wind tunnels. Coll. Aero. Cranfield Rep. 46. 1961. (221AAB)

Backstrom, M. (A) Hydraulic computing device for unsteady state heat transfer problems. (In Swedish.) Tidsskr. Varme, Ventilation, 19:113. 1948. (311B)

―――. (B) Kältetechnik. Verlag Brown Karlsruhe, p. 502. 1953. (311B)

―――, Juhasz, S., Liebaut, A., and Hooper, F. Hydraulic analogy for counterflow and reverse flow heat exchangers. International Analogy Computation Meeting, Bruxelles. 1955. (322AB)

Baer, D. H., Schlinger, W. G., Berry, V. J., and Sage, B. H. Temperature distribution in the wake of a heated sphere. Trans. ASME 75:407-414. 1953. (313A)

Baehr, H. D., and Schubert, F. Determination of the efficiency of quadratic plate-type fins by means of an electrical analogy method. (In German.) Kältetechnik 11:320-325. 1959. (313A)

Baes, L. See Piccard, A. (441AA)

Bagrinovskii, A. D. Elektricheskoe modelirovanie rudnichnykh ventilyatsionnykh setei. Gornyi Zhurnal 132:62-67. 1957. (122BB)

Bailey, R. See Liebmann, G. (Appendix II E)

Baird, E. G. See Einstein, H. A. (211AB)

Baird, R. C., and Bechtold, I. C. The dynamics of pulsative flow through sharp-edged restrictions with special reference to orifice-metering. Trans. ASME 74:1381-1387. 1952. (121AB)

Baker, H. D. See Paschkis, V. (311A)

Balachowsky, G. Note on the mapping of electric fields. (In French.) Bul. Soc. Fr. Elec. (6) 6:181-186. 1946. (612)

―――, and Tirroloy. Utilisation of the electric tank in different problems concerning high-voltage equipment. C. I. G. R. E., Paris, Paper 132. 1950. (612)

Balanin, V. V. Approximate method of calculating the output of water filtering through a coffer-dam. (In Russian.) Trudi Leningr. in-ta inzh. vod transp. n. 22, 119-132. 1955. (221AAA)

Ball, W. E., Jr. See Borden, A., and Shelton, G. L., Jr. (221AAA)

Ballet, M. Determination des Frequences Propres de Vibrations de Torsion des Linges d'Arbres. Revue Generale de Mecanique 34:429-434. 1950. (Also, Engineer's Digest 12:96-98. 1951) (Also, Bul. de l'Assn. Techn. Maritime et Aeronautique 48. 1949) (532C)

REFERENCES

Balloffet, A. Applications of the Hele-Shaw apparatus. (In Spanish.) Cienc. y Tecn. 114:191-206. 1950. (221AF)

Balmford, J. A. See Kayan, C. F. (122AC)

Bammert, K. See Hahneman, H. (221AAB)

Barbier, M. Study of anharmonic oscillations by means of an electromechanical analog. (In French.) Ind. Atom. 2:109-112. 1958. (651D)

Barker, C. L. Use of McIlroy analyzer on water distribution systems. J. Am. Water Works Assn. 50:15-20. 1958. (122AB)

Barker, J. R. Conducting analogs of a magnetic field. Am. J. Phys. 28:139-144. 1960. (632)

Barkhausen, H., and von Bruck, J. Shape of the electrical fields in electronic tubes, measured in the tank. (In German.) ETZ 54:175-177. 1933. (612)

Baron, F. Circuit analysis of laterally-loaded continuous frames. Proc., ASCE 83 (ST1 n. 1147):1-32. 1957. Discussion. (ST5 n. 1382):33-36. 1957. (472AC)

──, and Michalos, J. P. Laterally loaded plane structures and structures curved in space. Proc. ASCE n. 51. 1951. (Also, Trans. ASCE 117:279-311. 1952.) (Also, Engineering Report 15, Iowa Engineering Experiment Station, 1953.) (472AC)

Barron, R. A. Viscous flow tube model. Proc. 2nd Int. Conf. Soil Mech. Found. Eng. 3:209. 1948. (221AG)

Barta, J. (A) A Csarara's Problémájának újabb Analógiája. Math. es term est. Budapest. 54:496-506. 1936. (441AB)

──. (B) Die Darstellung ebener Potential-Stromungen Mittelseiner Elastischen Scheibe. Ing. Arch. 6:396-402. 1935. (454)

Baruch, J. J., and Lang, H. C. Analogue for use in loudspeaker design work. Inst. Radio Engrs. Trans. of Professional Group on Audio AU-1:8-13. 1953. (524)

Basu, S. K. On the determination of transient response of linear systems. J. Sci. Indust. Res. India. 18B:93-96. 1959. (Appendix II D1)

Bateman, H. Notes on a differential equation which occurs in the two-dimensional motion of a compressible fluid and the associated variational problems. Proc. Roy. Soc. London (A) 125:598-618, Sect. 3. 1929. (211AC)

Batzel, S., and Schmidt, W. Untersuchengen ueber die Wetterverzweigung unter Tage. Gueckauf 88:471-479. 1952. (122BB)

Baud, R. V. Fillet profiles for constant stress. Product Eng. 5:133. 1934. (454)

Bauer, B. B. (A) Equivalent circuit analysis of mechano-acoustic structures. Inst. Radio Engrs. Trans. of Prof. Group on Audio, v. AU-2:112-120. 1954. (121BBA)

──. (B) Transformer analogs of diaphragms. J. Acous. Soc. Am. 23:680-683. 1951. (121BBA)

──. (C) Transformer couplings for equivalent network synthesis. J. Acous. Soc. Am. 25:837-840. 1953. (121BBA) (511B)

Baumann, H. Determination of temperature distribution in gas-turbine rotor bodies and cylinders by electrolytic tank method. Engrs. Digest 15:139-141. 1954. (313A)

Bautz, W. See Thum, A. (A) - (441BB), (B) - (441BB), (C) - (441BB), (D) - (441BB).

Bazjanac, Davorin. Untersuchungen mit Hilfe der elektrischen Analogie uber den Einfluk der Luftstrahlbegrenzung in Windkanalen auf Tragflugelmessungen. Institut fur Aerodynamik, Zurich. 1943. (221AAA)

Beaver, W. L. Flux plotting analog for an axially symmetric potential field. J. Appl. Phys. 28:579-582. 1957. (632)

Bechtold, I. C. See Baird, R. C. (121AB)

REFERENCES

Becker, E. Analogies between surges and shock waves. (In German.) Ing. Arch. 21:42-54. 1953. (211AA)

Bedingfield, C., Jr., and Drew, T. Analogy between heat transfer and mass transfer (a psychrometric study). Ind. and Engr. Chem. 42:1164-1173. 1950. (322C)

Beggs, C. W. Distribution analysis; Public Service Electric and Gas Company. Gas Age 117:32-37. 1956. (122BAB)

Behrbohm, and Pinl. Zeits. f. Angew. Math. u. Mech. v. 21:193-203. 1941. (211AC)

Belgaumkar, B. M. (A) Application of electrical analogy to the analysis of hydrodynamic journal bearings. J. Sci. Engr. Res., India 3:385-392. 1959. (222A)

———. (B) Characteristics of series; electrical circuit used as an analogue for for mechanical vibration system. Proc. 2nd Congr. Theor. Appl. Mech., New Delhi, India; Indian Soc. Theor. Appl. Mech., Indian Inst. Tech., Kharagpur, 199-208. 1956. (511B)

———. (C) Limitations of voltage resonance in the electrical models of mechanical vibration systems. Proc. First Congress of Theor. and Appl. Mech., Indian Inst. of Tech., Kharagpur, 261-282. November 1-2, 1955. (511B)

Benedict, R. P. Liebmann network approximations to one-dimensional transient heat conduction problems. Mech. Engr. 81(3):106. 1959. (311A)

———, and Meyer, C. A. Electrolytic tank analog for studying fluid flow fields within turbomachinery. Mech. Engr. 80:112. 1958. (221AAA)

Bennett, R. R. See McCann, G. D. (531A)

Benscoter, S. U., and MacNeal, R. H. (A) Analysis of straight multicell wing on Cal. Tech. analog computer. NACA. TN 3113. (524)

———, and MacNeal, R. H. (B) Equivalent plate theory for a straight multicell wing. NACA. TN 2786. (524)

———, and MacNeal, R. H. (C) Introduction to electrical circuit analogies for beam analysis. U. S. NACA Tech. Note 2785. 1952. (524)

———. See MacNeal, R. H. (A), (524); (B) (524).

Benton, G. S. See Li, W. H., and Bock, P. (221AAA)

Bergman, E. O. See Gilkey, H. J. (441AA)

Bernard, J. J., and Siestrunck, R. Sur l'utilisation d'approximations successives dans la détermination de certains potentiels aérodynamiques. La Recherche Aéronautique, 17. 1950. (221AAA)

Berry, F. R., Jr. Electrical analog solution of certain non-linear problems in vibrations and elastic stability. Proc. Soc. Exp. Stress Anal. 13(1):1-12. 1955. (421B) (511C)

Berry, V. J. See Baer, D. H., Schlinger, W. G., and Sage, B. H. (313A)

———. See Schlinger, W. G., Mason, J. L., and Sage, B. H. (322B)

Bertram, S. Calculation of the resonant properties of electrical cavities. Proc. Inst. Radio Engrs. 42:579-585. 1954. (642)

Beuken, C. L. (A) Die Warmestromung durch die Ecken von Ofenwandunger. Waerme-und Kältetchnik 29:1. (n. 7). 1937. (313A)

———. (B) Economish Technish Tijdschrift. Mastricht, Netherlands. n. 1. 1937. (313A)

———, and Hamaker, H. J. De Warmteverliezen van electrische Ovens bij Periodiek Bedrijf. Economish Technisch Tijdschrift, 19. 1940. (311A)

———. See Paschkis, V. (313A)

Bey, S. L. See Hurst, H. E. (221AE)

Biezeno, C. B. See Boiten, R. G. (441AA)

———, and Koch, J. J. Uber einige Beispiele zur elektrischen Spannungsbestimmung. Ing. Arch. 4:384-393. 1933. (441AD) (455)

———, and Rademaker, J. Het Experimenteel Bespalen van de Schuffspannungsverdeeling in de Divarsdoorsnede van een Gewrongen Prismatische Staaf. De Ingenieur. 46:185-197. 1931. (441AA)

REFERENCES

Billington, N. S. (A) Electrical models for heating problems. J. Instn. Heating and Vent. Engrs. 18:247-261. 1950. (311A) (313A)
———. (B) Floor-panel heating — Some design data. J. Instn. Heating and Vent. Engrs. 21:256-265. 1953. (313A)
———. (C) Heat loss in slab floors. Heating and Vent. 50:89-90. 1953. (313A)
Binder, R. C. Fluid flow analogy for torsional stresses. Mach. Des. 10:45-46. 1938. (441AF)
Binnie, A. M., and Hooker, S. G. The flow under gravity of an incompressible and inviscid fluid through a constriction in a horizontal channel. Proc. Roy. Soc. 159:592-608. 1937. (211AB)
Biot, M. A. Contribution a la technique photo-elastique. Ann. Soc. Scientifique Bruxelles (B) 53:13-15. 1933. (452) (455)
———, and Smits, H. Etude photo-elastique des tensions de contraction dans un barrage. Bul. Tech. n. 4, p. 10. Bul. Tech. de l'Union des Ing. Sortis des Ecoles Speciales de Louvain. 1933. (452)
———, and Wianko, T. Electric network model for flexure torsion flutter. GALCIT Publication 195. 1941. (524) (532C)
———. See von Karman, T. (511A)
Birkebak, R. See Eckert, E. R. G., Hartnett, J. P., and Irvine, T. F. (312D)
Bisplinghoff, B. L., Pian, T. H. H., and Levy, L. I. Mechanical analyzer for computing transient stresses in airplane structures. J. Appl. Mech. 17:310-314. 1950. (490)
Bitterly, J. G. See Orlin, W. J., and Linder, N. J. (211AB)
Black, J., and Mediratta, O. P. Supersonic flow investigations with a "hydraulic analogy" water channel. Aero. Quart. 2:227-253. 1951. (211AB)
Blake, G. G. Mechanical model analogous to oscillatory electrical circuit. Engr. 181:535-536. 1946. (621)
Blass, E., and Wesser, U. Direct current network for an analogy model of refrigeration machinery and equipment. (In German.) Kältetechnik 11:326-331. 1959. (313A)
Bloch, A. (A) Electromechanical analogies and their use for analysis of mechanical and electromechanical systems. J. Instn. Elec. Engr. 92:157-169. 1945. (Pt. I-General) (511A)
———. (B) New approach to dynamics systems with gyroscopic coupling terms. London, Edinburgh and Dublin Phil. Mag. and J. Sci. 35:315-334. 1944. (511A)
Bluhm, J. I., and Flanagan, J. H. A procedure for the elastic stress analysis of threaded connections including the use of an electrical analogue. Proc. Soc. Exp. Stress Anal. 15(1):85-100. 1957. (422D)
Bock, P. See Li, W. H., and Benton, G. S. (221AAA)
Boelter, L. M. K., Martinelli, R. C., and Jonassen, F. Remarks on the analogy between heat transfer and momentum transfer. Trans. ASME 63:447-455. 1941. (322C)
Bömelburg, H. (A) Comment on "Wedge pressure coefficients in transonic flow by hydraulic analogy." J. Aero. Sci. 22:731-732. 1955. (211AB)
———. (B) The practical use of the hydraulic analogy in quantitative form for special problems in gas dynamics. (In German.) Mitt. Max-Planck-Inst. Strömungsforschung n. 10, 1954. (211AB)
Boisard, P. Solution of the transient flow problem of slightly compressible fluid in a porous medium by means of resistance network. (In French.) Rev. Inst. Fr. Petrole et Ann. Comb. Liquid 14:1107-1146. 1959. (212A)
Boiten, R. G., and Biezeno, C. B. Sterkte en stijfheid van een op wringing belaste, door een diepe spiesleuf verzwakte as van cirkelvormige dwarsdoorsnede. De Ingenieur 57:0-1. 1945. (441AA)
Bondar', N. G. (A) The electrical analogy of the oscillations of the static and

dynamic rigidities of bar frames. (In Russian.) Issledovania po teorii sooruzhenii, Moscow, Gosstroiizdat n. 7, 549-574. 1957. (524)

Bondar', N. G. (B) Solution of dynamic problems of rod systems by means of electric model. (In Russian.) Inzhener. Sbornik, Akad. Nauk SSSR 16:87-108. 1953. (524)

Boothroyd, A. R. Design of electric wave filters with aid of electrolytic tank. Proc. Instn. Elec. Engrs. 98:426-492. 1951. (640B)

———, Cherry, E. C., and Makar, R. An electrolytic tank for the measurement of steady-state response, transient response, and allied properties of networks. Proc. Instn. Elec. Engrs. 96(1):163-177. 1949. (Appendix II D1)

———. See Makar, R., and Cherry, E. C. (Appendix II D1)

Borden, A., Shelton, G. L., Jr., and Ball, W. E., Jr. An electrolytic tank developed for obtaining velocity and pressure distributions about hydrodynamic forms. David W. Taylor Mod. Basin Rep. 824. 1953. (221AAA)

Bordoni, G. Analogie electtrico-meccaniche. Alta Frequenza 9:133-161. 1940. (511A)

Boreli, M. (A) On the effect of an impermeable partial core on the flow through a dam. C. R. Acad. Sci. Paris 235:646-648. 1952. (221AAA)

———. (B) Use of a drain above an impermeable core in a dam. C. R. Acad. Sci. Paris 235:785-786. 1952. (221AAA)

Boscher, J. (A) Application of electric networks for the calculation of deformation of elastic plates. (In French.) C. R. Acad. Sci. Paris 238:1189-1192. 1954. (462)

———. (B) Resolution par analogie electrique d'equations aus derivees partielles du quatrieme ordre intervenant dans divers problemes d'elasticite. France. Ministere de l'Air — Publications Scientifiques et Techniques 348. 1958. (457)

———. (C) Sur la détermination numérique de fonctions, biharmoniques par un procédé analogique de réseaux superposés. C. R. Acad. Sci. Paris 236:44-46. 1953. (457)

———, and Malavard, L. Détermination analogique des fonctions biharmonique. Bul. 8, Société Francaise des Mécaniciens. 1953. (457)

———. See Malavard, L. (422D)

Boston, O. W. See Neubauer, T. P. (441AA)

Botset, H. G. The electrolytic model and its application to the study of recovery problems. Trans., AIME (Petroleum Div.) 165:15. 1946. (221AAA)

———. See Wyckoff, R. D. (221AAA)

———. See Wyckoff, R. D., and Muskat, M. (221AAA)

Boussinesq, J. Etude nouvelle sur l'equilibre et la mouvement des corps solides elastiques dont certaines dimensions sont trés petites par rapport a d'autres. J. Math. Pure et Appl. (2), 16. 1871. (441AG)

Bowman-Manifold, M., and Nicoll, F. H. Electrolytic field-plotting trough for circularly-symmetric systems. Nature 142:39. 1938. (612)

Boyd, J. E. Strength of materials. p. 153. 1917. (422C)

Boyer, H. M. McIlroy pipeline network analyzer. Instruments & Automation 29:928-929. 1956. (122AB)

Bradfield, K. N. E., Hooker, S. G., and Southwell, R. V. Conformal transformation with the aid of an electrical tank. Proc. Roy. Soc. London, 159(A):315-346. 1937. (441AD)

Bradley, C. B., and Ernst, C. E. Analyzing heat flow in cyclic furnace operation. Mech. Eng. 65:125-129. 1943. (311A)

———, Ernst, C. E., and Paschkis, V. Economic thickness of thermal insulation for intermittent operation. Trans., ASME 67:93-100. 1945. (311A)

Brahtz, J. H. A. Pressures due to percolating water and their influence upon

stresses in hydraulic structures. Second Congress on Large Dams, Washington 5:43-71. 1936. (221AC)
Bray, J. W. Electrical analyzer for rigid frameworks. Structural Engr. 35:297-311. 1957. Discussion, 36:202-206. 1958. (472AB)
Breckenridge, J. R. See Woodcock, A. H., and Thwaites, A. L. (311A) (321A) (330)
Brokmeier, K. H. Possibilities of analog study in heat transfer research. (In German.) VDE — Fachberichte 15. 1951. (313A)
Bromberg, R., and Martin, W. Theory and development of a thermal analyzer. Univ. of Calif. Los Angeles. 1947. (313A)
Broude, B. Apparatus for the study of streamlined bodies when flying at supersonic speeds. (In Russian.) Grazhd. Aviatsiya 10:27-28. 1955. (211AB)
Brough, H. W. See Jenkins, R., and Sage, B. H. (322C)
Brouwer, G. Electrical analog of eddy-current-limited domain-boundary motion in ferromagnetics. J. Appl. Phys. 26:1297-1301. 1955. (632)
———, and Van Der Meer, S. A network analog of a statically loaded two-dimensional frame. Proc. Soc. Exp. Stress Anal. 15(2):35-42. 1958. (472AB)
Brower, W. B., Jr. (A) The application of the electric-tank analogy to two- and three-dimensional problems in linearized aerodynamic theory. Rensselaer Polytech. Inst. (Research Division) TR AE5506. 1955. (221AAB)
———. (B) A linearized theory for the normal force on closed bodies of revolutions. (In English.) 9th Congrés Intern. Mécan. Appl., Univ. Bruxelles, 1, Chap. 3 - Subsonic:439-448. 1957. (221AAA)
———, and De Rienzo, P. The design of a basic electric-tank analogy installation. Rensselaer Polytech. Inst. Aero. Engr. AF 18(600) - 499. 1955. (Appendix II D1)
Brown, K. J. See Bycroft, G. N., and Murphy, M. J. (511A)
Bruce, W. A. An electrical device for analyzing oil-reservoir behavior. Trans. AIME 151:112. 1943. (212A) (221AAA)
———. See Horner, W. L. (221AAA)
Bruckmayer, F. (A) Elektrisches Modellmessverfahren für Waermebruecken im Kuehlraumbau. VDI Zeit. 88:270-272. 1944. (313A)
———. (B) Elektrische Modellversuch zur Loesung Waermetechnischer Aufgaben. Archiv für Waermewirtschaft 20:23-25. 1939. (313A)
Brüche, E., and Recknagel, A. An electrical device for analyzing oil-reservoir behavior. Z. Tech. Physik 17:126. 1936. (651)
Bruman, J. R. (A) Additional comments on "A study of transonic gas dynamics by the hydraulic analogy." J. Aero. Sci. 20:359. 1953. (211AB)
———. (B) Application of the water channel-compressible gas analogy. North American Aviation Inc., Rep. n. NA-47-87. 1947. (211AB)
Bryant, R. A. A. (A) Hydraulic analogy as a distorted dissimilar model. J. Aero. Sci. 23:282-283. 1956. (211AB)
———. (B) The hydraulic analogy for external flow. J. Aero/Space Sci. 27:148-149. 1960. (211AB)
———. (C) One-dimensional and two-dimensional gas dynamics analogies. Australian J. Applied Science 7:296-313. 1956. (211AB)
———. (D) The size of aerofoil models for quantitative hydraulic analogy research. J. Roy. Aero. Soc. 60:208-209. 1956. (211AB)
———. (E) Surface contamination in hydraulic analogy research. J. Aero. Sci. 23:1056-1057. 1956. (211AB)
———. (F) Transonic flow hydraulic analogy. J. Aero. Sci. 23:90-91. 1956. (211AB)
———. (G) Use of the hydraulic analogy for inside problems. J. Aero/Space Sci. 25:536. 1958. (211AB)

———, and Grant, J. N. G. Two-dimensional bow shock wave detachment distances. J. Roy. Aero. Soc. 61:424-426. 1957. (211AB)
Bublikov, E. I. Determining temperature distribution in walls by means of an electro-thermal analogy. (In Russian.) Teploenergetika 3:10-13. 1956. (313A)
Buchberg, H. (A) Cooling load from thermal network solutions. Heating-Piping 29:131-136. 1957. (311A)
———. (B) Electric analogue prediction of thermal behavior of inhabitable enclosure. Heating, Piping, and Air Cond. 27:131-138. 1955. (311A)
———. (C) Electric analogue studies of single walls. Heating, Piping, and Air Cond. 27:125-130. 1955. (311A)
Budrin, D. V. A hydrostatic integrator for solving the differential equation of thermal conductivity with varying temperature relationships of the thermophysical properties — Coefficient of thermal conductivity and heat capacity. (In Russian.) Trudi. Uralsk. politekhn. in-ta n. 53:22-41. 1955. (311B)
Burgers, J. M. First report on viscosity and plasticity. Acad. Sci. Amsterdam. 1935. (491)
Burke, S. P., and Parry, V. F. An electrical analogy for the determination of gas flow under steady conditions in gas-bearing strata. Am. Chem. Soc. Chicago, Illinois. 1933. (221AAA)
Burnand, G. Study of thermal behavior of structures by electrical analogy. Brit. J. Appl. Phys. 3:50-53. 1952. (311A)
Busemann, A. The hodograph method for gas dynamics. (In German.) ZAMM 17:73-79. 1937. (211C)
Bush, V. (A) Gimbal stabilization. J. Franklin Inst. 188:199-215. 1919. (511A)
———. (B) Structural analysis by electric circuit analogies. J. Franklin Inst. 217:289-329. 1934. (471A) (471B) (472AA) (511A)
Bycroft, G. N., and Murphy, M. J., and Brown, K. J. Electrical analog for earthquake yield spectra. Proc. ASCE 85 (EM 4 n. 2197):43-64. 1959. (511A)
———. See Murphy, M. J., and Harrison, L. W. (511A)
Cadambe, V., and Kaul, R. K. (A) The application of membrane analogy for the determination of torsional rigidity of non-circular solid shafts. J. Sci. Indus. Res. India 13B:455-461. 1954. (441AA)
———, and Kaul, R. K. (B) Deflection of clamped plates by two-step membrane analogue. J. Roy. Aero. Soc. 59:358-360. 1955. (463)
———, and Tewari, S. G. Application of electrical analogy for the solution of problems in elasticity. J. Sci. Indus. Res. India 15B:107-111. 1956. (431)
Cahen, G. Study of bending and torsional vibrations by means of electromagnetic analogies. Application to frameworks of ships and aircraft. Bul. Assn. Tech. Marit. Aero. 49:459-473. 1950. (524) (532C)
———. Calculators solve flow problems. Am. Gas Assn. Monthly 33:19-37. 1951. (Also, Gas Age 108:17-18, 66. 1951.) (Also, Gas 27:42-43. 1951.) (122BAC)
Caldwell, D. H. See Babbitt, H. E. (221AAB)
Calhoun, J. C., Jr. (A) Analogy between electrical and fluid flow. Oil and Gas J. 47:301. 1949. (221AAA)
———. (B) Equipressure and flow lines. Oil and Gas J. 47:107. 1949. (221AAA)
Callaghan, E. E. Analogy between mass and heat transfer with turbulent flow. NACA Tech. Note 3045. 1953. (322C)
Calvert, T. W. G. The determination of stress concentrations with an electrolytic tank model. Brit. J. Appl. Phys. 12:184-188. 1961. (455)
Camarasescu, N. See Patraulea, N. N. (221AAA)
Cambel, A. B. See Schneider, P. J. (A), (Appendix II A 1b); (B), (312A)
Camp, T. R. Hydraulics of distribution systems; Some recent developments in

REFERENCES

methods of analysis; Use of network analyzers and hydraulic models. J. N. E. Water Works Assn. 57:334-352. 1943. (122AA) (122AB)

Camp, T. R., and Hazen, H. L. Hydraulic analysis of water distribution systems by means of an electric network analyzer. J. N. E. Water Works Assn. 48:383. 1934. (122AA)

Campbell, F. B. See Lane, E. W., and Price, W. H. (221AAA)

Campbell, W. E. Two electrical analogies for the pressure distribution on a lifting surface. Nat. Res. Coun. Canada Mech. Eng. Rep. MA - 219. 1949. (221AAA)

Canac, F. Propagation, interference, reflection, absorption, diffusion of sonic and ultrasonic waves visualized by the strioscopic method. (In French.) Acustica 4:320-328. 1954. (644)

Canning, T. N. A simple mechanical analogue for studying the dynamic stability of aircraft nonlinear moment characteristics. U.S. NACA Tech. Note 3125. 1954. (511C)

Carne, E. B. See Guile, A. E. (311A)

Carr, L. H. A. The determination of field potentials near corners. Bul. Elec. Eng. Ed. 13:17-21. 1954. (612)

Carter, D. S. An electrical method for determining journal-bearing characteristics. Trans. ASME (J. Appl. Mech.) 74:A114-118. 1952. Discussion 75:440. 1953. (222A)

Carter, G. K. Numerical and network-analyzer solutions of the equivalent circuits of the elastic field. Trans. ASME 66:A162-167. 1944. (455)

―――. See Concordia, C. (221D)

―――, and Kron, G. (A) Network analyzer solution of an equivalent circuit for elastic structures. J. Franklin Inst. 238:443-452. 1944. (472AB)

―――, and Kron, G. (B) Network analyzer tests of equivalent circuits of vibrating polyatomic molecules. J. Chem. Physics 14:32-34. 1946. (710)

―――. and Kron, G. (C) Numerical and network-analyzer tests of an equivalent circuit for compressible fluid flow. J. Aero. Sci. 12:232-234. 1945. (211C) (221D)

―――. See Kron, G. (720)

Carter, W. J., and Oliphint, J. B. Torsion of a circular shaft with diametrically opposite flat sides. J. Appl. Mech. 19:249-251. 1952. (441AE)

Castagno, Aldo. (A) Determinazione del campo di moto attorno ad una schiera de profili alari mediante la vasca elettrica. Atti dell' Accademia delle Scienze di Torino, Gennaio. 1954. (221AAA)

―――. (B) Sulle trasformazioni conformi eseguite sperimentalmente con la vasca elettrica. Atti dell' Accademia delle Sci. di Torino, Luglio. 1953. (221AAA)

Castles, W., Jr., Durham, H. L., Jr., and Kevorkian, J. Normal component of induced velocity for entire field of a uniformly loaded lifting rotor with highly swept wake as determined by electromagnetic analog. NACA TN4238. 1958. (221C)

Caughey, T. K., and Hudson, D. E. A response spectrum analyzer for transient loading studies. Proc. Soc. Exp. Stress Anal. 13(1):199-206. 1955. (511C)

Chadeisson, R. Use of the electrolytic-tank method to represent any pumping test including many boundary conditions. (In French.) Vol. II, 208-212. (221AAA)

Chaffee, E. L. Mechanical model of coupled electrical instruments. Rev. Sci. Instr. 6:231-238. 1935. (622B)

Chapman, F. Laclede's new network analyzer starts attacking distribution layout problems. Gas 29:27-29. 1953. (122BAB)

Chapylgin. On gas jets (1904). Trans. Brown Univ., Providence. 1944. (See Supplementary Note 2.) (211AC)

REFERENCES

Chattarjee, P. N., and Dutt, D. Uses of an electric potential analogue computer to solve some elasticity problems. Proc. 3rd Congr. Theor. Appl. Mech., Bangalore, India; Indian Soc. Theor. Appl. Mech., Indian Inst. Technol., Kharagpar, 329-340. 1957. (492)

Chavasse. Quelques réflexions sur les bruits et les vibrations. Société des Ingenieurs de l'Automobile - J. 18:22-g. 1945. (121BBB)

Cheers, F., and Raymer, W. G. Tests in the NPL electric tank on a 4:1 axisymmetrical diffuser having a discontinuity in the wall velocity. Aero. Res. Coun. Lond. Rep. Mem. 9505. 1946. (221AAB)

─────, Raymer, W. G., and Fowler, R. G. Preliminary tests on electric potential flow apparatus. Gr. Brit. Aero. Res. Coun. R. and M. 2205. 1945. (221AAB)

Chegolin, P. M. (A) Application of the method of electromechanical analogies for the solution of the problems of complex deflections of rod systems. (In Russian.) Electric Modelling of Beams and Frameworks, Taganrog, 22-27. 1956. (421B)

─────. (B) Investigation of the frequency properties of beams and frames by the method of electric modelling. (In Russian.) Electric Modelling of Beams and Frames, Taganrog, 72-98. 1956. (524)

Cherry, E. C. Analogies between vibrations of elastic membranes and electromagnetic fields in guides and cavities. Proc. Inst. Elec. Engrs. (Pt. III— Radio and Communication Eng.) 96:346-358. 1949. (511A)

─────. See Boothroyd, A. R. (Appendix II D1)

─────. See Makar, R., and Boothroyd, H. R. (Appendix II D1)

Childs, E. C. The water table, equipotentials and streamlines in drained land. Soil Sci. Pt. I-56:317-330, 1943. Pt. II-59:313-327, 1945. Pt. III-59:405-415, 1945. Pt. IV-62:183-192, 1946. (221AAA)

Chilton, E. G., and Handley, L. R. Pulsations in gas-compressor systems. Trans. ASME 74:931-941. 1952. (121BAB)

Chow, C. K. See McIlroy, M. S. (122BAB)

Cima, R. M., and London, A. L. Transient response of a two-fluid counterflow heat exchanger, the gas-turbine regenerator. Trans. ASME 80:1169-1175. 1958. (321A)

Clark, A. V. Simplified method for study of two-dimensional transient heat flow using resistance paper. Mech. Engr. 79:583. 1957. (311A)

Clark, J. See Juhasz, S. (321B)

Clark, S. K., and Hess, R. L. Transient thermal stresses by an analogy. J. Appl. Mech. 25:627-628. 1958. (459)

Clarke, L. N. The effect of the number of sections on the accuracy of a particular RC electrical analogue. Australian J. Appl. Sci. 3:119-124. 1952. (311A)

Clement, P. R., and Johnson, W. C. Distributed electrical analog for waveguides of arbitrary cross section. Proc. Inst. Radio Engrs. 43:89-92. 1955. (642)

Clennon, J. P., and Dawson, J. K. Gas distribution problems solved by electric network calculators. Proc. Am. Gas Assn., p. 389-414. 1951. (122BAA) (122BAC)

Coet, P. Note sur un traceur de courbes d'isogradient électrique semi-automatique. Revue HF 4:85-88. 1958. (Appendix II D1)

Cohn, S. B. (A) Determination of aperture parameters by electrolytic tank measurements. Proc. IRE 39:1416-1421. 1951. (642)

─────. (B) The electric and magnetic constants of metallic delay media containing obstacles of arbitrary shape and thickness. J. Appl. Phys. 22:628-634. 1951. (612) (632) (642)

─────. (C) Electrolytic tank measurements for microwave metallic delay-lens media. J. Appl. Phys. 21:674-680. 1950. (642)

REFERENCES

Coker, E. G. Elasticity and plasticity (Thomas Hawksley Lecture). Proc. Inst. Mech. Engr., Vol. I. 1926. (441AA) (442A)

Colburn, A. P. A method for correlating forced convection heat transfer data and a comparison with fluid friction. AIChE Trans. 29:174. 1933. (322C)

Colin, E. C., Jr. A flow analogy for torsion. M. S. thesis, Univ. of Illinois. 1947. (492)

Columbia gas system installs analyzer to solve distribution problems. Gas Age 112:28-29. 1953. (122BAB)

Collins, J. D., and Jones, E. L. A study of the application of electric calculating boards to the solution of hydraulic problems arising in the design of water distribution systems. B. S. thesis, M. I. T. 1933. (122AA)

―――, and Jones, E. L. Complex heat transfer solved by electrical analogy. Chem. and Met. Engr. 50:111-113, 125. 1943. (311A)

Comier, J. J. See Nougaro, J., and Gruat, J. (221AAA)

Concordia, C. (A) Network analyzer determination of torsional oscillations of ship drives. Trans. Soc. Nav. Arch. Marine Engr. 51:51. 1943. (Also, Marine Engr. 48:158-159. 1943.) (531A)

―――. (B) Network and differential analyzer solution of torsional oscillation problems involving non-linear springs. Trans. ASME 67:A 43. 1945. (531A)

―――, and Carter, G. K. D.C. network-analyzer determination of fluid-flow pattern in centrifugal impeller. Trans. ASME 69:A 113-118. 1947. (221D)

―――. See Whinnery, J. R., Kron, G., and Ridgway, W. (641)

Conference on control of underseepage. U.S. Corps of Engineers, U.S. Army, Cincinnati, Ohio. June 13-14, 1944. (221AAB)

Cook, A. C. An experimental investigation to determine the effectiveness of using radio vacuum tube filaments as non-linear elements for an electric network analyzer for hydraulic networks. M. S. thesis, M. I. T. 1941. (122AB)

Coyle, M. B. (A) Air-flow analogy for solution of transient heat conduction problems. Brit. J. Appl. Phys. 2:12-17. 1951. (311C)

―――. (B) The solution of transient heat conduction problems by air-flow analogy. Proc. Gen. Discussion on Heat Transfer (London), 265-267. 1951. (311C)

Cramp, W. Measurement of capacity by analogy. World Power 2:267-273. 1924. (612)

Crandall, S. M. Lifting-surface-theory results for thin elliptic wings of aspect ratio 3 with chordwise loadings corresponding to 0.5-chord plain flap and to parabolic-arc camber. U.S. NACA Tech. Note 1064. 1946. (221CA)

―――. See Swanson, R. S. (221CA)

―――. See Swanson, R. S., and Miller, S. (221CA)

Cranz, H. (A) Die Experimentelle Bestimmung der Airyschen Spannungsfunktion mit Hilfe des Plattengleichnisses. Ing. Arch. 10:159-166. 1939. (451)

―――. (B) Experimentelle Losung von Torsionsaufgaben. Ing. Arch. 4:506-509. 1933. (441AD)

―――. (C) Modellversuche zur Losung von Aufgaben des eben Potentials. Ing. Arch. 7:432-438. 1936. (441AD)

Crausse, E., and Poirier, Y. Analogy study of infiltration in anisotropic media with the aid of conducting paper analogy. (In French.) C. R. Acad. Sci. Paris 243:475-477. 1956. (221AAA)

Cremer, L., and Klotter, K. New aspects of the electro-mechanical analogies. (In German.) Ing. Arch. 28:27-38. 1959. (621)

Criner, H. E., and McCann, G. D. Rails on elastic foundations under the influence of high-speed travelling loads. Trans. ASME 75:13-22. 1953. (422A)

―――, McCann, G. D., and Warren, C. E. New device for solution of transient

REFERENCES

vibration problems by method of electrical-mechanical analogy. Trans. ASME 67:A 135. 1945. (511A) (511B)

Criner, H. E. See McCann, G. D. (A), (511A), (511B); (B), (511A), (511B)
―――― . See McCann, G. D., and Warren, C. E. (531A)

Cross, H. The column analogy. Univ. of Ill. Eng. Exp. Sta., Bul. 215. 1930. (422A)

Crossley, H. E., Jr. See Gilmore, F. R., and Plesset, M. S. (211AB)

Cuenod. Influence des phenomenes de coup de helier sur le reglage de la vitesse des turbines hydrauliques. Houille Blanche 4:163-182. 1949. (121AB)

Culver, R. (A) Simple electrical network analogue for solution of rectangular beam stability equation. Australian J. Appl. Sci. 4:371-379. 1953. (421B)
―――― . (B) Solution of "simple" non-linear pipe-line networks. Civ. Eng. (London) 49:1185-1188, 1314-1316; 50:86-87. 1954. 1955. (122AA)
―――― . (C) Use of extrapolation techniques with electrical network analogue solution. Brit. J. Appl. Phys. 3:376-378. 1952. (Appendix II E)

Cushman, P. A. Shearing stresses in torsion and bending by the membrane analogy. ASME Advance Paper for the June 23-25, 1932 Meeting. (Unpublished.) (431) (432) (441AA) (441AB)

Dachler, R. Grundwasserstromung. Wien. 1936. (221AAA)

Dadda, L. (A) Direct determination in an electrolytic tank of the gradient in an electric field. (In Italian.) Energia Elett. 26:469. 1949. (612)
―――― . (B) Electrolytic analog for the study of harmonic fields in axial symmetry. (In Italian.) Energia Elett. 30:12. 1953. (632)

Dahlin, E. B. Equipotential plotting table. Rev. Sci. Instruments 25:951-953. 1954. (Appendix II 2D)

Damewood, G. Electrical analog helps solve design problems. Oil and Gas J. 54:98. 1956. See also Pipe Line Ind. 5:30-33. 1956. (121BAA)

Danforth, C. E. See Poritsky, H., and Sells, B. E. (211AAC) (211AAD)

Dantu. A new method for determination of stresses in plane elasticity. Ann. Ponts Chauss. 122:375-405. 1952. (451)

d'Arcy, D. F. See Kelk, G. F., and Misener, W. S. (221AAA)

Darlington, S. Potential analog method of network synthesis. Bell System Tech. J. 30:315-365. 1951. (621)

Daum, C. R. See Glover, R. E., and Herbert, D. H. (111)

Davey, P. E., and Spooner, J. C. Application of an electric analogue to domestic refrigerator cabinets. Refrig. Eng. 60:1300-1303, 1332. 1952. (313A)

Dawson, J. K. See Clennon, J. P. (122BAA) (122BAC)

Dean, L. E., and Shurley, L. A. Analysis of regenerative cooling in rocket thrust chambers. Jet Propulsion 28:104-110. 1958. (313A)

Deards, S. R. See Babister, A. W., Marshall, W. S. D., Lilley, G. M., and Sills, E. C. (221AAB)

de Crombrugghe, O., and Remacle, J. Ventilation miniére calcul de réseaux maillés. Annales des Mines de Belgique, n. 10, 876-896. 1958. (122BB)

de Haller, P. Application of electrical analogy to investigation of cascades. Sulzer Tech. Rev. n. 3/4. 1947. (Also, Bul. Tech. de la Suisse Romande 73:25-30. 1947.) (221AAA)

de Laclemandiere, J. L'electroanalyseur de flux calorifique. Metallurgie et Construction Mecanique 80:11, 13-15, 17, 19. 1948. (311A)

de Laclemandiere, L. Experimental study of the transmission of heat in the transient state by aid of electric and thermal analogies. (In French.) Communication to Asso. Ing. Chauffage et de Ventilation de France et aux Eleves du Cours Sup. de Chauffage Ind. de l'Office de Repartition du Charbon. 1946. (311A)

REFERENCES

De la Marre, P. H. Modèles analogiques électriques à trois dimensions pour l'étude des écoulements de filtration à surface libre. Houille Blanche 11:193-205. 1956. (221AAA)

Den Hartog, J. P. Experimentelle losung des ebenen spannungsproblems. Zeits. f. Angew. Math. u. Mech. 11:156. 1931. (452)

_____, and McGivern, J. G. On the hydrodynamic analogy of torsion. Trans. ASME 57:A-46-48. 1935. (441AF)

de Packh, D. C. A resistor network for the approximate solution of the Laplace equation. Rev. of Sci. Instr. 18:798. 1947. (Appendix II E)

De Rienzo, P. See Brower, W. B., Jr. (Appendix II D1)

de Sénarmout, H. (A) Memoire sur la conductibilite des substances cristallises pour la chaleur. Compte Rendus. 25:459-461. 1847. (611)

_____. (B) Second memoire sul la conductibilite des corps cristallises pour la chaler. Compte Rendus. 25:708-710. 1847. (610A)

Design and operation of an hydraulic analog computer for studies of freezing and thawing in soils. Arctic Construction and Frost Effects Lab., New England Div., Corps of Engineers, U. S. Army, Boston. 1956. (311B)

Deutler, H. Zur Versuchmassigen Losung von Torsionsaufgaben mit Hilfe des Seifen-Hautgleichnisses. Ing. Arch. 9:280-282. 1939. (Also, Jahrbuch Deutsche Luft. 2:366-367. 1938.) (441AB)

Diemer, G. See Knol, K. S. (643)

Diesendruck, L. See Katzoff, S., Gardner, C. S., and Eisenstadt, B. J. (221D)

Diggle, H., and Hartill, E. R. (A) Electrolytic tank in engineering design. Metropolitan-Vickers Gaz. 26:106-119. 1955. (Appendix II D1)

_____, and Hartill, E. R. (B) Some applications of electrolytic tank to engineering design problems. Proc. Instn. Elec. Engrs. 101 Part 2 (Power Eng.): 349-364. 1954. Discussion, 102, Part B (Radio & Electronic Eng.):447-452. 1955. (Appendix II D1)

Dix, C. H. The numerical computation of Cagniard's integrals. Geophysics 23:198-222. 1958. (Appendix II D1)

Dobie, W. B., and Gent, A. R. Accuracy of determination of the elastic torsional properties of non-circular sections using relaxation methods and the membrane analogy. Struct. Engr. 30:203-212. 1952. (441AA)

Doerfel, R. Die Losung der Fragen der Zahnflankenbernhrung. Zeits. V. D. I. 69:149-154. 1925. (454)

Dolcetta, A. Experimental equipment for the solution of hydrodynamic problems by the method of electric analogy. (In French.) Energia Elett. 26:461-468. 1948. (221AAA)

Douglas, J. F. H. The reluctance of some irregular magnetic fields. Trans. AIEE. 34:1067-1134. 1915. (631)

_____, and Kane, E. W. Potential gradient and flux density—Their measurement by an improved method in irregular electrostatic and magnetic fields. Trans. AIEE 43:982-988. 1924. (612) (632)

_____, and Voith, R. J. Inductance of A-C magnets from simple models. Trans. AIEE 78, Pt. I:562-568. 1959. (632)

Dove, D. B. Simple electrical analogue for the solution of linear simultaneous equations. J. Sci. Instr. 36:474-475. 1959. (Appendix II E)

Drew, T. See Bedingfield, C., Jr. (322C)

Drucker, D. C., and Frocht, M. M. Equivalence of photoelastic scattering patterns and membrane contours for torsion. Proc. Soc. Exp. Stress Anal. 5(2):34-41. 1948. (441AA)

Druzhinin, N. I. The method of electro-hydrodynamic analogy and its application in filtration analysis. (In Russian.) Moscow-Leningrad, Gosenergoizdat, 356 pp. 1956. (221AAA)

Duffy, F. L. See Stephenson, R. E., and Eaton, J. R. (122BAA)

Duncan, J. P. Geared marine turbine drives. Engineering 179:404-409. 1955. (531A)

Duquenne, R. On the analogy calculation for lifting surfaces. C. R. Acad. Sci. Paris 234:2150-2152. 1952. (221AAA)

──, and Grandjean, C. (A) Application de l'analogie rhéoélectrique au calcul de trois ailes de meme forme en plan. ONERA N. T. 9/1292. A. 1953. (221AAA)

──, and Grandjean, C. (B) Calcul d'effets de volets par analogie rhéoélectrique. ONERA P. V. n. 11/1292.A. 1954. (221AAA)

──, and Grandjean, C. (C) Determination of the effect of thickness of symmetrical wings upon the zero lift with aid of electrical analog. (In French.) C. R. Acad. Sci. Paris 238:1564-1566. 1954. (221AAA)

──. See Malavard, L. (221AAA)

──. See Malavard, L., Enselme, M., and Grandjean, C. (221AAA)

Durham, H. L., Jr. See Castles, W., Jr., and Kevorkian, J. (221C)

Dusinberre, G. M. A note on the "implicit" method for finite-difference heat-transfer calculations. Trans. ASME, J. Heat Transf. 83C:94-95. 1961. (Appendix II E)

Dutt, D. See Chattarjee, P. N. (492)

Eaton, J. R. See Klein, E. O. P., and Tonloukian, Y. S. (311A)

──. See Stephenson, R. E., and Duffy, F. L. (122BAA)

Eckert, E. R. G., Hartnett, J. P., Irvine, T. F., and Birkebak, R. Electrical analog for determining temperature distribution in electrical components. Trans. AIEE 78, Pt. II (Applications and Industry):5-10. 1959. (312D)

Edamoto, I. (A) An electrical method for solving the torsion problem of a cylindrical body. Tech. Rep. Tohoku Univ. 21:51-75. 1956. (441AD)

──. (B) An electrical method for solving the torsional problem of a cylindrical body. Proc. 1st Japan. Nat. Congr. Appl. Mech. 1951. Nat. Committee for Theor. Appl. Mech., 215-218. 1952. (441AD)

──. (C) An electrolyte tank method of solution of the torsion problem of a cylindrical body. Sci. Rep. Ritu (B) 35:25-28. 1954. (441AD)

──, and Kanayama, M. Experimental study of torsion problems by means of a special electrolytic tank, Parts I, II, and III: Circular cylinder having a lengthwise radial cut; twist drill; measurement of moment. Tech. Rep. Tohoku Univ. 23:1-20. 1958. (441AD)

Efros, D. A. Construction of streamline flow by the method of electric modeling. (In Russian.) Nauk USSR Ser. Teckh. Nauk, 9:1061-1068. 1947. (221AAA)

Egorov, P. M. Eksperimental 'noe issledovanie potentsial' nikh poley posredstvom konformo preobrazovannickh modeley. Elektrichestvo 3:6-13. 1954. (Appendix II D1)

Ehrenfried, A. D. A device for rapid and automatic recording of electrostatic fields. Amer. J. Phys. 12:371-372. 1944. (Appendix II D1)

Ehret, L. See Hahneman, H. W. (632)

Einsporn, E. Ebenheit. Zeits. f. Instrumentenkunde 57:265-285. 1937. (451)

Einstein, H. A., and Baird, E. G. Progress reports on the analogy between surface shock waves on liquids and shocks in compressible gases. Hydrodynamic Laboratories, C. I. T. 1946; 1947. (211AB)

──, and Harder, J. A. Electric analog model of a tidal estuary. Proc. ASCE 85 (WW3 2173):153-165. 1959; Discussion 86 (WW1 2419):161. 1960; Reply 87 (WW1 2750):169. 1961. (111)

Einstein, P. A. Factors limiting accuracy of electrolytic plotting tanks. Brit. J. Applied Physics. 2(2):49-55. 1951. (Appendix II D1)

Eisenstadt, B. J. See Katzoff, S., Gardner, C. S., and Diesendruck, L. (221D)

REFERENCES

Electrical analogy method for investigation of transient heat flow problems. Glass Ind. 27:290. 1946. (311A)

Electrical analyzer solves heat-flow problems. Power Plant Engr. 45:62-63. August, 1941. (Also, Power 85:76-78, 170, 172, 174. 1941.) (311A)

Electrical computer solves wing flutter problems. Aviation 45:111. 1946. (524) (532C)

Electrical quantities; a mechanical analogy. Elec. Rev. 124:603-604. 1939. (621)

Electrical waves in slow motion; apparatus designed and built by C. F. Wagner. Power Plant Engr. 43:760, 809. 1939. (621)

Electrolytic model developed by Gulf Corp. expected to increase usable oil resources. Nat. Petr. News 37:38. October 17, 1945; 37:50. 1945. (221AAA)

Electronic brain reports for work; McIlroy pipeline-network analyzer. Am. Gas Assn. 35:18-19. 1953. (122AA)

Ellerbrock, H. H., Jr., Schum, E. F., and Nachtigall, A. J. Use of electric analogs for calculation of temperature distribution of cooled turbine blades. U. S. NACA TN 3060. 1953. (311A) (313A)

Emanueli, L. Experimental research on the distribution of the gradient of potential of three-phase conductors. (In Italian.) L'Elettrotecnica 8:573-580. 1921. (612)

Emersleben, O. The period of a revolving pendulum as an analog to the potential of a circle. (In German.) Zeit. angew. Math. u. Mech. 29:279-282. 1949. (614)

Engle, M. E., and Newton, R. E. Experimental solutions of shear-lag problems by an electrical analogy. Rep. No. R-149, Curtis-Wright Corp. Airplane Div. (St. Louis). 1944. (482)

Engleman, F. Verdrehung von Staben mit Einseitigringformigem Querschnitt. Forsch. Ing. Wes. 6:146-154. 1935. (441AB)

Enselme, M. (A) Calcul par analogies électriques d'ailes minces supportant une répartition de pression donnée. ONERA N. T. 5/1101. 1955. (221AAA)

———. (B) Two electrical analogies for the solution of wings. (In French.) Rech. Aéro. 64:35-42. 1958. (221AAA)

———. See Malavard, L., Duquenne, R., and Grandjean, C. (221AAA)

Ericksson, B. E. An experimental study of heat transmission in the surface layers of the Skagastöl glacier. (In English.) Kyltekn. Tidskr., Stockholm 18:21-30. 1959. (311B)

Ernst, C. E. See Bradley, C. B. (311A)

———. See Bradley, C. B., and Paschkis, V. (311A)

Ertaud, A. L'optique electronique. Ed. L. de Broglie (Paris: Rev. d'Optique) 105. 1946. (651D)

Eschenbrenner, G. P. See McKeon, J. T. (311A)

Estorff, W. Die Ausmessung der Electrostatischer Felder von Isolatoren nach dem Electrolytverfahren. Elek. Zeit. 39:53-55, 62-67, 76-78. 1918. (612)

Experimental determination of fluctuating heat flow. Engr. Digest 1:50-51. 1940. (Condensed from Sulzer Mitteilungen.) (311A)

Experimental determination of non-permanent heat flow. Swizer Tech. Review. 14-15. 1939. (311A)

Eyres, N. R. See Jackson, R., Sarjant, R. J., Wagstaff, J. B., Hartree, D. R., and Ingham, J. (311A)

Fagin, I. Electrical analogs of rigid body suspension systems. M. S. thesis, Ohio Univ. 1948. (511B)

Falkner, V. M., and Gandy, R. W. Notes on the use of an electric tank for the solution of wing loading problems. Aero. Res. Coun. London. 10,022. October, 1946. (221AAA)

Fant, C. G. M. See Stevens, K. N., and Kasowski, S. (121BBC)

REFERENCES

Farberov, I. L. See Pitin, R. N., and Ponnik, Yu A. (211C)
Farnsworth, S. W. See Fortescue, C. L. (612)
Farr, H. K., and Wilson, F. R. Some applications of the electrolytic field analyzer. Trans. AIEE, Vol. 70, 1301-1309. 1951. (Appendix II D1)
Fedorovich, V. See Warshavsky, L. A. (511A)
Fenn, F. H. (A) Application of equivalent currents to slope deflection analysis of structures. La. Univ. Engr. Exp. Sta. Bul., Series 20. 1950. (472AA) (492)
───. (B) Application of equivalent voltages to slope deflection analysis of structures. La. Univ. Engr. Exp. Sta. Bul., Series 22. 1951. (472AA) (492)
Ferrari, C. (A) Experimental determination of aerodynamic fields in two and three dimensions by means of their analogies with electric fields. (In Italian.) Aerotecnica 10:453-469. 1930. (221AAB)
───. (B) Sull 'analogia fra i campi elettrici e i campi aerodinamici. Atti dell 'Accademia d. Scienze di Torino v. LXIV. 1929. (221AAA)
Filchakov, P. F. (A) Electromodeling of seepage problems in heterogeneous soils. (In Russian.) Doklady Akad. Nauk SSSR. 66:593-596. 1949. (221AAA)
───. (B) A simulation of circulation problems with breakaway of the streamlines. (In Ukrainian.) Dopovidi Akad. Nauk USSR (5) 440-443. 1955. (221D)
───, and Panchishin, V. I. Apparatus for filtration study based on electric analogy principle. (In Russian.) Gidrotekh. Stroit. 22:39-40. 1953. (221AAA)
Finn, R. S. See Katzoff, S. (221AAC)
Finzi-Contini, B. Un modello elettrico per lo studio dei campi termici dei sistemi de riscaldamento a turbi immersi nelle muratur. Il Politecnico 85:291. 1937. (313A)
Firestone, F. A. (A) The mobility method of computing the vibration of linear mechanical and acoustical systems, mechanical-electrical analogies. J. Appl. Phys. 9:373. 1938. (511B)
───. (B) A new analogy between mechanical and electrical systems. Acous. Soc. of Amer. J., 4:249-267. 1933. (511B)
───. (C) Twixt earth and sky with rod and tube: Mobility and classical impedance analogies. J. Acous. Soc. Amer. 28:1117-1153. 1956. (511B)
Flanagan, J. H. See Bluhm, J. I. (422D)
Flowers, J. V. The development of experimental techniques for the study of compressible flow by the hydraulic analogy. M. S. thesis. Univ. of Calif., Berkeley. 1951. (211AB)
Fluid analog studies air-conditioning loads. Prod. Engr. 26:201. 1955. (311B)
Fok, T. D. Y., and Au, T. On the solution of rigid frames by the column analogy. Proc. ASCE 85 (ST1 n. 1914):103-112. 1959. (422A)
Föppl, A. Uber die Torsion von Runden Staben mit Veranderlichen Durchmesser. Sitzungsberichter Bayerischen Akademie der Wissenschaften, München 35:249-304. 1905. (441BB)
───, and Föppl, L. Drang und Zwang. (2nd edition.) I:248. 1924. (451)
Föppl, L. Ein Erganzung des Prandtlschen Seifhautgleichnisses zur Torsion. Zeits. f. Angew. Math u. Mech. 15:37-40. 1935. (441AB)
───. See Foppl, A. (451)
Ford, R. L. Electrical analogues for heat exchangers. Proc. Instn. E. E. 103, Part B:65-82. 1956. (321A)
Forster, R. Experimentelle Losung von Randivertaugaben der Gleichung $\Delta u = 0$. Arch. für Eletct. 2:175-181. 1914. (612)
Fortescue, C. L., and Farnsworth, S. W. Air as an insulator when in the presence of insulating bodies of higher specific inductive capacity. Trans. AIEE 32:893-906. 1913. (612)
Foskett, L. W. See Linsley, R. K., and Kohler, M. A. (112A)

REFERENCES

Fossa, D. Geometric representation of plane laplacian fields by methods of Hele-Shaw and Lehmann. (In Italian.) Energia Elett. 29:650-657. 1952. (221AF)

Foster, G. C., and Lodge, O. On the flow of electricity in a uniform plane conducting surface. Proc., Phys. Soc. London 1:113-149, 193-208. 1876. (Appendix II D2)

Fothergill, C. A. An improved blotter model for analog studies. J. Petr. Technol. 9:55-56. 1957. (221AAA)

Fowler, R. G. See Cheers, F., and Raymer, W. G. (221AAB)

Fox. See Allen, Motz, and Southwell. (643)

Francken, J. C. Resistance network, simple and accurate aid to solution of potential problems. Philips Tech. Rev. 21:10-23. 1959/1960. (Appendix II E)

Franks, C. V. See Humphreys, C. M., Nottage, H. B., Huebscher, R. G., Schutrum, L. F., and Locklin, D. W. (313A)

Freberg, C. R. Solution of vibration problems by use of electrical models. Purdue Univ. Eng. Bul., Res. Series 92. 1944. (511A)

———. See Kemler, E. N. (511A)

Frevert, R. K. Development of a three dimensional electric analogue with application to field measurement of soil permeability below the water table. Ph. D. thesis, Iowa State Univ. 1948. (221AB)

Fridman, M. M. Solution of the general problem of bending of a thin isotropic elastic plate supported along one edge. Prikl. Mat. Mekh. 16:429-436. 1952. (464)

Friedmann, N. E. (A) Analog methods for study of transient heat flow in solids with temperature-dependent thermal properties. J. Appl. Phys. 26:129-130. 1955. (311A)

———. (B) Truncation error in semi-discrete analog of heat equation. J. Math. and Phys. 35:299-308. 1956. (Appendix II E)

———, Yamamoto, Y., and Rosenthal, D. Solution of torsional problems with aid of electrical conducting sheet analogy. Proc. Soc. Exp. Stress Anal. 13(2):1-6. 1956. (441AD)

Friend, W. L. See Metzner, A. B. (322C)

Fritzsche, G. Technische Netzwerkeigenschaften-funktionen-theoretische gesehen. Dresden. Technische. Hochschule-Wissenschaftliche Zeit. 7:341-354. 1957-1958. (Appendix II E)

Frocht, M. M. See Drucker, D. C. (441AA)

Fromm, H. Shear-elasticity of liquids. (In German.) Technik. F:125-126. 1952. (491)

Frost, A. D. Analog circuit representation for wall panels. J. Acous. Soc. Amer. 28:1285-1291. 1956. (540)

Fulop, W. (A) Rubber membrane and solution of Laplace's equation. Brit. J. Appl. Phys. 6:21-23. 1955. (Appendix II A1b)

———. (B) Some further results on rubber membrane theory and Laplace's equation. J. Sci. Instruments 34:453-454. 1957. (Appendix II A1b)

Gabor, D. Mechanical tracer for electron trajectories. Nature (London) 139:373. 1937. (651D)

Gagarina, A. A. Application of electrical analogy to the calculations for a square plate with a square orifice. (In Russian.) Investigations on the Theory of Constructions, n. 7, Moscow, Gosstroiizdat, 533-547. 1957. (458)

Gair, F. C. Unifying design principle for resistance network analogue. Brit. J. Appl. Phys. 10:166-172. 1959. (Appendix II E)

Gandy, R. W. See Falkner, V. M. (221AAA)

Gardner, C. S., and LaHatte, J. A., Jr. Determination of induced velocity in front of inclined propeller by magnetic-analogy method. U. S. NACA Wartime Rep. L-154. (Adv. Restr. Rep. L6A05b. 1946.) (221AD)

REFERENCES

———. See Katzoff, S., Diesendruck, L., and Eisenstadt, B. J. (221D)
Gardner. See Hazen, and Schurig. (Appendix II E)
Garner, F. H., Jenson, V. G., and Keey, R. B. Flow pattern around spheres and the Reynolds analogy. Trans. AIChE 37:191-197. 1959. (322C)
Gast, T. Demonstrationen zum zaehelastischen verhalten hochpolymerer stoffe. Zeit. für Angewandte Physik 7:82-86. 1955. (491)
Gates, R. G. See Kayan, C. F. (A), (313A); (B), (313A)
Gaugain. Ann. de Chemie et de Physik, 64(3):200. 1862. (Appendix II D1)
Gay, N. R. See Mackey, C. O. (A), (321B), (330); (B), (321B), (330)
Gehlshøj, B. Electromechanical and electroacoustical analogies and their use in computations and diagrams of oscillating systems. Ingen. Vidensk. Skr., n. 1. 1947. (511A) (121BBB)
Gelfand, R., Shinn, B. J., and Tuteur, F. B. Automatic field plotter. Trans. Am. Inst. Elec. Engrs. 74 (1, Communications and Electronics):73-78. 1955. (Appendix II D1)
Gelissen, H. C. J. H. Development of the Beukenmodde for analysis of unsteady-state heat transfer. Ind. Heating 15:402-414. 1948. (311A)
Gent, A. R. See Dobie, W. B. (441AA)
Gerber, H., and Ackert, J. Experimental methods for determination of potential-flow maps. (In German.) Escher Wyss Mitt. 6:171-176. 1928. (221AAA)
Gerber, S. and Pilod, P. Proving the existence of two types of seepage of a fluid through a dam with vertical walls, by means of electric analogy. (In French.) C. R. Acad. Sci., Paris 249:2006-2008. 1959. (221AAA)
Germain, P. (A) Application of a rheographic tank in the study of systems of axial symmetry. (In French.) Rev. Gen. Elec. 64:263. 1955. (632)
———. (B) Application of the rheoelectric method for calculating flow over conical bodies of small angle. (In French.) C. R. Acad. Sci. Paris 226: 1126-1127. 1948. (221AAB)
———. (C) La théorie générale des mouvements coniques et ses applications à l'aerodynamique supersonique. (voir par. 3, 1.3.2, Utilization des analogies electriques. Publication ONERA 34. 1949.) (211AAA)
———. (D) Un procédé analogique pour la résolution de l'equation de Laplace. Bulletin Technique 5 de l'A. I. Br. (Université Libre de Bruxelles.) 1952. (Appendix II D1)
———. See Malavard, L., and Siestrunck, R. (221AAA)
Geyer, J. C. See Suryaprakasam, M. V., and Reid, G. W. (A), (122AA), (122AB); (B), (122AA)
Gibbings, J. C. The design of an annular entry to a circular duct. Aero. Quart. 10:361-372. 1959. (221AAB)
Gilbert, E. G. See Gilbert, E. O. (312D)
Gilbert, E. O., and Gilbert, E. G. A capacitively coupled field mapper for two-dimensional distributed-source problems. Trans. AIEE. 72:345-349. 1953. (312D)
Gilkey, H. J., and Bergman, E. O. Supplementary methods of stress analysis. Civil Eng. 2:97-101, 376-377. 1932. (441AA)
Gilmore, F. R., Plesset, M. S., and Crossley, H. E., Jr. The analogy between hydraulic jumps in liquids and shock waves in gases. J. Appl. Phys. 21:243-249. 1950. (211AB)
Glover, R. E., Herbert, D. H., and Daum, C. R. Solution of hydraulic problem by analog computer. Proc. ASCE Sep. n. 134. 1952. (111)
Godsey, F. W., Jr. Two methods of mapping flux lines. Elec. Engr. 54:1032-1036. 1935. (632)
Gohar, M. K. Experimental determinations of the electric field and equipotential surfaces using the heat-conduction analogy. J. Appl. Phys. 25:805-807. 1954. (611)

REFERENCES

Goldenberg, H. Calculation of transient heating in single-core cable dielectric. Engineer 197:779-780. 1954. (311A)

Goldsworthy, R. A. Two-dimensional rotational flow at high mach number past thin aerofoils. Quart. J. Mech. and Appl. Math. 5:54-63. 1952. (211B)

Goodier, J. N. An analogy between the slow motions of a viscous fluid in two dimensions and systems of plane stress. Phil. Mag. Series 7 17:554-576. 1934. (222B)

―――. See Timoshenko, S. (441AG)

Goos, and Haenchen. Ann. Phys. (6) 1:333. 1947. (121BBD)

Goran, L. A. Minimum energy solution and an electrical analogy for the stress distribution in stiffened shells. J. Aero. Sci. 18:407-416. 1951. (481)

Grad, E. M. See Liebmann, G. (632)

Graham, R. E. Linear servo theory. Bell System Tech. Univ. 25:616-651. 1946. (511A)

Grandjean, C. See Duquenne, R. (A), (221AAA); (B), (221AAA), (C), (221AAA)

―――. See Malavard, L., Duquenne, R., and Enselme, M. (221AAA)

Grant, J. N. G. See Bryant, R. A. A. (211AB)

Grave, D. F. H. See Wiles, G. G. (313A)

Gray, R. B. Experimental smoke and electromagnetic analog study of induced flow field about a model rotor in steady flight within ground effect. NASA TN D-458. 1960. (221CA)

Green, J. E. The pressure surge in oil pipelines. Proc. Third World Petr. Cong., The Hague 9:7-21. 1951. (121AAB)

Green, L. Nonsteady flow of gas through a porous wall. C. I. T. Jet Propulsion Lab. Prog. Rep. 4-109. 1949. (212B)

Green, L., Jr., and Wilts, C. H. Nonsteady flow of gas through a porous wall. Proc. First U.S. Nat. Cong. Appl. Mech. 777. (212B)

Green, P. E. Automatic plotting of electrostatic fields. Rev. Sci. Instr. 19:646-653. 1948. (612)

Greene, C. E. Michigan Technic. 1869. (422C)

Greenhill, A. G. Hydromechanics. Encyclopedia Britannica, Cambridge Univ. Press, New York (Eleventh Edition) 14:115-135. 1910. (441AE)

Greenland, R. V., and Holden, W. N. Electrolytic plotting tank. S. African Mech. Engr. 9:318-328. 1960. (Appendix II D1)

Greenwood, A. Simple mechanical analogue for solving certain power system stability problems. Elec. Eng. 73:879-884. 1954. (621)

Griffith, A. A. The use of soap films in solving stress problems. Proc. First Int. Cong. Appl. Mech., Delft 39-42. 1924. (441AA)

―――, and Taylor, G. I. (A) The application of soap films to the determination of the torsion and flexure of hollow shafts. Gr. Brit. Adv. Comm. Aero, R. and M. (392). 1918. (432) (441AA)

―――, and Taylor, G. I. (B) The determination of torsional stiffness and strength of cylindrical bars of any shape. Gr. Brit. Adv. Comm. Aero. R. and M. (334). 1917-18. (441AA)

―――, and Taylor, G. I. (C) The problem of flexure and its solution by the soap-film method. Gr. Brit. Adv. Comm. Aero, R. and M. (399). 1917. (Also, Gr. Brit. Adv. Comm. Aero., T. R. 1917-1918.) (432)

―――, and Taylor, G. I. (D) The use of soap films in solving torsion problems. Gr. Brit. Adv. Comm. Aero. R. and M. (333). 1917. (441AA)

―――, and Taylor, G. I. (E) The use of soap films in solving torsion problems. Proc. Instn. Mech. Engrs. 755-809. 1917. (441AA) (441AB)

Griscom, G. B. A mechanical analogy of the problem of transmission stability. The Electric J. 23:230-235. 1926. (621)

Gross, D. See Robertson, A. F. (313A)

REFERENCES

Gross, B., and Pelyer, H. Some aspects of relation between relaxation spectrum and creep spectrum. New mathematical relation concerning delta function. Brit. Elec. and Allied Ind. Res. Assn. - Tech. Report L/T264. 1951. (491)

Gross, W. A., and Soroka, W. W. (A) Electrical analogy for dynamical systems containing broken-linear unsymmetrical elasticity. Proc. First U. S. Nat. Cong. Appl. Mech. 133. (531A)

———, and Soroka, W. W. (B) Network representation of elastic problems in cylindrical coordinates. Proc. Soc. Exp. Stress Anal. 11(2):45-58. 1954. (455)

Grossman, K. H. Stroemungen durch Schaufelgitter. Schweiz Bauzeitung 66:429-430. 1948. (221D)

Grossmann, G. Experimentelle Durchfuehrung einer neuen hydrodynamischen Analogie für das Torsionsproblem. Ingenieur-Archiv 25:381-388. 1957. (441AH)

Gruat, J. Studies of constant cross-section surge tanks with the aid of electric analogies. (In French.) C. R. Acad. Sci. Paris 240:2120-2122. 1955. (121AB)

———. See Nougaro, J., and Comier, J. J. (221AAA)

Guébhard, A. (A) Figuration electrochimique des lignes equipotentielles sur des portions quelconques du plan. Jour. de Phys. (2nd series) 1:205-222. 1882. (612)

———. (B) Sur la figuration electrochimique des systemes equipotentiel. Jour. de Phys. (2nd series) 1:483-492. 1882. (612)

Guenther, E. Forsch. Geb. Ing. 11, 76. 1940. (221AG)

Guerra, G. The Ricci analogy, and the experimental stress analysis of elastic frames. (In Italian.) Aerotecnica 35:11-16. 1955. (472AF)

Guha, S. K., and Ram, Gurdas. The development of electrical analogy method in India. J. Centr. Board of Irrig. 7:290. 1950. (221AAA)

Guile, A. E., and Carne, E. B. Analysis of an analogue solution applied to the heat conduction problem in a cartridge fuse. Elec. Eng. 73:224. 1954. (311A)

Gustafson, P. N. Statical analogue for natural vibrations. Proc. Soc. Exp. Stress Anal. 11(1):147. 1953. (511A)

Habib, P., and Sabarly, F. Study of water circulation in a permeable soil by means of three-dimensional electrical analogy. (In French.) Vol. II, 250-254. (221AAC)

Hacques, G. (A) Analogie rhéoélectrique en domaine illimite de la fonction associée d'un potentiel harmonique à symétrie axiale. Assn. Internationale pour le Calcul Analogique — Annales 2:57-62. 1960. (Appendix II D1)

———. (B) Application of the electrolytic analogy to the problem of ring airfoils. (In French.) Publ. Sci. Tech. Air, France 358. 1960. (221AAA)

———. (C) Calculation of the effect of conicity of an annular rotational wing in axial flow with help of the method of rheoelectrical analogy. (In French.) C. R. Acad. Sci., Paris 245:1510-1520. 1957. (221AAA)

———. (D) Calculation of the effect of incidence on an annular lifting surface by the method of rheoelectric analogy. (In French.) C. R. Acad. Sci. Paris 245:2476-2479. 1957. (221AAA)

———. See Malavard, L. (221AAA)

Haefeli, R. J. See Zangar, C. N. (221D)

Hagg, A. C., and Warner, P. C. Oil whip of flexible rotors. Trans. ASME 75:1339-1344. 1953. (524)

Hague, B. Determining the distribution of electrical and magnetic fields. Elec. 102:185-187, 315-317. 1929. (631)

Hahn, E. P. (A) Experimental solution of hydrodynamic equations. Eng. 123:178. 1927. (221AAA) (221AB) (221B) (221CB)

Hahn, E. P. (B) Methode experimentale pour la resolution des equations au mouvement des fluides. Rev. Gen. Elec. 21:485-489. 1927. (221AAA) (221AB) (221B) (221CB)

Hahneman, H., and Bammert, K. Tests in an electrolytic tank on a ramming and non-ramming air intake for a gas turbine power plant. M. of S. Volkenrode Mon. 11:2; R and T. 182. 1946. (221AAB)

Hahneman, H. W., and Ehret, L. Mapping of axial symmetrical potential fields in a new electrolytic tank. (In German.) Forschung Ing. Wes. 20:141-144, 171-177. 1954. (632)

Hahnle, W. Die Darstellung Elektromechanischer Gebilde durch Rein elektrische Schaltbilder. Wiss. Veroff. a. d. Siemens - Konzern., Vol. 11. (Julius Springer, Berlin. 1932.) (511B)

Hamaker, H. J. See Bueken, C. L. (311A)

Haenchen. See Goos. (121BBD)

Handley, L. R. See Chilton, E. G. (121BAB)

Hansen, I. E. Calculation of iron losses in dynamo electric machinery. AIEE Trans. 28, (II):993-1001. 1909. (631)

Hansen, V. E. Complicated well problems solved by membrane analogy. Trans. Amer. Geo. Union 33:912-916. 1952; Discussion 34:951-952. 1953. (221AC)

Hansen, W. W., and Lundstrom, C. C. Experimental determination of impedance functions by the use of an electrolytic tank. Proc. IRE 33:528-534. 1945. (Appendix II D1)

Hansson. See Merbt. (221D)

Harder, J. A. See Einstein, H. A. (111)

Hardie, A. M. Mechanical synthesis of the amplitude — modulated wave. Wireless Eng. 29:326-333. 1952. (621)

Hargest, T. J. (A) An electric tank for the determination of theoretical velocity distributions. Aero. Res. Coun. London Rep. Mem. 2699. 1949. Publ. 1952. (221AAA)

———. (B) National Gas Turbine Res. Establishment Memo. M. 48. 1949. (221AAA)

Harleman, D. R. F., and Ippen, A. T. The range of application of the hydraulic analogy in transonic and supersonic aerodynamics. "Memoires sur la mecanique des Fluides." Publ. Sci. Tech. Min. Air. Paris 91-112. 1954. (211AB)

———. See Ippen, A. T. (211AB)

Harrison, H. C. See Maxfield, J. P. (511A)

Harrison, L. W. See Murphy, M. J., and Bycroft, G. N. (511A)

Hartill, E. R. (A) The analogy between the flow of current in the electrolytic tank and the electric and electromagnetic field. Bul. Elec. Engng. Ed. 8:42-54. 1952. (612)

———. (B) Electrolytic tank and its application to engineering design. Metropolitan Vickers Gaz., v. 24 (394):147-154. 1952. (Appendix II D1)

———. (C) The initial impulse-voltage distribution on a transformer winding. Bul. Elec. Eng. Ed. 10:45-52. 1953. (612)

———. (D) Use of deep electrolytic tank for solution of field problems in engineering. Elec. Energy 2:118-124, 184-187. 1958. (Appendix II D1)

———, McQueen, J. G., and Robson, P. N. Deep electrolytic tank for solution of 2- and 3-dimensional field problems in engineering. Proc. Instn. Elec. Engrs. 104A:401-411. 1957. (Appendix II D1)

———. See Diggle, H. (A), (Appendix II D1); (B), (Appendix II D1)

Hartnett, J. P. See Eckert, E. R. G., Irvine, T. F., and Birkebak, R. (312D)

Hartree, D. R. See Jackson, R., Sarjant, R. J., Wagstaff, J. B., Eyres, N. R., Ingham, J. (311A)

REFERENCES

Harza, L. F. Uplift and seepage under dams on sand. Trans. ASCE 100:1352-1385. 1935. (221AAA)

Hassan, M. M. Use of the three-dimensional electrical analogy in the design of conduit contractions. Ph. D. thesis, Univ. of Iowa. 1948. (221AAC)

———. See Rouse, H. (221AAC)

Hatamura, M. See Sunatani, C., and Matuyama, T. (441AA) (441AB)

Hauger, H. H., Jr. Intermittent heating of airfoils for ice protection, utilizing hot air. Trans. ASME 76:287-298. 1954. (311A)

Haupt, L. M. (A) Pipe line fluid network calculator. Petr. Engr. 24:D33-34. 1952. (122AA) (122AB)

———. (B) Solution of hydraulic flow distribution problems through use of the network calculator. Texas Eng. Exp. Sta. Res. Rep. 18. 1950. (122AA)

———. (C) Water distribution problems solved by network calculators. Proc. ASCE. 84(PL1 n. 1577):1-6. 1958; Discussion 85(PL1 n. 1920):49-50. 1959. (122AB)

Hay, A. See Hele-Shaw, H. S. (631)

———. See Hele-Shaw, H. S., and Powell, P. H. (631)

Hay, G. E. See Prager, W. (472AD) (472AE)

Hay, N., and Markland, E. The determination of the discharge over weirs by the electrolytic tank. Proc. Instn. Civ. Engrs. 10:59-86. 1958. (221AAA)

Hay, N. See Markland, E. (221AAA)

Haythornthwaite, B. See Morris, R. E. (221AB)

Hazen, H. L. See Camp, T. R. (122AA)

Hazen, Schurig, and Gardner. The M. I. T. network analyzer. Trans. AIEE 49:1102-1113. 1930. (Appendix II E)

Heat transfer problems solved by electrical analyzer. Prod. Eng. 14:442-446. 1943. (311A)

Heat-transfer problems solved with roomful of r-c networks; Paschkis heat and mass flow analyzer. Electronics 16:180. 1943. (311A)

Hebert, D. J. Hydrostatic uplift pressures under dams on pervious earth foundation. U.S. Dept. of Interior. Bureau of Reclamation, Memo. 384. (221AAA)

Hedgecock, J. L. A study of detached shock waves by the water channel-compressible gas analogy. M. S. thesis, Univ. of Calif., Berkeley. 1950. (211AB)

Heindlbofer, K. and Larsen, B. M. Electric analogue of flow of heat in regenerator system. Trans. Am. Inst. Min. Met. Engr. (Iron and Steel Div.) 162: 15-36. 1945. (321A)

Heisler, M. P. See Paschkis, V. (A), (311A); (B), (311A); (C), (313A)

Hele-Shaw, H. S. (A) Investigation of the nature of surface resistance of water and of streamline motion under certain experimental conditions. Second Paper, Trans. Inst. Naval Arch. 40:21-40. 1898. (221AF)

———. (B) Streamline motion of a viscous film. Brit. Assoc. Adv. Sci. Report of 68th Meeting: 136-142. 1898. (221AF)

———, and Hay, A. Lines of induction in a magnetic field. Phil. Trans. Roy. Soc. (A) 195:303-327. 1901. (631)

———, Hay, A., and Powell, P. H. Hydrodynamical and electromagnetic investigations regarding the magnetic flux distribution in toothed-core armatures. J. Instn. Elec. Engrs. 34:21-53. 1905. (Also, Electrician 54:213-215, 307. 1904.) (631)

Henderson, E. N. See Sharp, J. M. (121BAA)

Hepp, G. Measurements of potential by means of the electrolytic tank. Philips Tech. Rev. 4:223-230. 1939. (Appendix II D1)

Herbert, D. H. See Glover, R. E., and Daum, C. R. (111)

Hess, J. J., Jr. See Linvill, J. G. (311A)

Hess, R. L. See Clark, S. K. (459)

REFERENCES

Hetenyi, M. I. (A) Handbook of experimental stress analysis. Wiley, New York. 1950. (431) (432) (433) (441AA) (451)

――――. (B) On similarities between stress and flow patterns. J. Appl. Phys. 12:592-595. 1941. (454)

Higgins, T. J. (A) Analogic experimental methods in stress analysis as exemplified by Saint-Venant's torsion problem. Proc. Soc. Exp. Stress Anal. 2(2):17-27. 1944. (441AA) (441AB) (441AE) (441AF) (441AG)

――――. (B) Stress analysis of shafting exemplified by Saint-Venant's torsion problem. Proc. Soc. Exp. Stress Anal. 3(1):94-101. 1945. (441BB)

――――. See Swenson, G. W., Jr. (Appendix II E)

――――. See Teasdale, R. D. (612)

Hill, B. F. See Torgeson, W. L., and Kitchar, A. F. (313A)

Hill, R. On Inoue's hydrodynamical analogy for the state of stress in a plastic solid. J. Mech. Phys. Solids 2:110-116. 1954. (454)

Himpan, J. (A) An improved electrolytic tank. (In German.) ETZ 63:349. 1942. (612)

――――. (B) A modified electrolytic tank for mapping of potential fields. (In German.) Die Telefunkenröhre 17:198. 1939. (612)

Hines, L. E. Cranfield cycling-pressure maintenance unit gas well flow rates from electric models. World Oil 130:174. 1950. (221AAA)

Hinsley, F. B. See Scott, D. R. (122BB)

Hiramatsu, Y. (A) Ermittlung der Staerke von Wetterstroemen in Grubensetternetyen nach Formeln für den Elektrischen Strom. Glückauf 89:355-359. 1953. (122BB)

――――. (B) Finding air quantities in mine ventilation networks by making use of calculation of electric circuit. Kyoto Univ. Faculty Eng. Memoirs 15:71-78. 1953. (122BB)

Hoff, N. J., and Libby, P. A. Recommendations for numerical solution of reinforced-panel and fuselage-ring problems. U.S. NACA Rep. 934. 1949. (483)

Hohenemser, K. Experimentelle Losung Ebener Potentialaufgaben. Fors. a. d. Gebiete d. Ing. 2:370-371. 1931. (441AD)

Holden, W. N. See Greenland, R. V. (Appendix II D1)

Hollway, D. L. (A) The determination of electron trajectories in the presence of space charge. Austral. J. Phys. 8:74-89. 1955. (651D)

――――. (B) An electrolytic-tank equipment for the determination of electron trajectories, potential and gradient. Proc. Instn. Elec. Engrs. 103B:155-160. 1956. (651D)

――――. (C) A method of tracing electron trajectories in crossed electric and magnetic fields. Proc. Instn. Elec. Engrs. 103B:161-163. 1956. (651D)

――――. See Murray, C. T. (Appendix II D2)

Hooker, S. G. See Binnie, A. M. (211AB)

――――. See Bradfield, K. N. E., and Southwell, R. V. (441AD)

Hooper, F. See Backstrom, M., Juhasz, S., and Liebaut, A. (322AB)

――――. See Juhasz, S. (A), (322AB); (B), (322AB); (C), (322AB)

Horner, W. L., and Bruce, W. A. Electrical-model studies of secondary recovery; aid in locating best relative positions of injection and production wells. Proc. Amer. Petr. Inst. (Sect. IV, Production) 24(4):190-198. 1943. (221AAA)

Horton, C. W. Loss mechanism for the Pierre shale. Geophysics 24:667-680. 1959. (491)

Housner, G. W., and McCann, G. D. The analysis of strong motion earthquake records with the electric analog computer. Bul. Seis. Soc. Amer. 39:47-56. 1949. (511A)

Houstoun, J. See Ackroyd, R. T., Lynn, J. W., and Mann, E. (311A)

REFERENCES

How to trace ground water flow. Pub. Works. 85:75-76. 1954. (221AAA)

Huang, Y.-S. Column analogy for multi-connected rigid frames. Scientia Sinica 8:568-579. 1959. (472AC)

Huard de la Marre, P. (A) Résolution de problèmes d'infiltrations à surface libre au moyen d'analogies électriques. Ministère de l'Air — Publications Scientifiques et Techniques 340. 1958. (Translation, "Solution of infiltration problems with free surface with the aid of electrical analogies." (221AAC)

———. (B) Three dimensional electrical analogy for the study of seepage flow with free surface. (In French.) C. R. Acad. Sci. Paris 240:2203-2205. 1955. (221AAC)

Hubbard, P. G. Application of electrical analogy in fluid mechanics research. Rev. Sci. Instr. 20:802-807. 1949. (221AAB) (221AAC)

———, and Ling, S. C. Hydrodynamic problems in three dimensions. Proc. ASCE Sep. n. 143. 1952. (221AAC)

Hudson, D. E. See Caughey, T. K. (511C)

Hudson, G. E. See Probstein, R. F. (211AB)

Hudson, R. F. See Scott, D. R. (122BB)

Hübner, R. Aufbau und Arbeitsweise des elektrischen Wettermodells "Rheinelbe." Glückauf 91:705-714. 1955. (122BB)

Hübscher, R. G. See Humphreys, C. M., Nottage, H. B., Franks, C. V., Schutrum, L. F., and Locklin, D. W. (313A)

Huggins, W. H. A note on frequency transformations for use with the electrolytic tank. Proc. IRE 36:421-424. 1948. (Appendix II D1)

Hughes, W. F. See Stokey, W. F. (458)

Humphreys, C. M., Nottage, H. B., Franks, C. V., Hübscher, R. G. Schutrum, L. F., and Locklin, D. W. Laboratory studies on heat flow within a concrete panel. Heating, Piping, and Air-Cond. 22:109-115. 1950. (313A)

Hurst, H. E., and Bey, S. L. Flow-net construction simplified by using principle of magnetic field. Civ. Eng. 20:268-269. 1950. (221AE)

Hurst, W. Electrical models as an aid in visualizing flow in condensate reservoirs. Petr. Engr. 12(10):123. 1941. (221AAA)

———, and McCarty, G. M. The application of electrical models to the study of recycling operations in gas distillate fields. Amer. Petr. Inst. (Drilling and Production Practice.) 228-240. 1941. (221AAA)

Hutcheon, I. C. (A) Iterative analogue computer for use with resistance network analogues. Brit. J. Appl. Phys. 8:370-373. 1957. (Appendix II E)

———. (B) Simulation of non-linear field problems. Brit. Communications and Electronics 5:96-99. 1958. (Appendix II E)

Hutter, R. G. E. Electron beam deflection — applications of the small-angle deflection theory. J. Appl. Phys. 18:797-810. 1947. (612)

Ikeda, K. Soap film technique for solving torsion problems. Japan Sci. Rev. 2:113-118. 1951. (441AA)

Il'enko, O. V. (A) Design and modelling of certain three-dimensional frames by the use of electrical systems of substitution. (In Russian.) Electric Modelling of Beams and Frames. Taganrog, 42-49. (472AB)

———. (B) Determination of deflection and torsion moments in simple intercrossing and broken (in plane) beams. (In Russian.) Electrical Modelling of Beams and Frames. Taganrog, 50-58. 1956. (472AB)

Ingham, J. See Jackson, R., Sarjant, R. J., Wagstaff, J. B., Eyres, N. R., and Hartree, D. R. (311A)

Inoue, N. (A) Application of the theory of supersonic flow to the two-dimensional isostatical problem in the theory of elasticity. J. of the Phys. Soc. of Japan, 6:460-475. 1951. (454)

———. (B) Mechanics of perfectly plastic and pulverulent bodies studied as a kind of gas dynamics. (In Japanese.) Doshisha Eng. Rev. 2:85-108. 1952. (454)

Inoue, N. (C) New mechanical analogy for the flow of compressible fluid. J. Aero. Sci. 19:783-784. 1952. (211AAE)

Ip, S. K. An electrolytic tank as an analogue computing machine for factorizing high degree polynomials. Quart. J. Mech. Appl. Math. 10:369-384. 1957. (Appendix II D1)

Ippen, A. Soap film technique for solving torsion problems. Ph. D. thesis, C. I. T. 1936. (211AB)

Ippen, A. T. See Harleman, D. R. F. (211AB)

─────, and Harleman, D. R. F. Certain quantitative results of the hydraulic analogy to supersonic flow. Proc. Sec. Midwest. Conf. Fluid Mech. 219-230. Ohio State Univ. Press. 1952. (211AB)

Irvine, T. F. See Eckert, E. R. G., Hartnett, J. P., and Birkebak, R. (312D)

Isaachsen, I. Innere Vorgange in Stromenden Flussigkeiten und Gasen. Zeits. d. Ver. Deut. Ing. 55:428-431. 1911. (211AB)

Isakoff, S. E. Analysis of unsteady fluid flow using direct electrical analogs. Ind. and Eng. Chem. 47:413-421. 1955. (121BAB)

Ishiguro, S. A method of analysis for long-wave phenomena in the ocean, using electronic network models, part I: the earth's rotation ignored. Phil. Trans. Roy. Soc. Lond. (A)251:303-340. 1959. (113)

Isobe, T., and Nihei, T. Automatic plotting of equipotential lines. Tokyo University, Aeronautical Research Inst. 24, Report 338:211-221. 1958. (Appendix II D1)

Itaya, S., and Tomita, Y. A study of high-speed gas flow by the hydraulic analogy. (1st Report) (In Japanese.) Trans. Japan Soc. Mech. Engrs. 21:18-21. 1955. (211AB)

Jackson, R., Sarjant, R. J., Wagstaff, J. B., Eyres, N. R., Hartree, D. R., and Ingham, J. Variable heat flow in steel. J. Iron and Steel Inst. 150:211-267. 1944. (311A)

Jacobsen, L. S. (A) Torsional stress concentrations in shafts of circular cross section and variable diameter. Trans. ASME 47:619. 1925. (441BA) (441BB)

─────. (B) Torsional stresses in shafts having grooves or fillets. Trans. ASME 57:A154-155. 1935. (441BA) (441BB)

Janssen, E. Flow past a flat plate at low Reynolds numbers. J. Fluid Mech. 3:329-343. 1958. (222B)

Janssen, J. E. See McNall, P. E., Jr. (A), (312D); (B), (312D)

Janzen, H. B. See Nobles, M. A. (221AAA)

Jasper, N. H. (A) Ship vibration problems. J. Am. Soc. Naval Engrs. 65:23-28. 1953. (540)

─────; (B) Statistical approach to longitudinal strength design of ships. J. Amer. Soc. Naval Engrs. 62:565-573. 1950. (492)

Jefferson, H. Notes on electro-mechanical equivalence. Wireless Engr. 21:563-570. 1944. (511A)

Jenkins, R., Brough, H. W., and Sage, B. H. Prediction of temperature distribution in turbulent flow; application of the analog computer. Ind. and Eng. Chem. 43:2483-2486. 1951. (322C)

Jennings, S. T. The solution of the aircraft wing flutter problem by means of electric circuit analysis. Ph. D. thesis, Ohio Univ. 1947. (524) (531B)

Jensen, J. L. An electric analogue for hydraulic network analysis. M. S. thesis, Iowa State Univ. 1951. (122AA) (122AB)

Jensen, V. P. Experimental determination of non-linear distribution of stresses by the slab analogy, with application to Hoover Dam. M. S. thesis, Univ. of Ill. (451)

Jenson, V. G. See Garner, F. H., and Keey, R. B. (322C)

Johner, W. See Piccard, A. (441AA)

REFERENCES

Johnson, B. G. Torsional rigidity of structural sections. Civ. Eng. 5:698-701. 1935. (441AA)

Johnson, H. A. Seepage forces in a gravity dam by electrical analogy. Proc. ASCE 81(757):1-16. 1955. Discussion. 82(SM2 942):7-9. 1956. Reply. 82(SM4 1095):3-4. 1956. (211AAA) (211AAC)

Johnson, H. P. Seepage through loessal earth dams. M. S. thesis, Iowa State Univ. 1950. (221AAA)

Johnson, W. C., and Alley, R. E., Jr. An electrical analogy method for the solution of differential equations. Report 3, ONR Contract N6 ori-105, Task Order VI, 1948. (Appendix II E)

_____. See Clement, P. R. (642)

Jonassen, F. See Boelter, L. M. K., and Martinelli, R. C. (322C)

Jones, D. A. Anisotropic field plotting in electrolytic tank. Rev. Sci. Instr. 30:577-578. 1959. (Appendix II D1)

Jones, E. L. See Collins, J. D. (122AA) (311A)

Jones, W. P. Note on lifting plane theory with special reference to Falkner's approximate method and a proposed electrical device for measuring downwash distributions. Aero. Res. Counc. Lond. Rep. Mem. 2225:1-3. 1946. (221CA)

Joukovsky, N. I. Flow of water in an open channel and flow of gases in pipes. II. An analogy between the motion of a heavy liquid in a narrow channel and the high velocity of gas in a pipe. U.S.S.R. Sci. Tech. Dept. 63, Moscow. 1925. (211AB)

Jouquet, E. Some problems in general hydrodynamics. J. de Mathematiques Pures et Appliquées (8) 3.1.:3-13. 1920. (211AB)

Juhasz, S. Hydraulic analogy for transient crossflow heat exchangers. Mech. Engr. 80:107. 1958. (321B)

_____, and Clark, J. Hydraulic analogy for transient conditions in heat exchangers. IXe Congrès International de Mécanique Appliquée, Actes II, Universite Bruxelles, 495-505. 1957. (321B)

_____, and Hooper, F. C. (A) Hydraulic analog for studying steady-state heat exchangers. Ind. and Eng. Chem. 45:1359-1362. 1953. (322AB)

_____, and Hooper, F. C. (B) Hydraulic analogy applied to crossflow heat exchangers. Proc. Second U.S. Nat. Cong. Appl. Mech. 1954. (322AB)

_____, and Hooper, F. C. (C) Hydraulic analogy for multipass-crossflow heat exchangers. Mech. Engr. 77:262. 1955. (322AB)

_____. See Backstrom, M., Liebaut, A., and Hooper, F. (322AB)

Kanayama, M. See Edamoto, I. (441AD)

Kane, E. W. See Douglas, J. F. H. (612) (632)

Karapetian, V. V. Determination of the pressure of an interference-affected wave on a vertical structure by means of electro-hydrodynamic analogies. (In Russian.) Trudin. -i. in-ta. osnovanye i fundamentov. 25:62-84. 1955. (211C)

Karplus, W. J. (A) Electric circuit theory approach to finite difference stability. Trans. AIEE 77(Pt. 1):210-213. 1958. (Appendix II E)

_____. (B) Simulation of field problems. Instruments and control systems 32:253-254. 1959. (311A)

_____, and Allder, J. R. Atmospheric turbulent diffusion from infinite line sources: an electric analog solution. J. Meteor. 13:583-586. 1956. (221D)

Kasowski, S. See Stevens, K. N., and Fant, C. G. M. (121BBC)

Katz, D. L. See Knudsen, J. G. (322C)

Katzoff, S., and Finn, R. S. Determination of jet boundary corrections to cowling-flap-outlet pressures by electrical analogy method. U.S. NACA Wartime Rep. L-240. (Adv. Restr. Rep. 4B23. 1944.) (221AAC)

Katzoff, S., Gardner, C. S., Diesendruck, L., and Eisenstadt, B. J. Linear theory of boundary effects in open wind tunnels with finite jet lengths. U.S. NACA Rep. 976. (221D)

Kaul, R. K. See Cadambe, V. (A), (441AA); (B), (463)

Kayan, C. F. (A) Effect of floor slab on building structure temperatures and heat flow. Heating, Piping, and Air-Cond. 19:103-111. 1947. (313A)

―――. (B) Electric analogger studies on panels with imbedded tubes. Heating, Piping, and Air-Cond. 22:123-130. 1950. (313A)

―――. (C) Electrical analogger analysis of cooled-structure complex. Heating, Piping, and Air-Cond. 27:143-148. 1955. (322AA)

―――. (D) Electrical analogger application to heat pump process. Heating, Piping, and Air-Cond. 25:123-127. 1953. (322AA)

―――. (E) Electrical analogy prediction of heat exchanger performance. (In English.) 9th Congrès. Intern. Mécanique Appl., Univ. Bruxelles, 2:484-494. 1957. (322AA)

―――. (F) Electrical geometrical analogue for complex heat flow. Trans. ASME 67:713-718. 1945. (313A)

―――. (G) Fin heat transfer by geometrical electrical analogy. Ind. and Eng. Chem. 40:1044-1049. 1948. (313A)

―――. (H) Heat flow and temperature analysis of complex structures through application of electrical resistance concept. 6th Int. Cong. for Appl. Mech., Paris. 1946. (313A)

―――. (J) Heat exchanger analysis by electrical analogy studies. Proc. Gen. Discussion on Heat Transfer. 227-231. 1951. (313A) (322AA)

―――. (K) Heat transfer equipment analysis by resistance concept. Refrig. Eng. 59:1195-1199. 1240-1241. 1951. (322AA)

―――. (L) Heat transfer temperature patterns of a multicomponent structure by comparative methods. Trans. ASME 71:9-16. 1949. (313A)

―――. (M) Performance prediction for a process heat-and-power complex by resistance concept. Trans. ASME 80:547-554. 1958. (322AA)

―――. (N) Refrigerating-plant performance characteristics by electrical-analog analysis. Trans. ASME 77:543-548. 1955. (322AA)

―――. (O) Temperature distribution in complex wall structures by geometrical electrical analogue. Refrig. Eng. 49:113-116. 1945. (313A)

―――. (P) Temperature patterns and heat transfer for a wall containing a submerged metal member. Refrig. Eng. 51:533-537. 1946. (313A)

―――. (Q) Temperatures and heat flow for a concrete slab with imbedded pipes. Refrig. Eng. 54:143-151. 1947. (313A)

―――, and Balmford, J. A. Analysis of process-fluid-flow network by electrical analogy. Trans. ASME 79:1957-1962. 1957. (122AC)

―――, and Gates, R. G. (A) Influence of insulation on moisture condensation aspects of a steel-framed cold-storage warehouse structure. Refrig. Eng. 66:39-44. 1958. (313A)

―――, and Gates, R. G. (B) Temperature distribution in fins and other projections, including those of building structures, by several procedures. Trans. ASME 80:1599-1607. 1958. (313A)

―――, and McCague, J. A. Transient refrigeration loads as related to energy-flow concepts. J. ASHRAE 1:77-82. 1959. (311A)

Keey, R. B. See Garner, F. H., and Jenson, V. G. (322C)

Kelk, G. F., Misener, W. S., and d'Arcy, D. F. An electrolytic tank for potential plotting. A. V. Roe Canada Ltd., Gas Turbine Engineering Division, rep. 17, 20. 1947, 1948. (221AAA)

Kelton, F. C. Electrolytic-model study of cycling in the grapeland field, Houston County, Texas. Proc. Amer. Petr. Inst. 24(4):199-202. 1943. (221AAA)

REFERENCES

Kemble, E. C. Fundamental principles of quantum mechanics. McGraw-Hill Co., N. Y. and London. 1937. (651A)

Kemler, E. N. (A) An experimental method of determining sucker rod loads. Trans. AIME 118:89-98. 1936. (512B)

———. (B) Models and analogues for solving design problems. Prod. Engr. 13:660-661. 1942. (511A)

———. (C) Pumping problems study; electronic analyzer seen as valuable aid. Oil and Gas J. 50:49-50. 1951. (512B)

———, and Freberg, C. R. Electrical models in the teaching and application of vibration theory. J. Eng. Educ. 34:274-281. 1943. (511A)

Kennedy, A. J. (A) Design of comprehensive computer for handling complex creep problems. Engineer 200:2-4, 34-35. 1955. (491)

———. (B) Physical and computing significance of electrical analogue of creep and recovery. Brit. J. Appl. Phys. 6:49-58. 1955. (491)

Kennedy, P. A., and Kent, G. Electrolytic tank, design and applications. Rev. Sci. Instruments 27:916-927. 1956. (Appendix II D1)

Kennedy, R. J. See Miller, W. (612)

Kennelly, A. E., and Whiting, S. E. On an approximate measurement, by electrolytic means, of the electrostatic capacity between a vertical metallic cylinder and the ground. Elec. World 48:1239-1241. 1906. (612)

Kent, G. See Kennedy, P. A. (Appendix II D1)

Kettenacker, L. Untersuchung Mechanischer Schwinggebil de Mittels Elektrischer Ersatzschaltungen. Forschung auf dem Gebiete des Ingenieurwesens-Ausgabe A. 5:67-71. 1934. (511B)

Kettleborough, C. F. (A) An analogue study of the temperature distribution in cooled gas-turbine blades. Brit. J. Appl. Phys. 6:174-176. 1955. (313A)

———. (B) A simple electrical network analogy for the solution of the stepped thrust bearing. Austral. J. Appl. Sci. 6:32-37. 1955. (222A)

Kevorkian, J. See Castles, W., Jr., and Durham, H. L., Jr. (221C)

Key, F. A., and Lamb, W. G. P. Non-linear resistance-capacitance circuit. Electronic Eng. 27:446-448. 1955. (121BC)

Kharlamov, A. A. Simulation by modelling of the aerodynamic reactions of an oscillating wing in plane flow. (In Russian.) Nauchn. Dokladi Vyssh. Shkoly. Fiz.-Matem. Nauk n. 1, 149-154. 1958. (524) (532)

Kihara, H., and Masubuchi, K. Theoretical studies on the residual welding stress. Rep. Transport, tech. Res. Inst. 6. 1953. (492)

Kingery, D. S. Fluid network analyzer for solving mine ventilation distribution problems. Missouri Univ. School of Mines and Met. Bul. (Tech. Series) 95:1-18. 1958. (122BB)

Kingsbury, Albert. On problems in the theory of fluid-film lubrication, with an experimental method of solution. Trans. ASME 53:59. 1931. (222A)

Kirchhoff, G. Über den Durchgang eines elektrischen Stromes durch eine Ebene, insbesonder durch eine Kreisförmige. Ann. D. Phys. u. Chem., 64:497-514. 1845. (Appendix II D2)

Kitchar, A. F. See Torgeson, W. L., and Hill, B. F. (313A)

Klein, E. O. P., Touloukian, Y. S., and Eaton, J. R. Limits of accuracy of electrical analog circuits used in solution of transient heat conduction problems. ASME - Paper 52-A65. 1952. (311A)

Kleynen, P. H. J. A. The motion of an electron in two-dimensional electrostatic fields. Philips Tech. Rev. 2:338. 1937. (651C)

Klotter, K. Analogies between electrical and mechanical oscillators. Ing. Arch. 18:291-301. 1950. (511A)

———. See Cremer, L. (621)

Knol, K. S., and Diemer, G. A model for studying electromagnetic waves in rectangular wave guides. Philips Tech. Rev. 11:156-163. 1949. (643)

REFERENCES

Knudsen, J. G., and Katz, D. L. Fluid dynamics and heat transfer. Mich. Univ. Eng. Research Inst. -Bul. 37:243. 1953. (322C)

Knuth, E. L., and Kumm, E. L. Application of hydraulic analog method to one-dimensional transient heat flow. Jet Propulsion 26:649-654, 659. 1956. (311B)

Koch, J. J. See Biezeno, C. B. (441AD) (455)

Koffman, J. L. Vehicle development. Automobile Engr. 46:97-102. 1956. (511B)

Kogan, L. G. See Vyalov, S. S. (311A)

Kohler, K. Multistory frames with symmetrical structure, loaded by transverse forces. (In German.) Stahlbau 22:274-276. 1953. (472AB)

Kohler, M. A. Application of electronic flow routing analog. Proc. ASCE, Sep. 135. 1952; Discussion D-135. 1953. (112A)

―――. See Linsley, R. K., and Foskett, L. W. (112A)

―――. See Linsley, R. K., and Paulhus, J. L. H. (112A)

Kohler, R. W. See Perry, H. A., Jr., and Vierling, D. E. (122AA)

Kopf, E., and Weber, E. Verfahren zur Ermittlung der Torsionsbeanspruchung Mittels Membranmodell. Zeit. d. ver. Deutscher Ing. 78:913-914. (2). 1934. (441AA)

Kopper, J. M. See McCann, G. D. (511A)

Korobov, M. A. See Mashovets, V. P. (312D)

Kostiuk, A. G., and Sokolov, V. S. About electric modelling of temperature field arising in turbine rotors. (In Russian.) Teploenergetika 10:22-27. 1959. (313A)

Kourim, G. Circuit diagrams for analog simulating heat flow in nuclear reactors. (In German.) Archiv fur Technisches Messen Und Industrielle Messtechnik, Leaflet 263. 1957. (313A)

Kozdoba, L. A. The application of resistance networks to the solution of non-stationary heat conductivity problems. (In Russian.) Inzhener. Fiz. Zh. 3:72-79. 1960. (313A)

Kraichnan, R. H. Electromagnetic analogy to sound propagation in moving media. J. Acous. Soc. Amer. 27:527-530. 1955. (121BBF)

Kroeger, C. V. Role of the computer in distribution design. Am. Gas J. 182:14-16. 1955. (122BAB)

Kron, G. (A) Electric circuit models for the vibration spectrum of polyatomic molecules. J. Chem. Phys. 14:19-31. 1946. (710)

―――. (B) Electric circuit models of partial differential equations. Elec. Eng. 67:672-684. 1948. (221D) (456) (641)

―――. (C) Electric circuit models of the Schroedinger equation. Physical Revue 67:39-43. 1945. (720)

―――. (D) Equivalent circuits for numerical solution of critical speeds of flexible shafts. Trans. ASME 68:A109-116. 1946. (492) (524)

―――. (E) Equivalent circuits for oscillating systems and the Riemann-Christoffel curvature tensor. Trans. AIEE 62:25-31. 1943. (511C)

―――. (F) Equivalent circuits of compressible and incompressible fluid-flow fields. J. Aero. Sci. 12:221-231. 1945. (211C) (221D)

―――. (G) Equivalent circuits of the elastic field. Trans. ASME 66:A149-161. 1944. (456)

―――. (H) Equivalent circuit of the field equations of Maxwell-I. Proc. IRE 32:289-299. 1944. (641)

―――. (J) Equivalent circuits to represent the electromagnetic field equations. Phys. Rev. 64:126-128. 1943. (641)

―――. (K) Generalization of calculus of finite differences to non-uniformly spaced variables. Trans. AIEE 77 (Pt. 1):535-539. 1958. (Appendix II E)

REFERENCES

———. (L) Numerical solutions of ordinary and partial differential equations by means of equivalent circuits. J. Appl. Phys. 16:172-186. 1945. (221D) (456) (641)

———. (M) Set of principles to interconnect solutions of physical systems. J. Appl. Phys. 24:965-980. 1953. (Appendix II E)

———. (N) Tensorial analysis and equivalent circuits of elastic structures. J. Franklin Inst. 238:399-442. 1944. (471B) (472AB) (492)

———, and Carter, G. K. A-C network analyzer study of the Schrodinger equation. Phys. Rev. 67:44-49. 1945. (720)

———. See Carter, G. K. (A), (472AB); (B), (710); (C), (211C), (221D)

———. See Prebus, A. F., and Zlotowsky, I. (652)

———. See Whinnery, J. R., Concordia, C., and Ridgway, W. (641)

Kron, L. C. See Simond, A. W. (612)

Krueger, R. F. See Vogel, L. C. (311A)

Kuehni, H. P., and Lorraine, R. G. A new A-C network analyzer. Trans. AIEE 57:67. 1938. (Appendix II E)

Kumm, E. L. See Knuth, E. L. (311B)

Kunin, I. A. A solution of a certain category of problems by modelling on an electrolytic trough. (In Russian.) Izv. Sibirsk. Otd. Akad. Nauk SSSR 7:53-61. 1958. (Appendix II D1)

Kuros, G. R. An electric experimental method for determination of bending moments of statically-indeterminate frames. (In German.) Bauingenieur 29:10-15. 1954. (472AB)

Kuttruff, H. Optical model experiments of a stationary sound field; diffusivity in large rooms. (In German.) Acustica 8:330-336. 1958. (121BBE)

Kvitka, A. L., Agarev, V. A., and Umanski, E. S. The solution of the axially-symmetrical problem of the theory of elasticity by electrical analogy for the case of the presence of centrifugal forces and temperature fields. (In Russian.) Izv. Kievsk. Politekhn. In-ta 19:455-461. 1956. (492)

Ladner, A. W. Marconi Rev. n. 9, 10, 11. 1929. (621)

LaHatte, J. A., Jr. See Gardner, C. S. (221AD)

Laitone, E. V. (A) Developments in gas dynamics by the hydraulic analogy. "Memoirs sur la mécanique des Fluides." Publ. Sci. Tech. Min. Air, Paris 203-217. 1954. (211AB)

———. (B) An experimental investigation of transonic and accelerated supersonic flow by the hydraulic analogy. Univ. Cal. Dept. Eng. Tech. Rep. HE-116-315. 1950. (211AB)

———. (C) A rational discussion of the hydraulic analogy. J. Aero. Sci. 20:59. 1953. (211AB)

———. (D) A study of transonic gas dynamics by the hydraulic analogy. J. Aero. Sci. 19:265-272. 1952. (211AB)

———, and Nielsen, H. Transonic flow past wedge profiles by hydraulic analogy. J. Aero. Sci. 21:498-499. 1954. (211AB)

———, and Stout, J. E. Mach reflections in two-dimensional diffusers from hydraulic analogy experiments. Jet Propulsion 28:257-259. 1958. (211AB)

Lamb, H. Lehrbuch der Hydrodynamik. Deutsche Ausgabe, Zweite Auflage. G. B. Teubner, Leipzig und Berlin. 1931. (211AB)

Lamb, W. G. P. See Key, F. A. (121BC)

Landahl, M. T., and Stark, V. J. E. Electrical analogy for solving oscillating surface problem for incompressible nonviscid flow. Stockholm, Kungl Tekniska Hogskolan — Inst. for Flygteknik (Roy. Inst. Technology - Div. Aeronautics) — Tech. Note 34. 1953. (221AAA)

Landau, H. G. (A) High order accuracy in the solution of partial differential equations by resistor networks. J. Appl. Mech. 25:17-20. 1958. (Appendix II E)

REFERENCES

Landau, H. G. (B) A simple procedure for improved accuracy in the resistor-network solution of Laplace's and Poisson's equation. J. Appl. Mech. 24:93-97. Discussion 639-640. App. II E. 1957. (Appendix II E)

Landau, J. See Prochazka, J., and Standart, G. (322AB)

Landis, F., and Zupnik, T. Effectiveness of stub-fins as determined by teledeltos paper analog. Mech. Engr. 79:960. 1957. (313A)

Lane, E. W., Campbell, F. B., and Price, W. H. The flow net and the electrical analogy. Civil Eng. 4:510. 1934. (221AAA)

Lang, H. C. See Baruch, J. J. (524)

Lang, R., and Petrick, E. N. Application of electrical analog theory in preliminary design of air-cooled turbines. Mech. Engr. 81:92. 1959. (313A)

Langmuir, D. B. (A) An automatic plotter for electron trajectories. RCA Rev. 11:143-154. 1950. (651D)

―――. (B) Automatic plotting of electron trajectories. Nature (London) 139: 1066-1067. 1937. (651D)

Langmuir, E., Adams, E. Q., and Mickle, G. S. Flow of heat through furnace walls: the shape factor. Trans. Am. Electrochem. Soc. 24:53. 1913. (313A)

Larmor, J. The influence of flaws and air cavities on the strength of materials. Phil. Mag. (5) 33:70-78. 1892. (441AF)

Larras, J. Means of solution of the problem of rippling by an electro-analogic method. (In French.) Ann. Ponts. Chauss. 13:207-225. 1936. (221AAA)

Larsen, B. M. See Heindlbofer, K. (321A)

Laurent, G. J. Design investigations for an electric network analyzer for hydraulic networks. M. S. thesis, M.I.T. 1940. (122AB)

Lawson, D. I., and McGuire, J. H. The solution of transient heat-flow problems by analogous electrical networks. Proc. Instn. of Mech. Engrs. 167:(A)275-287. 1953. (311A)

Lazaryan, V. A. On the electrical simulation of transient conditions of the motion of a bar. (In Russian.) Trudi Dnepropetrovsk. in-ta inzh. zh.-d. transp. 25:84-123. 1956. (512B) (532C)

Lazzarino, O. The analogy between hydrodynamic and electromagnetic fields. Atti Accad. Gioenica Sci. Nat. Catania, Mem. 12:1-16. 1923. (211C) (221D)

Leclerc, A. Deflection of a liquid jet by a perpendicular boundary. M. S. thesis, State Univ. of Iowa. 1948. (221AAB)

Le Corbeiller, P., and Yeung, Y. Duality in mechanics. J. Acous. Soc. Amer. 24:643-648. 1952. (511A) (511B)

Lee, B. D. Potentiometric-model studies of fluid flow in petroleum reservoirs. Trans. AIME (Petr. Dev. and Tech.) 174:41-66. 1948. (221AAA)

Lefkovits, H. C. See Matthews, C. S. (221AF)

Legendre, R. Hydraulic analogy for the study of hypersonic flow. (In French.) C. R. Acad. Sci. Paris 250(23):3771-3772. 1960. (211AB)

Legras, J., and Malavard, L. Electric tank study of problems of the lifting surface. (In French.) Proc. 6th Int. Congr. Appl. Mech., Paris. 1946. (221AAA)

Leibowitz, R. C. Natural modes and frequencies of vertical vibration of a beam with an attached sprung mass. David W. Taylor Mod. Basin Rep. 1125. 1958. (524)

Leicester, R. Resistance analogue of beams and arches. J. Inst. Engrs., Australia 32:99-103. 1960. (422D)

Leopold, C. S. Hydraulic analogue for the solution of problems of thermal storage, radiation, convection and conduction. Heating, Piping, and Air-Cond. 20:105-111. 1948. (311B)

Levi, F. Magnetic leakage evaluated with an electrolytic tank. Electronics 21:178. 1948. (632)

Levin, A. M. Hydraulic jump and separation of boundary layers. (In Russian.) Dokladi Akad. Nauk SSSR (N.S.) 96:1121-1124. 1954. (211AB)

Levy, L. I. See Bisplinghoff, B. L., and Pian, T. H. H. (490)

Lewis, W. T. O. Potential flow apparatus. Aeronautics. 1946. (221AAB)

─── , and Newman, E. J. (A) The development of a diffuser outline having specific velocity distribution characteristics by means of the potential flow apparatus. Bristol Aeroplane Co. Rep. KR76/45. 1945. (221AAB)

─── , and Newman, E. J. (B) Investigations into the flow over cowlings for radial aero engines using potential flow apparatus. Bristol Aeroplane Co. Rep. KR66/44. 1944. (221AAB)

Li, W. H., Bock, P., and Benton, G. S. New formula for flow into partially penetrating wells in aquifers. Trans. Amer. Geo. Union 35:805-812. 1954. (221AAA)

Libby, P. A. See Hoff, N. J. (483)

Liebaut, A. See Backstrom, M., Juhasz, S., and Hooper, F. (322AB)

Liebmann, G. (A) The effect of pole piece saturation in magnetic electron lenses. Proc. Phys. Soc. (London) B 66:448. 1953. (632)

─── . (B) A method for the mapping of vector potential distributions in axially symmetrical systems. Phil. Mag. 41:1143. 1950. (633)

─── . (C) New electrical analogue method for solution of transient heat conduction problems. Trans. ASME 78:655-663. 1956. (311A)

─── . (D) Note on the resistance-network analogue solution of field problems of spherical symmetry. Brit. J. Appl. Phys. 5:412. 1954. (Appendix II E)

─── . (E) Precise solution of partial differential equations by resistance networks. Nature, London 164:149. 1949. (Appendix II E)

─── . (F) A resistance-network analogue method for solving plane stress problems. Associated Electrical Industries, Research Report A.271. 1953. (457)

─── . (G) A resistance-network analogue method for solving plane stress problems. Nature (London) 172:78. 1953. (457)

─── . (H) Resistance-network analogues with unequal meshes or subdivided meshes. Brit. J. Appl. Phys. 5:362-366. 1954. (Appendix II E)

─── . (J) Solution of partial differential equations with a resistance network analogue. Brit. J. Appl. Phys. 1:92-103. 1950. (Appendix II E)

─── . (K) Solution of plane stress problems by electrical analogue method. Brit. J. Appl. Phys. 6:145-157. 1955. (457)

─── . (L) Solution of some nuclear reactor problems by resistance-network analogue method. J. Nuclear Energy 2:213-225. 1956. (730)

─── . (M) Solution of transient heat-transfer problems by the resistance-network analog method. Trans. ASME 78:1267-1272. 1956. (311A)

─── . (N) The solution of waveguide and cavity-resonator problems with the resistance-network analogue. Proc. Inst. Elec. Engrs. 99(Pt. IV):260-272. 1952. (641)

─── , and Bailey, R. An improved experimental iteration method for use with resistance-network analogs. Brit. J. Appl. Phys. 5:32-35. 1954. (Appendix II E)

─── , and Grad, E. M. Imaging properties of a series of magnetic electron lenses. Proc. Phys. Soc. (London) B 64:956-971. 1951. (632)

Light, L. Extension of the conducting sheet analogy to externally pressurized gas bearings. Trans. ASME (J. Basic Engrg.) 83D:209-212. 1961. (222A)

Ligtenberg, F. K. The Moiré method—a new experimental method for the determination of moments in small slab models. Proc. Soc. Exp. Stress Anal. 12(2):83-98. 1955. (451)

Lilley, G. M. See Babister, A. W., Marshall, W. S. D., Sills, E. C., and Deards, S. R. (221AAB)

REFERENCES

Linaweaver, F. P., Jr. See Appleyard, V. A. (122AB)

Lindner, N. J. See Orlin, W. J., and Bitterly, J. G. (211AB)

Ling, S. C. See Hubbard, P. G. (221AAC)

Linsley, R. K., Foskett, L. W., and Kohler, M. A. Electronic device speeds flood forecasting. Eng. News Rec. 141:64-66. 1948. (112A)

____, Kohler, M. A., and Paulhus, J. L. H. Applied hydrology. McGraw-Hill, New York. 537-541. 1949. (112A)

Linvill, J. G., and Hess, J. J., Jr. Studying thermal behavior of houses; thyratrons and pentode provide electronic control for electrical model. Electronics 17:117-119. 1944. (311A)

Litman, B. Damping effect of d-c marine propulsion motors on vibrations produced in drive shafts by large propellers. Trans. AIEE 64:31. 1945. (531A)

Litvinishin, G. See Olshak, V. (491)

Litvinov, M. M. Determination of steady temperature fields in the blades and disks of turbines by an electric analogue method. (In Russian.) Izv. Akad. Nauk SSSR, Otd. Tekh. Nauk 5:15-22. 1956. (313A)

Locanthe, B. N. See McCann, G. D., and Wilts, C. N. (Appendix II E)

Locklin, D. W. See Humphreys, C. M., Nottage, H. B., Franks, C. V., Huebscher, R. G., and Schutrum, L. F. (313A)

Lodge, O. See Foster, G. C. (Appendix II D2)

Loeb, A. M. Electric analog study of hydrostatic bearings. J. Franklin Inst. 263:450-452. 1957. (222A)

____, and Rippel, H. C. Determination of optimum proportions for hydrostatic bearings through the use of the electric analog field plotter. J. Franklin Inst. 265:342-344. 1958. (222A)

Loh, W. H. T. (A) Hydraulic analogue for one-dimensional unsteady gas dynamics. J. Franklin Inst. 269:43-55. 1960. (211AB)

____. (B) Hydraulic analogy for two-dimensional and one-dimensional flows. J. Aero Space Sci. 26:389-391. 1959. (211AB)

____. (C) A study of the dynamics of the induction and exhaust systems of a four-stroke engine by a hydraulic analogy. Sc. D. thesis, M. I. T. 1946. (121BAB) (211AB)

London, A. L. See Cima, R. M. (321A)

Lorraine, R. G. See Kuehni, H. P. (Appendix II E)

Love, A. A treatise on mathematical theory of elasticity. University Press, Cambridge. (3rd edition) 343. 1920. (451)

Lukyanov, V. S. Hydraulic apparatus for engineering computations. (In Russian, English Translation USA Corps of Engineers, New England Division.) Izv. Akad. Nauk SSSR Otd. Tekh. Nauk 2:53. (311B)

Lundstrom, C. C. See Hansen, W. W. (Appendix II D1)

Luthra, H. R., and Vaidhianathan, V. I. Uplift pressures under weirs with three sheet piles. Proc. Ind. Acad. Sci. 4:491-502. Sect. A, 4. 1936. (221AAA)

Luthra, S. D. L. (A) Effect of impervious strata on pressures under weir and testing stability of existing weir by electrical method. J. Central Board of Irrigation and Power (India) 12:102-117. 1955. (221AAA)

____. (B) Electrical method for testing design of inlet curves for canal sluices. J. of Central Board of Irrigation and Power (India) 12:611-615. 1955. (221AAA)

____, and Ram, G. (A) Electrical analogy applied to determine effect of impervious layer on potential distribution below weir founded in pervious media. (India.) J. Central Board of Irrigation and Power 10:245-252. 1953. (221AAA)

____, and Ram, G. (B) Electrical analogy applied to study seepage into drain

REFERENCES

tubes in stratified soil. J. Central Board of Irrigation and Power (India) 11:398-405. 1954. (221AAA)

Luu, T. S. Deviation of a main jet by an auxiliary jet slightly inclined in direction by the help of a rheoelectrical analogy. (In French.) C. R. Acad. Sci. Paris 246:55-58. 1958. (221AAA)

Lyle, E. S. E. Analysis of gable frames by column analogy method. Concrete and Constr. Eng. 51:389-394. 1956. (422A)

Lynn, J. W. See Ackroyd, R. T., Houstoun, J., and Mann, E. (311A)

McAdams, W. H. Heat transmission. 2nd ed. McGraw-Hill, New York. 1942. (322C)

McAllister, J. F., Jr. Equivalent circuits of the electromagnetic field. G. E. Review. 47:9-14. 1944. (641)

McCague, J. A. See Kayan, C. F. (311A)

McCann, G. D. Electric analogies for mechanical structures. ISA J. 3:161-165. 1956. (422D) (472AB)

──, and Bennett, R. R. Vibration of multifrequency systems during acceleration through critical speeds. Trans. ASME 71:375. 1949. (531A)

──, and Criner, H. E. (A) Mechanical problems solved electrically. Westinghouse Engr. 6:48-56. 1946. (511A) (511B)

──, and Criner, H. E. (B) Solving complex problems by electrical analogy. Mach. Des. 17:137-142; 18:129-132, 157. 1945, 1946. (511A) (511B)

──, and Kopper, J. M. Generalized vibration analysis by means of the mechanical transients analyzer. Trans. ASME. 69:A127-134. 1947. (511A)

──, and MacNeal, R. H. Beam vibration analysis with the electric analog computer. J. Appl. Mech. 17:13. 1950. (422D) (521) (522) (532A) (532B)

──, Warren, C. E., and Criner, H. E. Determination of transient shaft torques in turbine generators by means of the electro-mechanical analogy. Trans. AIEE 64:51. 1945. (531A)

──, and Wilts, C. H. Application of electric-analog computers to heat-transfer and fluid-flow problems. Paper presented at the First Heat Transfer and Fluid Mechanics Institute, Los Angeles, Calif. June 21-23, 1948. (Also, Trans. ASME 71:247-258. 1949.) (211AAD) (313A)

──, Wilts, C. H., and Locanthi, B. N. Electronic technique applied to analogue methods of computation. Proc. IRE 37:954-961. 1949. (Appendix II E)

──. See Criner, H. E. (422A)

──. See Criner, H. E., and Warren, C. E. (511A) (511B)

──. See Housner, G. W. (511A)

──. See MacNeal, R. H., and Wilts, C. H. (521) (532C)

McCarty, G. M. See Hurst, W. (221AAA)

McCormick, B. W., Jr. The application of an electromagnetic analogy to the determination of induced camber correction for wide-bladed propellers. Heat Transfer Fluid Mech. Inst., Stanford Univ. Press 111-124. 1952. (221CA)

McDonald, D. (A) Electrolytic analogue in design of high-voltage power transformers. Proc. Inst. Elec. Engrs. 100, Pt. 2 (Power Eng.):145-166. 1953. Discussion, 176-183. 102 Pt. A(Power Eng.):89-93. 1955. (612)

──. (B) Stenographic record of discussions: utilisation of the electric tank. C.I.G.R.E., Paris 1:265. 1950. (612)

McElroy, G. E. Network analyzer for solving mine-ventilation-distribution problems. U. S. Bur. Mines — Information Cir. 7704. 1954. (122BB)

McGhee, J. See Mossop, I. A. (730)

McGivern, J. G., and Supper, H. L. A membrane analogy supplementing photoelasticity. Trans. ASME 56:601. 1934. (411)

──. See Den Hartog. (441AF)

McGuire, J. H. See Lawson, D. I. (311A)

McKeon, J. T., and Eschenbrenner, G. P. Thermal analysis and design of intermediate heads in pressure vessels. Mech. Engr. 80:135. 1958. (311A)

McIlroy, M. S. (A) Analyzing pipeline networks; non-linear electrical analogy process. Heating, Piping, and Air Cond. 24:100-103. 1952. (122AA) (122AB)

———. (B) Direct reading electric analogue computer helps solve distribution problems. Gas Age 110:31-35. 1952. (122AB)

———. (C) Direct-reading electric analyzer for pipeline networks. J. Amer. Water Works Assn. 42:347-366. 1950. (122AA) (122AB)

———. (D) Gas pipe networks analyzed by direct-reading electric analogue computer. Proc. Amer. Gas Assn. 34:433-447. 1952. (122BAB)

———. (E) How to make water-line networks talk; electric analyzer. Amer. City 66:112-113. 1951. (122AB)

———. (F) Nonlinear electrical analogy for pipe networks. Proc. ASCE, Sep. 139. 1952. (122AB)

———. (G) Studies of fluid network analysis and the basic design of an electric analyzer for fluid distribution systems. Sc. D. thesis, M. I. T. 1946. (122AB)

———. (H) There's an easy way out of your pipe-network woes; pipeline network analyzer at Binghampton, N. Y. Eng. News Record 147:35-36. 1951. (122AB)

———. (J) Water-distribution systems studied by complete electrical analogy. J. N. E. Water Works Assn. 65:299-318. 1951. (122AA) (122AB)

———, and Chow, C. K. Steam distribution systems analyzed by nonlinear electrical analogy method. Proc. Nat. Dist. Heating Assn. 43. 1952. (122BAB)

The McIlroy pipeline network analyzer. The Standard Electric Time Company, Springfield, Massachusetts, Bul. 183. (122AB)

McMahan, K. D. Fluidynamic control of fluid flow. Proc. 5th Int. Cong. for Appl. Mech. 617-624. 1938. (221D)

McNall, P. E., Jr., and Janssen, J. E. (A) An electrolytic analog applied to heat conduction within a transistor. Trans. ASME 78:1181-1186. 1956. (312D)

———, and Janssen, J. E. (B) An electrolytic analog applied to the solution of a thermal conduction problem. ASME Preprint 54-SA-45. 1954. (312D)

McPherson, M. B., and Radziul, J. V. Water distribution design and McIlroy network analyzer. Proc. ASCE 84(HY2 1588):1-15. 1958. (122AB)

McQueen, J. G. See Hartill, E. R., and Robson, P. N. (Appendix II D1)

Maas, W. (A) Electrical analogue for mine ventilation. Colliery Eng. 28:24. 1951. (122BB)

———. (B) An electrical analogue for mine ventilation and its application to ventilation planning. Geologie en Mijnbouw, the Hague, Holland 117. 1950. (122BB)

Mach, E. (A) Mitteilungen über einfache Vorlesungsversuche. Cärl's Rep. für Exp. Phys. 6:8-12. 1870. (612)

———. (B) Photography of projectile phenomena in air. Sitzungberichte der Wiener Akademie 95:164. 1887. (211AB)

Mackey, C. O., and Gay, N. R. (A) Cooling load from sunlit glass. Trans. ASHVE 58:321. 1952. (321B) (330)

———, and Gay, N. R. (B) Cooling loads from sunlit glass and wall. Heating, Piping, and Air Cond. 26:123-128. 1954. (321B) (330)

Mackie, A. G. One-dimensional unsteady motion of a gas initially at rest and the dam-break problem. Proc. Camb. Phil. Soc. 50:131-138. 1954. (211AB)

MacNeal, R. H. (A) An asymmetrical finite difference network. Quart. Appl. Math. 11:295-310. 1953. (Appendix II E)

REFERENCES

——. (B) Electrical analogies for stiffened shells with flexible rings. NACA Tech. Note 3280. 1954. (484)

——. (C) The equivalent circuits of shells used in airframe construction. Proc. Joint IRE, AIEE, ACM Western Computer Conf. 1953. (484)

——. (D) Solution of elastic plate problems by electrical analogies. Trans. ASME 73:59-67. 1951. Discussion, 427-428. (461)

——. (E) The solution of partial differential equations by means of electrical networks. Ph. D. thesis, CIT, Pasadena. 1949. (Appendix II E)

——. (F) Theoretical and experimental effects of sweep upon the stresses and deflections distribution in aircraft wings of high solidity, Part IV. Application of the analog computer to the elastic plate problem. U. S. Defense Research Laboratory Publication AETR 5761-4. 1949. (461)

——. (G) Vibrations of composite systems. ARDC Report 4, Project R-354-30-1. 1954. (540)

——, and Benscoter, S. U. (A) Analysis of multicell delta wings on Cal. Tech. analog computer. NACA Tech. Note 3114. (524)

——, and Benscoter, S. U. (B) Analysis of sweptback wings on Cal. Tech. analog computer. NACA 3115. (524)

——, McCann, G. D., and Wilts, C. H. The solution of aeroelastic problems by means of electrical analogies. J. Aero. Sci. 18:777-789. 1951. (521) (532C)

——. See Benscoter, S. U. (A), (524); (B), (524)

——. See McCann, G. D. (422D) (521) (522) (532A) (532B)

——. See Russell, W. T. (433)

——. See Shields, J. H. (421B) (422D) (461) (472AB)

Magnusson, P. C. (A) Transients in coupled inductance-capacitance circuits analyzed in terms of a rolling-ball analogue. Trans. AIEE 69:1525. 1950. (622A)

——. (B) Transients in coupled inductance-capacitance circuits analyzed in terms of a rolling-ball analogue. Elec. Eng. 69:1079. 1950. (622A)

Maher, F. J. See Pletta, D. H. (441AA)

Makar, R., Boothroyd, H. R., and Cherry, E. C. An electrolytic tank for exploring potential field distributions. Nature (London) 161:845-846. 1948. (Appendix II D1)

——. See Boothroyd, A. R., and Cherry, E. C. (Appendix II D1)

Makinson. A mechanical analogy for transverse electric waves in a guide of rectangular section. J. of Sci. Instr. 24:189. 1947. (643)

Malavard, L. (A) Apercu sur la méthode d'analogie rhéoélectrique. Publ. Scint. et Techn. du Ministere de l'Air 261. 1950. (221AAA)

——. (B) Application des analogies electriques a la solution de quelques problèmes de l'hydrodynamique. Publications Scientifiques et Techniques du Ministere de l'Air 57. 1934. (221AAA)

——. (C) Application of electric analogies to some problems in aerodynamics. (In French.) Rev. Tech. Asso. Ing. Aéro. 1938. (221AAA)

——. (D) Applications aerodynamique du calcul experimental analogique. Commun. au Congres National de l'Aviation Francaise. 1945. (221AAA)

——. (E) Calculateur d'ailes et reseau de resistances lineaire pouvant remplacer, dans certaines questions, le bassin electrique. Comptes-Rendus de l'Academie des Sciences. July, 1945. (221AAA)

——. (F) Contribution à l'étude théorique du soufflage au bord de fuite d'un profil d'aile (résolution par analogie rhéoélectric). ONERA NT 4/1727.A. 1954. (221AAA)

——. (G) Electroanalogic method for the calculation of wall corrections. (In French.) C.R. Acad. Sci. Paris 204:1052-1054. 1937. (221AAA)

——. (H) Electrolytic plotting tank. High Speed Aerodynamics and Jet Propulsion IX:322-340. Princeton University Press. 1954. (221AAA)

Malavard, L. (J) Etude de quelques problèmes techniques relevant de la theorie des ailes. Application a leur solution de la methode rheoelectrique. Publ. Scient. et Techn. du Ministere de l'Air, 1953. 1939. (221AAA)

_____. (K) Experimental realization of the electric analogy of M. J. Larras for the analytical problem of rippling. (In French.) Ann. Ponts Chauss. 14:226-231. 1936. (221AAA)

_____. (L) L'analogie electrique comme methode auxiliarre de la photo-elasticite. C. R. Acad. Sci., Paris 206:38-39. 1938. (455)

_____. (M) On a new technique in experimental solutions by rheoelectric analogies. La Rech. Aero., 20:61-68. 1951. (221AAA)

_____. (N) On the fundamental problem concerning airfoils of finite aspect ratio. (In French.) C. R. Acad. Sci. Paris 195:733-736. 1932. (221AAA)

_____. (O) Pour le calcul des effets du fuselage et des fuseaux-moteurs sur la répartition en envergure des efforts aerodynamiques. C. R. Acad. Sci., Paris 215:266-268. 1942. (221AAA)

_____. (P) Recent developments in the method of the rheoelectric analogy applied to aerodynamics. J. Aero. Sci., 24:321-331. 1957. (221AAA)

_____. (Q) Some recent applications of electroanalogic methods. (In French.) Proc. Colloq. Methodes de Calcul, 55-71. 1949. (211AAA)

_____. (R) Sur la solution rheoelectrique de questions de representation conforme et application a la theorie des profils d'Ailes. C. R. Acad. Sci., Paris 218:106-108. 1944. (221AAA)

_____. (S) Sur une nouvelle technique dans le calcul expérimental par analogies rhéoélectriques. La Rech. Aero. 20. 1951. (221AAA)

_____. (T) The technique of electrical analogies. Translation of the Aeronautics Dept., Rensselaer Polytechnic Inst. 1953. (221AAA)

_____. (U) The use of rheo-electrical analogies in certain aerodynamical problems. J. Roy. Aero. Soc. 51:739-753, 753-756. 1947. (221AAA)

_____, and Boscher, J. Electric analogs for studying the flexure of beams. (In French.) C. R. Acad. Sci. Paris 236:1130-1133. 1953. (422D)

_____, and Duquenne, R. Study of lifting surfaces by rheoelectric analogies. La Rech. Aero. 23:3-12. 1951. (221AAA)

_____, Duquenne, R., Enselme, M., and Grandjean, C. Properties of delta wings with swept or straight trailing edges calculated by the electric tank method. (In French.) ONERA NT 25, 1955. (221AAA)

_____, Germain, P., and Siestrunck, R. Calculation of the distribution of velocity over the cross sections of grills of blades (cascades). (In French.) La Rech. Aero. 4. 1948. (221AAA)

_____, and Hacques, G. Problems in ring airfoils, solved with rheoelectric analogies. (In French.) Advances in Aeronautical Sciences, v. 2, (Proc. of the First International Congress in the Aeronautical Sciences, Madrid, September 8-13, 1958), Pergamon Press, 771-776. 1959. (221AAA)

_____, and Miroux, J. Electrical analogies for heat transfer problems. Engrs. Digest 13:416-420. 1952. (313A)

_____, and Siestrunck, R. (A) Diverses methodes analogiques pour le calcul des grilles d'aubes. Proc. 6th Int. Congr. Appl. Mech., London. September, 1946. (221AAA)

_____, and Siestrunck, R. (B) Sur une methode d'etude des grilles rectilignes indefinies de profils quelconques. Communication au Congres National de l'Aviation Francaise. May, 1945. (221AAA)

_____. See Boscher, J. (457)

_____. See Legras, J. (221AAA)

_____. See Peres, J. (A), (221AAA); (B), (221AAA); (C), (221AAA); (D), (221AAA); (E), (441AD), (633); (F), (221AAA); (G), (221AAA)

REFERENCES

——. See Peres, J., and Romani, L. (221AAA)
Mandelstam, L. Pendlmodell zur Demonstration electrischer gekoppelter Kreise. Jahrbuch der Drahtlosen Telegraphie und Telephonie 4:515. 1911. (622B)
Manley, R. G. Electro-mechanical analogy in oscillation theory. J. Roy. Aero. Soc. 47:22-26. 1943. (511A)
Mann, E. See Ackroyd, R. T., Houstoun, J., and Lynn, J. W. (311A)
Many, Abraham, and Meiboom, Saul. An electrical network for determining the eigenvalues and eigenvectors of a real symmetrical matrix. Rev. Sci. Instr. 18:831-836. 1947. (Appendix II E)
March, H. See Trayer, G. W. (441AA)
Marchal, G. H. Tables de calcul electriques. Bul. Société Belge des Electriciens 64:84-92. 1948. (121BBB)
Marchet, P. Détermination des lignes de jet dans les mouvements plans et de révolution. ONERA NT D-359-A. 1949. (221AAB)
Marin, J. Evaluating torsional stresses by membrane analogy. Mach. Des. 15:118. 1943. (441AA)
Markland, E., and Hay, N. Potential flow tank. Engineering, 173:292-294. 1952. (221AAA)
——. See Hay, N. (221AAA)
Marshall, D. L., and Olivery, L. R. Some uses and limitations of model studies in cycling. Trans. Amer. Inst. Min. Metall. Engrs. 174:67-87. 1948. (221AAA)
Marshall, W. S. D. See Babister, A. W., Lilley, G. M., Sills, E. C., and Deards, S. R. (221AAB)
Martin, W. See Bromberg, R. (313A)
Martinelli, A. Stresses and strains in plane systems. (In Italian.) Annali dei Lavori Publicci 67:218-243. 1929. (451)
Martinelli, R. C. (A) Further remarks on the analogy between heat and momentum transfer. 6th Int. Cong. for Appl. Mech., Paris. 1946. (332C)
——. (B) Heat transfer to molten metals. Trans. ASME 69:947-959. 1947. (322C)
——. See Boelter, L. M. K., and Jonassen, F. (322C)
Marvaud, J. (A) Traceur automatique de trajectoires électroniques. C. R. Acad. Sci., Paris 226:476-478. 1948. (651D)
——. (B) Traceur automatique de trajectoires électroniques et son adaptation à la determination des lignes de courant dans un bassin électrique. C. R. Acad. Sci., Paris 234:45-47. 1952. (651D)
Mashovets, V. P., and Korobov, M. A. Conditions of electric simulation of a thermal field with internal heat sources. Soviet Phys. - Tech. Phys. 3:1952-1957. 1959. (Translation of Zh. Tekh. Fiz. 28:2124-2129, 1958, by Amer. Inst. Phys., Inc., New York.) (312D)
Mason, J. L. See Schlinger, W. G., Berry, V. J., and Sage, B. H. (322B)
Masten, J. K. Electrical energy in balance. Power Plant Engr. 44:54-55. 1940. (621)
Masubuchi, K. See Kihara, H. (492)
Matsunaga, S. (A) Mach number at the diffuser throat of an ejector according to the hydraulic analogy. J. Aero. Sci. 24:918-919. 1957. (211AB)
——. (B) On thermal resistance of water droplet on metallic surface. Jet Propulsion 28:126-127. 1958. (313A)
——. (C) A proposal on the electric tank method for study of a symmetrical airfoil with a rear suction slot and a retractable flap. J. Roy. Aero. Soc. 58:434-435. 1954. (221AAA)
——. (D) A study of the adiabatic exponent of the "hydrodynamical gas" by statistical thermodynamics. J. Aero./Space Sci. 26:527-528. 1959. (211AB)

REFERENCES

Mattarolo, L. Electric analogy and method of quadra-poles in heat-transfer problems. (In Italian.) Termotecnica 6:355-363. 1952. (313A)

Matthews, C. S., and Lefkovits, H. C. Studies on pressure distribution in bounded reservoirs at steady state. J. of Petr. Technol. (Petr. Trans. AIME 204) 7:182-189. 1955. (221AF)

Matthews, C. W. The design, operation, and uses of the water channel as an instrument for the investigation of compressible flow phenomena. U.S. NACA Tech. Note 2008. 1950 (211AB)

Matuyama, T. See Sunatani, C., and Hatamura, M. (441AA) (441AB)

Maxfield, J. P., and Harrison, H. C. Methods of high quality recording. Trans. AIEE 45:334. 1926. (511A)

Maytin, Iury L. Pipeline-network problems on the McIlroy analyzer. Washington State Inst. of Tech., Bul. 227, Pullman. 1955. (122AB)

Mechanical circuits; electrical analogy is not foolproof. Wireless World 65:570-574. 1959. (511B)

Mediratta, O. P. See Black, J. (211AB)

Meiboom, Saul. See Many, Abraham. (Appendix II E)

Merbt, and Hansson, Proceedings, aeroelastics colloquium. (In German.) Gottingen, April 16-17, 1957. Mitt. Max-Planck Inst. Stromungsforschung 8. 1958. (221D)

Messerle, H. K. Electronic high speed simulation of hydraulic problems. J. Instn. Engrs. Australia 25:35-41. 1953. (112A)

Metzner, A. B., and Friend, W. L. Theoretical analogies between heat, mass, and momentum transfer and modifications for fluids of high Prandtl or Schmidt numbers. Canadian J. Chem. Engrg. 36:235-240. 1958. (322C)

Meyer, C. A. See Benedict, R. P. (221AAA)

Meyer, H. I., and Searcy, D. F. Analog study of water coning. J. Petr. Technol. 8:61-64. 1956. (221AF)

Meyer, H., and Tank, F. Uber ein verbessertes elektrisches verfahren zur auswertung der gleichung $\nabla^2 \emptyset = 0$ und seine anwendung bei photoelastischen untersuchungen. Helvetica Physica Acta 8:315-317. 1935. (441AD) (455)

Michalos, J. P. See Baron, F. (472AC)

Michell, A. G. M. The lubrication of plane surfaces. Zeitschrift fur Mathematik u. Physics. 1905. (222A)

Mickelson, J. K. Automatic equipment and techniques for field mapping. General Electric Review, 52:19-23. 1949. (Appendix II D1)

Mickle, G. S. See Langmuir, I., and Adams, E. Q. (313A)

Miles, A. J., and Stephenson, E. A. Pressure distribution in oil and gas reservoirs by membrane analogy. Trans. AIME (Petroleum Development and Technology) 127:135. 1938. (211AD)

———. See Wilson, L. H. (313B)

Miles, J. W. (A) An analogy among torsional rigidity, rotating fluid inertia, and self-inductance for an infinite cylinder. J. Aero. Sci. 13:377-380. 1946. (441C)

———. (B) Applications and limitations of mechanical-electrical analogies, new and old. Acous. Soc. of Amer. 14:183-192. 1943. (511A)

Miller, S. See Swanson, R. S., and Crandall, S. M. (221CA)

Miller, W., and Kennedy, R. J. Measurement of field distortion in free-air ionization chambers by analog method. J. Res. Nat. Bur. Stand. 55:291-297. 1955. (612)

Millstone, S. D. Electric analogies for hydraulic analysis. Mach. Des. 24:185-189. 1952; 25:166-170; 131-135. 1953. (121AAB) (121AB)

Mindlin, R. D. The analogy between multiply connected slabs and slices. Quart. Appl. Math. 4(3):279-290, 1946; 5(2):238, 1947. (451) (452)

REFERENCES

Miroux, J. (A) A new analog computer for transient conditions, for the study of certain phenomena with variable physical properties. (In French.) ONERA Publ. 81. 1955. (311A)

―――. (B) On the measurement of gradients in electric-current analogies. Rech. Aero. 29:9-20. 1952. (211AAA)

―――. (C) The potentialities of an analog apparatus for measuring gradients and for tracing fields of equal speeds directly. (In French.) ONERA 53:53-57. 1956. (221AAA)

―――. See Malavard, L. (313A)

Misener, W. S. See Kelk, G. F., and d'Arcy, D. F. (221AAA)

Mizushina, T. Analogy between fluid friction and heat transfer in annuli. Proc. General Discussion on Heat Transfer, 191-192. 1951. (322C)

Model studies of penstocks and outlet works. 2:107-118. U.S. Bureau of Reclamation, Denver. 1938. (221AAA)

Models and analogies for demonstrating electrical principles. The Engineer 142:167, 194, 228, 242, 273, 299, 339, 354, 380, 408, 447, 464, 503, 518, 548, 574, 602, 628, 655. 1926. (621)

Moeller, F. (A) Ausmessung von Waermefeldern durch Abbildung in Elektrischen Modellen. Archive für Technisches Messen., 151:T6(V24-1). 1947. (313A)

―――. (B) Über einige Grundfragen bei Analogien. Elektrotechnische Zeit. 69:73-78. 1948. (313A)

―――. (C) Zur Behandlung Stationaerer Waermestroemungen Mittels elektrischer Abbilder. Elektrotechnik u. Maschinenbau 61:4-8. 1943. (313A)

Mohr, O. (A) Abhandlungen aus dem Gebiete der Technischen Mechanik. 2nd ed., 342-374. 1914. (422C)

―――. (B) Beitrag zur Theorie der Holz-und Eisenconstruktionen. Z. Arch. Ing.-Ver., Hanover. 1868. (422C)

Molchanov, E. I. Analysis of temperature field in gas-turbine rotor, using hydraulic integrator. Engrs. Digest 17:469-471. 1956. (311B)

Molnar, L. Determination of the laws of streaming of liquids by the aid of electrical analogies. (In Hungarian.) Hidrológiai Közloeny 37:306-317. 1957. (221D)

Moon and Oliphant. Current distributions near edges of discharge tube cathodes. Proc. Cambridge Phil. Soc. 25:461. 1929. (651C)

Moore, A. D. (A) Fields from fluid mappers. J. Appl. Phys. 20:790-804. 1949. (221AF) (312B) (313C) (613) (631)

―――. (B) Four electromagnetic propositions, with fluid mapper verification. Elec. Eng. 69:607-610. 1950. (613) (631)

―――. (C) The further development of fluid mappers. Trans. AIEE 69(II): 1615-1624. 1950. (221AF) (313C) (613) (631)

―――. (D) The hydrocal. Ind. and Engr. Chem. 28:704. 1936. (311B)

―――. (E) Soap-film and sand bed mapper techniques. Trans. ASME 72:291-298. 1950. (221AF) (312B) (313C) (613) (631)

Moore, R. C. McIlroy fluid network analyzer. Instruments and Automation 28:428-429. 1955. (122AB)

―――. See Werner, B. L. (122AB)

Morgan, F., Muskat, M., and Reed, D. W. Studies in lubrication: VI. Electrolytic models of full journal bearings. J. Appl. Phys. 11:141. 1940. (222A)

Morris, R. E., and Haythornthwaite, B. Water flow analogues for gas dynamics. Engineering 190:261-263. 1960. (211AB)

Morris, W. L. Analogical computing devices in the petroleum industry. Ind. and Eng. Chem. 43:2478-2483. 1951. (221AAA)

Morton, G. A. See Ramberg, E. G. (612) (651C)

Mossop, I. A., and McGhee, J. Use of analogue wax model methods for reactor calculations. United Kingdom Atomic Energy Authority, Sellafield, Cumb., England. (730)

Motyakov, V. I. (A) Construction of networks of flow of incompressible liquids for multilinked heterogeneous regions. (In Azerb.) Izv. Akad. Nauk Azerb.-SSR 3:19-37. 1957. (221AAA)

———. (B) Solution of inverse problems on stationary flow in underground hydraulics by means of electric grid models. (In Russian.) Zh. Tekh. Fiz. 27:364-367. 1957. (221AAA)

Motz, See Allen, Fox, and Southwell. (643)

Mozley, J. M. Predicting dynamics of concentric pipe heat exchangers. Indust. Engrg. Chem. 48:1035-1041. 1956. (321A)

Müller-Breslau, H. (A) Beitrag zur Theorie des Fachwerks. Zeitschrift des Architekten und Ingenieurvereines zu Hannover 31:418. 1885. (422C)

———. (B) Graphische statik. 11:99-120. 1892. (422C)

Müller, H. Das Beukenmodell, Grundlagen, Grenzen, und Andwendungsmöglichkeiten. Elektrowaerme Technik 2:51-54. 1951. (313A)

Müllner, F. Electrical mapping of magnetic vortex fields. (In German.) ETZ 50:1321-1323. 1929. (633)

Mumford, A. R., and Powell, E. M. Heat flux pattern in fin tubes under radiation. Trans. ASME 67:693-695. 1945. (313A)

Muncey, R. W. Calculation of temperature rise with intermittent heating. Australian J. Appl. Sci. 7:29-37. 1956. (311A)

Murphy, E. F. (A) Determination of natural frequencies of piping systems by an electrical analog. Ph. D. thesis, Illinois Inst. of Technology. Chicago. 1948. (524)

———. (B) Natural frequencies of compressor and engine manifolds. Petr. Engr. (Ref. annual.) 21:C60-66. 1949. (121BAA)

Murphy, Glenn. (A) The conjugate frame as a tool for evaluating deflections. Proc. Seventh Int. Cong. Appl. Mech., London 1(12):131. 1948. (422C) (472BA)

———. (B) Memorandum to chief design engineer, U.S. Bureau of Reclamation. 1931. (451)

———. (C) Similitude in engineering. Ronald Press Co., New York. 1950. (421C)

Murphy, M. J., Bycroft, G. N., and Harrison, L. W. Electrical analog for shear stresses in multi-storied buildings. Proc. World Conf. on Earthquake Engrg. Berkeley. 1956. (511A)

———. See Bycroft, G. N., and Brown, K. J. (511A)

Murray, C. T., and Hollway, D. L. A simple equipment for solving potential and other field problems. J. Sci. Instr. 32:481-483. 1955. (Appendix II D2)

Muskat, M. (A) Potential distribution in large cylindrical disks with partially penetrating electrodes. J. Appl. Phys. 2:329. 1932. (221AAA) (221AAB)

———. (B) The theory of potentiometric models. Petr. Technol. 11:1-6. 1948. Trans. Amer. Inst. Min. Metall. Engrs., Petr. Div. 179:216-221. 1949. (221AAA)

———, and Wyckoff, R. D. A theoretical analysis of waterflooding networks. Trans. AIME 107:62. 1934. (221AAA)

———. See Morgan, F., and Reed, D. W. (222A)

———. See Wyckoff, R. D., and Botset, H. G. (221AAA)

Musson-Genon, R. Ann. Télécomm. 2:298-320. 1947. (651D)

Nabor, G. W. See Ramey, H. J., Jr. (221AAA)

Nachtigall, A. J. See Ellerbrock, H. H., Jr., and Schum, E. F. (311A) (313A)

Nadai, A. (A) Der beginn des fliessvorganges in einem tordierten stab. Zeits. f. Angew. Math. u. Mech. 3:442. 1923. (442A)

———. (B) Plastic torsion. Trans. ASME 53:29. 1931. (442A)

REFERENCES

———. (C) Plasticity (Engineering Societies Monographs). McGraw-Hill, New York. 1931. (442A)

Nagao, S. Electrical analog solves reactor design problems. Nucleonics 16:88-90. 1958. (730)

Nakazawa, H. Electrical analogies with resistive paper for the torsion problems of bar. Proc. 6th Japan Nat. Congr. Appl. Mech., Univ. of Kyoto, 99-102. 1956. (441AD)

Nechayev, S. V. The simulation of a circulation flow by an electro-hydrodynamical analog in accordance with the Chapligin-Joukowsky postulate. (In Russian.) Trudi U. fimsk. aviats. in-ta 1:28-39. 1955. (221D)

Needs, S. J. Effects of side leakage in 120-degree centrally supported journal bearings. Trans. ASME 56:721. 1934. (222A)

Neel, C. B. (A) A procedure for the design of air-heated ice-prevention systems. NACA Tech. Note 3130. 1954. (322AA)

———. (B) An investigation utilizing an electric analogue of cyclic de-icing of hollow steel propellers with internal electric heaters. NACA Tech. Note 3025. 1953. (311A)

Negoro, S. (A) On a method of solution of torsion and bending problems of a straight beam with axial hole. Trans. Soc. Mech. Eng. Tokyo 4:6-9. 1938. (433) (441AD)

———. (B) On a method of solving torsion and bending problems of a bar with axial hole. Tohoku Univ. Tech. Rep. 12:517-528. 1938. (433) (441AD)

———. (C) The use of the electric bath in solving torsion and bending problems. Trans. Soc. Mech. Eng. Tokyo 2:S-38, 112-116. 1936. (433) (441AD)

Nemenyi, P. (A) Stromlinien and hauptspannungstrajectorien. Zeits. f. Angew. Math. u. Mech. 13:364-366. 1933. (454)

———. (B) Über spannungsfelder, die nut bekannten stromungsfeldern isomorph sind. Proc. Third Int. Cong. for Appl. Mech. (Stockholm) 2:155-166. 1930. (454)

Neubauer, T. P., and Boston, O. W. Torsional stress analysis of twist-drill sections by membrane analogy. Trans. ASME 69:897-902. 1947. (441AA)

Neuber, H. (A) Der ebene Stromlinienspannungszustandmit lastfreiem Rand. Ing. Arch. 6:325-334. 1935. (454)

———. (B) Die Randbedingungen der Hauptspannungs-stromlinien. Zeits. f. Angew. Math. u. Mech. 15:374-475. 1935. (454)

New electrical method for mass and heat transfer problems developed at Columbia University by Victor Paschkis. News Edition 19:740. 1941. (311A)

New Heat-transfer research tool. Mech. Eng. 64:95-96. 1942. (311A)

Newman E. J. See Lewis, W. T. O. (A), (221AAB); (B), (221AAB)

Newman, P. C. Investigation of adiabatic colorimeter by heat transfer analogue. Fuel 35:295-302. 1956. (311A)

Newton, R. E. (A) Electrical analogy for shear lag. Rep. R-140, Curtiss-Wright Corp., Airplane Div. (St. Louis). 1944. (482)

———. (B) Electrical analogy for shear lag problems. Proc. Soc. Exp. Stress Anal. 2(2):71-80. 1945. (482)

———. See Engle, M. E. (482)

Nickle, C. A. Oscillographic solution of electro-mechanical systems. Trans. AIEE 44:844. 1925. (511A) (531A)

Nicoll, F. H. See Bowman-Manifold, M. (612)

Nielsen, H. See Laitone, E. V. (211AB)

Nihei, T. See Isobe, T. (Appendix II D1)

Nobles, M. A., and Janzen, H. B. Application of a resistance network for studying mobility ratio effects. J. Pet. Tech. 10:60-62. 1958. (221AAA)

Norbury, J. F., and Platt, A. A simple method of constructing duct models for the electrolytic tank. J. Roy Aero. Soc. 61:775-776. 1957. (Appendix II D1)

REFERENCES

Northrup, E. F. Some aspects of heat flow. Trans. Amer. Electrochem. Soc. 24:85-106. 1913. (313A)

Nottage, H. B. See Humphreys, C. M., Franks, C. V., Huebscher, R. G., Schutrum, L. F., and Locklin, D. W. (313A)

Nougaro, J., Gruat, J., and Comier, J. J. Principles of electric analogy for the study of coefficients of discharge in the infiltration of a dam. (In French.) C. R. Acad. Sci., Paris 246:1661-1664. 1958. (221AAA)

Nukiyama, S., and Tanasawara, Y. On an electric experiment upon the flow of heat axiel symmetric about an axis. J. Soc. Mech. Engr. Tokyo. 1930. (313A)

Oatley, C. W. See Sander, K. F., and Yates, J. G. (651D)

Okamoto, T. An experiment of location of detached shock wave ahead of wedge profile by hydraulic analogy. Bul. Tokyo Inst. Tech. (B) 1:1-5. 1956. (211AB)

Oliphant, J. B. See Moon. (651C)

Oliphint, J. B. See Carter, W. J. (441AE)

Olivery, L. R. See Marshall, D. L. (221AAA)

Olshak, V., and Litvinishin, G. The nonlinear phenomenon of liquid flow as a rheological simulator. (In Russian.) Bul. Polsk. Akad. Nauk. (IV) 2:71-76. 1954. (491)

Olson, H. F. Dynamical analogies. D. Van Nostrand Co., Inc., New York. 1943. (511A) (121BBA)

Olszak, W. Generalization of the elastic membrane analogy to problems of anisotropic systems. (In Polish, with French Summary.) Arch. Mech. Stos. 5:89-106. 1953. (441AA)

Oppenheim, A. K. (A) Radiation analysis by the network method. Trans. ASME 78:725-735. 1956. (330)

———. (B) Water-channel analog to high-velocity combustion. Trans. ASME 75:115. 1953. Discussion 439. (211AB)

Opsal, F. W. Analysis of two- and three-dimensional groundwater flow by electrical analogy. Trends. Engrg. Exp. Sta., Univ. Wash. 7:15-20, 32. 1955. (221AAC)

Orlin, W. J., Lindner, N. J., and Bitterly, J. G. Application of the analogy between water flow with a free surface and two-dimensional compressible gas flow. U. S. NACA Tech. Note 1185. 1947. (211AB)

Otis, D. R. Electric analogy to transient heat conduction in a medium with variable thermal properties. Proc. Nat. Simulation Conf., Dallas, Texas. Paper 3. 1956. (311A)

Otsuka, S. Latticed wing solution with the aid of electricity. Proc. 2nd Japan Nat. Congr. Appl. Mech. 207-211. 1952. (221AAA)

Packer, J. S. See Shen, D. W. C. (531A)

Palmer, P. J. (A) The determination of torsion constants for bulbs and fillets by means of an electrical potential analyzer. Aluminum Development Association, Research Report 22. 1953. (441AC)

———. (B) Lift on wings at sonic speeds by means of an electrical resistance analogue. J. Roy. Aero. Soc. 60:137-139. 1956. (221AAA)

———, and Redshaw, S. C. (A) Electrical resistance analogue with graded mesh. J. Sci. Instr. 34:407-409. 1957. (Appendix II E)

———, and Redshaw, S. C. (B) Experiments with electrical analogue for extension and flexure of flat plates. Aero. Quart. 6:13-30 (Pt. I.) 1955. (457)

Panchishin, V. I. See Filchakov, P. F. (221AAA)

Parker, R. J. Understanding and predicting permanent magnet performance by electrical analog methods. J. Appl. Phys. 29:409-410. 1958. (632)

Parry, V. F. See Burke, S. P. (221AAA)

Paschkis, V. (A) Combined geometric and network analog computer for transient heat flow. J. Heat Transfer 81C:144-148. 1959. (311A)

———. (B) Discussion of "Temperature distribution in complex wall structures by geometrical electrical analogue," by C. F. Kayan. Refrig. Eng. 49:113-116, 117, 139. 1945. (313A)

———. (C) Electric analogy method for the investigation of transient heat flow problems. Ind. Heating 9:1162-1170. 1942. (311A)

———. (D) Electric method for the solution of Laplace's equation. Proc. Soc. Exp. Stress Anal. 2(2):39-43. 1945. (311A) (455)

———. (E) Elektrisches Modell zur Verfolgung von Waermestrahlungsvorgaengen, insbesondere in elektrischen Oefen. Elektrotechnik und Maschinenbau 54:617-621. 1936. (311A)

———. (F) Establishment of cooling curves of welds by means of the electric analogy. Welding Journal, Welding Research Supplement 22:462-483s. 1943. (311A)

———. (G) Heat and mass flow analyzer — tool for heat research. Metal Prog. 52:813-818. 1947. (311A)

———. (H) Heat flow problems in foundry work. Am. Foundrymen's Assn. Preprint 44-46 for Meeting, April 25-28, 1944. (311A)

———. (J) Heat flow problems solved by electric circuits. Heating, Piping, and Air-Cond. 13:756-757. 1941. (311A)

———. (K) Heat flow through plastic materials. J. Soc. Plastics Engrs. 10:28-30. 1954. (313A)

———. (L) More heat flow. Electronics 18:380. 1945. (311A)

———. (M) New tool of heat-flow research in glass industry and ceramic industry at large. Bul. Am. Cer. Soc. 27:450-461. 1948. (311A)

———. (N) Periodic heat flow in building walls determined by electrical analogy method. ASHVE 48:75-90. 1942. (311A)

———. (O) Quenching of steel balls and rings; temperature-time-space relationships investigated on heat-and-mass-flow analyzer at Columbia University. Trans. Amer. Soc. for Metals 37:216-244. 1946. (311A)

———. (P) Temperature distribution on the surface of a brick wall with mortar joints. Refrig. Eng. 47:469-473. 1944. (313A)

———. (Q) Theoretical thermal studies of steel ingot solidification. Trans. Amer. Soc. for Metals. 38:117-147. 1947. (311A)

———, and Baker, H. D. A method for determining unsteady-state heat transfer by means of an electrical analogy. Trans. ASME 64:105-112. 1942. (311A)

———, and Beuken, C. L. Die Berechnung der Durchwarmungszeiten von Gutstuecken auf Grund der relativen Mindertemperatur. Elektrotechnik und Maschinenbau 56:98-100. 1938. (313A)

———, and Heisler, M. P. (A) The accuracy of lumping in an electrical circuit representing heat flow in cylindrical and spherical bodies. J. Appl. Phys. 17:246-254. 1946. (311A)

———, and Heisler, M. P. (B) The accuracy of measurements in lumped r-c cable circuits as used in the study of transient heat flow. Elec. Eng. 63:165-171. 1944. (311A)

———, and Heisler, M. P. (C) The influence of through metal on the heat loss from insulated walls. Trans. ASME 66:653-661. 1944. (313A)

———. See Avrami, M. (311A)

———. See Bradley, C. B., and Ernst, C. E. (311A)

———. See Weiner, J. H., and Salvadore, M. G. (441AC) (442B)

Paschoud, M. (A) Le probleme de la torsion et d'analogie hydrodynamiques de M. Boussinesq. Bl. Tech. de la Suisse Romande 51:277-281. 1925. (441AG)

———. (B) Compt. Rend. 179-451. 1924. (441AG)

Patigny, J. L'étude de la ventilation des mines par l'analogie electrique. Revue Universelle des Mines 14:381-416. 1958. (122BB)

Patraulea, N. N., and Camarasescu, N. On the increased lift obtained by placing the propellers on the upper side of the wing. (In Roumanian.) Studii Si Cercetari Mecan. Apl. 10:1013-1020. 1959. (221AAA)

Paulhus, J. L. H. See Linsley, R. K., and Kohler, M. A. (112A)

Pavlovsky, N. N. Motion of water under dams. First Congres des Grands Barrages, Stockholm, 4:179. 1924. (221AAA)

Pawley, M. Design of a mechanical analogy for the general linear electrical network with lumped parameters. J. Franklin Inst. 223:179-198. 1937. (511A) (621)

Paynter, H. M. Electrical analogies and electronic computers: surge and water hammer problems. Proc. ASCE Separate 146. 1942. (121AAA)

Pearce, C. A. R. Electrical conductivity and permittivity of mixtures, with special reference to emulsions of water in fuel oil. Brit. J. Applied Physics 6:113-120. 1955. (612)

Peattie, K. R. Conducting paper technique for construction of flow nets. Civ. Eng. (London) 51:62-64. 1956. (221AAA)

Peierls, R. E. Use of the electrolytic tank for magnetic problems. Nature (London) 158:831. 1946. (633)

──, and Skyrme, T. H. R. Tank model for magnetic problems of axial symmetry. Phil. Mag. 40:269-273. 1949. (633)

Pelyer, H. See Gross, B. (491)

Peplow, M. E. Electrical Times. 256. February 15, 1951. (612)

Peres, J. (A) La methods des analogies electriques en aerodynamique. Science Aerienne 5:1-4. 1936. (221AAA)

──. (B) Les méthodes d'analogies en mecanique appliquée. 5th Int. Congr. Appl. Mech., Cambridge, Massachusetts. 1938. (221AAA)

──, and Malavard, L. (A) Analogic realizations of jet flow. (In French.) C. R. Acad. Sci. Paris. 206:418-420. 1938. (221AAA)

──, and Malavard, L. (B) Application of the electric method to a problem concerning airfoils of finite aspect ratio. (In French.) C. R. Acad. Sci. Paris. 195:599-601. 1932. (221AAA)

──, and Malavard, L. (C) Application of the electric method to some questions concerning wing lift. (In French.) Sci. Aérienne 6:360. 1936. (221AAA)

──, and Malavard, L. (D) Application of the electric tank to some questions in the mechanics of fluid. (In French.) Asso. Tech. Marit. Aero. 1938. (221AAA)

──, and Malavard, L. (E) La method d'analogies rheographiques et rheometriques. Bul. Soc. Fran. d. Elecl. (5) 8:715-744. 1938. (441AD) (633)

──, and Malavard, L. (F) On electric analogies in hydrodynamics. (In French.) C. R. Acad. Sci. Paris. 134:1314-1316. 1932. (221AAA)

──, Malavard, L., and Romani, L. Probleme non lineaire de la theorie de l'aile. Application a la determination du maximum de portance. Actuellement a l'impression: Rapport Techn. du Groupement Francais pour les Recherches Aeronautiques. 20. 1946. (221AAA)

Perry, H. A., Jr., Vierling, D. E., and Kohler, R. W. Network flow analysis speeded by modified electrical analogy. Eng. News Rec. 143:19-23. 1949. (122AA)

Pestel, E. (A) A new flow analogy for torsion. (In German.) ZAMM 34:322. 1954. (441AH)

──. (B) A new hydrodynamic analogy of the torsion of prismatic rods. (In German.) Ing. Arch. 23:172-178. 1955. (441AH)

Petrick, E. N. See Lang, R. (313A)
Pian, T. H. H. See Bisplinghoff, B. L., and Levy, L. I. (490)
Piccard, A., and Baes, L. Mode experimental nouveau relatif a l'application des surfaces a courbure constante a la solution du probleme de la torsion des barres prismatiques. Proc. Second Int. Cong. Appl. Mech., Zurich 195-199. 1926. (441AA)
———, and Johner, W. Eine indirekte experimentelle Methode zur Bestimmung der Torsionspannungen in prismatischer Staben. Festschrift Prof. Dr. A. Stolda, 489-498. 1929. (441AA)
Pierce, J. R. Rectilinear electron flow in beans. J. Appl. Phys. 11:548-555. 1940. (651D)
Pierre, B. Influence of frames on insulation of cold storage chambers on board ships. Trans. Roy. Inst. Tech. Stockholm. 50. 1951. (313A)
Pilod, P. See Gerber, S. (221AAA)
Pinl. See Behrbohm. (211AC)
Pipes, L. A. (A) Analysis of longitudinal motions of trains by electrical analog. J. Appl. Phys. 13:780-786. 1942. (512B)
———. (B) Electrical circuit analysis of torsional oscillations. J. Appl. Phys. 14:352-362. 1943. (531A)
———. (C) Operational theory of longitudinal impact. J. Appl. Phys. 13:503-511. 1942. (512A)
Piquemal, J. (A) The investigation by electrical analogy methods of pressure surges set up in a penstock by a periodic disturbance at the downstream end. (In French.) Houille Blanche 13, B:767-774. 1958. (121AAA)
———. (B) Use of electric analogy for determination of water hammer overpressures. (In French.) C. R. Acad. Sci., Paris 245:1117-1119. 1957. (121AAA)
Pirard, A. Les problemes de torsion et les mesures de tensions par membranes minces. Rev. Univ. Min. 4:527-539. 1948. (441AA) (452)
Pitin, R. N., Ponnik, Yu. A., and Farberov, I. L. Application of the method of electrohydrodynamic analogy to the investigation of some problems of the underground gasification of coal. (In Russian.) Podzemn. Gazifik. Uglei 4:46-49. 1958. (211C)
Pizer, H. I., Wallis, R. A., and Warden, M. C. An electrolytic tank simulator. Aerodynamic Note 76, Council for Scientific and Industrial Research, Australia. 1948. (221AAA)
Platt, A., and Norbury, J. F. Temperature inequalities in the electrolytic tank. J. Roy. Aero. Soc. 62:456 (Tech. Note). 1958. (Appendix II D1)
Plesset, M. S. See Gilmore, F. R., and Crossley, H. E., Jr. (211AB)
Pletta, D. H., and Maher, F. J. The torsional properties of round-edged flat bars determined: I. Experimentally, II. Analytically. Va. Poly. Inst. Eng. Exp. Sta. Bul. 50. 1942. (441AA)
———. See Szebehely, V. G. (454)
Pocock, P. J. Calculation of wave drag of arbitrary slender body by means of electrical analogy tank. Can. Aeronautical J. 1:169-177, 168. 1955. (211AAA)
Podolsky, V. A. See Abramov, F. A. (122BB)
Poirier, Y. See Crausse, E. (221AAA)
Poisson-Quinton, Ph. Recherches theoriques et expérimentales sur le controle de la couche limite. Proc. 6th Int. Congr. Appl. Mech., London. 1946. (221AAA)
Polya, G. On the fundamental frequency of vibrating membranes and the torsional resistance of elastic rods. C. R. Acad. Sci., Paris 225:346-348. 1947. (524)
Ponnik, Yu. A. See Pitin, R. N., and Farberov, I. L. (211C)

Poritsky, H., Sells, B. E., and Danforth, C. E. Graphical, mechanical, and electrical aids for compressible fluid flow. J. Appl. Mech. 17:37-46. 1950. (211AAC) (211AAD)

Potseluiko, V. A., and Trofimenko, A. T. Investigation of the thermal field by the method of electrothermal analogy. (In Russian.) Investigations of the physical bases of the working processes of furnaces and ovens. Alma-Ata, Akad. Nauk. Kaz. SSR, 242-251. 1957. (311A)

Potter, N. L. Electrical analogue for heat flow problems in semi-conductors. Electronic Eng. 31:454-457. 1959. (311A)

Powell, E. M. See Mumford, A. R. (313A)

Powell, P. H. The air-gap correction coefficient. J. Instn. E. E. 40:228-234. 1908. (631)

——. See Hele-Shaw, H. S., and Hay, A. (631)

Prager, W., and Hay, G. E. On plate rigid frames loaded perpendicularly to their plane. Quart. Appl. Math. 1:49-60. 1943. (472AD) (472AE)

Prandtl, L. (A) Abriss der Lehre von der Flussigkeits und Gasbewegung. Handworterbuch der Naturwissenschaften, 4. 1913. (221AB)

——. (B) Bemerkung über den Wärmeübertragung im Rohr. Physikalische Zeitschrift 29:487-489. 1928. (322C)

——. (C) Bemerkung zu den Arbeiten von H. Cranz und H. Quest in band IV, seite 506 and 510. Ing. Archiv. 4:606. 1933. (441AA)

——. (D) Eine neue Darstellung der Torsionspannungen bei prismatischen Staben von beliebigem Querschnitt. Jahr. d. Deutschen Math. Ver. 19:31-36. 1904. (441AA)

——. (E) Zur Torsion von prismatischen Staben. Physikalische Zeitschrift 4:758. 1903. (441AA)

Prebus, A. F. See Zlotowski, I. (652)

——, Zlotowski, I., and Kron, G. The application of network analysis to some electron optical problems. Phys. Rev. 67:202. 1945. (Abstract) (652)

Preiswerk, E. (A) Application of the methods of gas dynamics to water flows with a free surface. Mitteilungen der Institut für Aerodynamik, 7, E.T.H. Zurich. 1938. (Translated as U. S. NACA Tech. Mem. 934, 935. 1940.) (211AB)

——. (B) Zweidimensionale stromung schiessenden Wassers. Schweizerische Bauzeitung 109:237. (20) 1937. (211AB)

Price, P. H., and Sarjant, R. J. Unsteady heat flow. Proc. Gen. Discussion on Heat Transfer. 281-283. 1951. (311A)

Price, W. H. See Lane, E. W., and Campbell, F. B. (221AAA)

Priddy, E. L. See Swanson, R. S. (221CA)

Probine, M. C. Electrical analogue method of predicting permeability of unsaturated porous materials. Brit. J. Appl. Phys. 9:144-148. 1958. (221AAA)

Probstein, R. F., and Hudson, G. E. A water analogue of the isentropic flow of compressible gases which have arbitrary ratios of specific heats. Project SQUID Report No. NYU-20-R. 1948. (211AB)

Prochazka, J., Landau, J., and Standart, G. Hydraulic analogue for studying steady-state heat exchangers. Brit. Chem. Eng. 5:242-247. 1960. (322AB)

Progress in Brown Boveri designs during 1941: Section VII, experimental work. Brown Boveri Review 29:76-77. 1942. (211AB)

Provasnik, F. Electricke modelly neustalenych tepelnych stavu elektrickych stroju. Elektrotechnicky Obzor 46:277-283. 1957. (311A)

Prudkovskii, G. P. (A) The use of an electrolytic tank as a computing device. Instruments and Experimental Techniques 3:423-426. 1960. (Translation of Pribory i Tekhnika Eksperimenta, USSR 3:77-79. 1959., by Instrument Society of America, Pittsburgh 22, Pa.) (Appendix II D1)

REFERENCES

———. (B) Use of electrolytic tank as computing device. Instruments and Experimental Techniques (English Translation of Pribory i Tekhnika Eksperimenta) 3:423-426. 1959. (Appendix II D1)

Przemieniecki, J. S. Electrical analogy for transient axisymmetrical heat flow. J. Aero. Sci. 24:922. 1957. (311A)

Puppini, U. Electric analogs for the study of the motion of seeping water. (In Italian.) Mem. Reale Accad. Sci. Bologna, Cl. Sci. Fis. 9(7):79-84. 1921-1922. Monitore Tecnico 28:209-214, 298. 1922. (221AAA)

Quest, H. Eine experimentelle losung des torsions problems. Ing. Arch. 4:510-520. 1933. (441AA)

Quill, J. S. The design of an electric analyzer for the solution of hydraulic problems. M. S. thesis, M. I. T. 1942. (122AB)

Quincke. Poggendorff's Annalen 97:382. 1856. (Appendix II D2)

Rademaker, J. See Biezeno, C. B. (441AA)

Radziul, J. V. See McPherson, M. B. (122AB)

Rajchman, J. Arch. des Sciences, Geneve 20(5):1938. (651C)

Ram, G., and Vaidhianathan, V. I. (A) Pressure under a flush floor with inclined sheet piles. Proc. Ind. Acad. Sci. 12:245-250 (A, 3). 1940. (221AAA)

———, and Vaidhianathan, V. I. (B) Uplift pressure and design of weirs with two sheet piles. Proc. Ind. Acad. Sci. 4:147-156 (A, 2). 1936. (221AAA)

———, Vaidhianathan, V. I., and Taylor, E. M. (A) The design of falls with reference to uplift pressure. Proc. Ind. Acad. Sci. 3:360-368 (A, 4). 1936. (221AAA)

———, Vaidhianathan, V. I., and Taylor, E. M. (B) Potential distribution in infinite conductors and uplift pressure on dams. Proc. Ind. Acad. Sci. 2:22-29 (A, 1). 1935. (221AAA)

———. See Guha, S. K. (221AAA)

———. See Luthra, S. D. L. (A), (221AAA); (B), (221AAA)

———. See Vaidhianathan, V. I. (A), (221AAA); (B), (221AAA)

———. See Vaidhianathan, V. I., and Taylor, E. M. (A), (221AAA); (B), (221AAA)

Ramachandran, A. Analogic experimental methods in heat transfer. Electrotechnics 23:110. (313A)

Ramberg, E. G., and Morton, G. A. Electron optics. J. Appl. Phys. 10:465. 1939. (612) (651C)

Ramey, H. J., Jr., and Nabor, G. W. A blotter-type electrolytic model determination of areal sweeps in oil recovery by in-situ combustion. Trans. Amer. Inst. Min. Metall. Engrs., Petr. Div. 201:35-39. 1954. (221AAA)

Ramo, S. See Whinnery, J. R. (641)

Randels, W. C. Comments on "A study of transonic gas dynamics by the hydraulic analogy." J. Aero. Sci. 19:572. 1952. (211AB)

Ranger, A. E. Electrical analogue for estimating die temperatures during wire drawing. J. Iron and Steel Inst. 185:383-388. 1957. (313A)

Rawcliffe, G. H. Mechanical analogies in electrical work. Electrical Review 123:121-122. 1938. (621)

Rawlings, B. Electrical analogue for plastic moment distribution. Civ. Eng. Trans. Inst. Engrs., Australia, CE 1:65-70. 1959. (492)

Rayleigh. On the flow of viscous fluids, especially in two-dimensions. Phil. Mag. Series 5, 36:354-372. 1893. (222B)

Raymer, W. G. See Cheers, F. (221AAB)

———. See Cheers, F., and Fowler, R. G. (221AAB)

Recknagel, A. See Brüche, E. (651)

Redshaw, S. C. (A) Electrical analogues for the solution of problems concerning the extension and flexure of flat elastic plates. Aero. Res. Counc. Rep. Mem. 15, 335. 1952. (457)

Redshaw, S. C. (B) An electrical potential analyzer. Proc. Instn. Mech. Eng. 159:55-62. 1948. (Appendix II E)
―――. (C) Electrical potential analyzer provides, in effect, an analogy to solution of Poisson's and Laplace's equations. Inst. Mech. Eng. J. and Proc. 159:55-62. (War emergency issue 38.) 1948. (Appendix II E)
―――. (D) The use of an electrical potential analyzer for the calculation of the pressures on lifting surfaces. Aero. Quart. 5:163-175. 1954. (211C)
―――. (E) Use of electrical analogue for solution of variety of torsion problems. Brit. J. Appl. Phys. 11:461-468. 1960. (441AC) (442B)
―――, and Rushton, K. R. An electrical analogue solution for the stresses near a crack or hole in a flat plate. J. Mech. Phys. Solids 8(3):173-186. 1960. (457)
―――. See Palmer, P. J. (A), (Appendix II E); (B), (457)
Reed, D. W. See Morgan, F., and Muskat, M. (222A)
―――. See Wyckoff, R. D. (221AAA)
Reich, H. J. A mechanical analogy for coupled electrical circuits. Rev. Sci. Inst. 3:287. 1932. (622B)
Reichardt, W. Topologisch entsprechende mechanische und elektrische schaltungen. Frequenz 13:278-286. 1959. (511B)
Reichenbach, G. S. Experimental measurement of metal-cutting temperature distributions. Trans. ASME 80:525-536. 1958. (313A)
Reichenbacher, H. Selbsttatige Ausmessung von Seifenhautmodellen. Ing. Arch. 7:257-272. 1936. (441AA)
Reid, G. W., and Wolfenson, L. B. How electric calculators solve water systems distribution problems. Pub. Works 84:78-79, 112, 114. 1953. (122AA)
―――. See Suryaprakasam, M. V., and Geyer, J. C. (A), (122AA), (122AB); (B), (122AA)
Reiniger, F. The study of thermal conductivity problems by means of the electrolytic tank. Philips Tech. Rev. 18:52-60. 1956-1957. (313A)
Reinius, E. On the stability of the upstream slope of earth dams. Victor Pettersons Bokindustriaktiebolag, Stockholm. 1948. (221AAA)
Relf, E. F. An electrical method of tracing stream lines for the two-dimensional motion of a perfect fluid. Phil. Mag. (6) 48:535-539. 1924. (221AB)
Reltov, B. F. Electrical analogy applied to three-dimensional study of percolation under dam built on pervious heterogeneous foundations. Second Cong. on Large Dams 5:73-85. 1936. (221AAC)
Remacle, J. See de Crombrugghe, O. (122BB)
Renard, G. Representation direct par analogie rheoelectrique des gradients de fonctions harmoniques en domaine plan limite ou illimite. France. Ministere de l'Air. Publications Scientifiques et Techniques. Tech. Note 78. 1958. (221AAA)
―――. Nouvelles applications du principe d'inversion dans le calcul analogique experimental. Assn. Internationale pour le Calcul Analogique — Annales 2:69-72. 1960. (Appendix II D1)
Revuz, J. (A) Families of profiles of blades for axial compressors. (In French.) Rech. Aero. 37:11-16. 1954. (221AAA)
―――. (B) Mise au point d'une methode de reduction des essais sur une aile munie de plaques de garde: Part I: Resolution par analogie electrique. ONERA NT 2/1358.A. 1950. Part II: Resultats des experiences. NT 4/1358.A. 1951. (221AAA)
―――. (C) Profil d'ailette pour compresseur axial. La Recherche Aeronautique 31. 1953. (221AAA)
―――. See Romani, L. (221AAA)

REFERENCES

Reynolds, O. On the extent and action of the heating surface for steam boilers. Proc. Manchester Library and Phil. Soc. 14:7-12. 1874. (322C)

Riabouchinsky, D. (A) Hydraulic analogy of the motion and resistance of a compressible fluid as an aid to aeronautical research. Reissner Anniv. Vol., J. W. Edwards, Ann Arbor 61-68. 1949. (211AB)

——. (B) On the hydraulic analogy to flow of a compressible liquid. C. R. Acad. Sci. Paris 195:998-999. 1932. (211AB)

——. (C) Quelques nouvelles remarques sur l'analogie hydraulique des mouvements d'un fluide compressible. C. R. Acad. Sci. Paris 199:632-634. 1934. (211AB)

——. (D) Recherche comparative sur l'aerodynamique des petites et des grandes vitesses. C. R. Acad. Sci. Paris 202:1725-1728. 1936. (211AB)

——. (E) Recherches sur l'amelioration des qualities aerodynamiques des profils d'ailes aux grandes vitesses. Publications Scientifiques et Techniques du Ministere de l'Air, Paris 108: Ch. 3. 1937. (211AB)

Rice, C. W. An experimental method of obtaining the solution of electrostatic problems, with notes on high voltages bushing design. Trans. AIEE 36:905-1051. 1917. (612)

Ridgway, W. See Whinnery, J. R., Concordia, C., and Kron, G. (641)

Richards, T. H. Analogy between the slow motion of a viscous fluid and the extension and flexure of plates: a geometric demonstration by means of moiré fringes. Brit. J. Appl. Phys. 11:244-254. 1960. (222B) (451)

Riesz, G. W., and Swain, B. J. Structural analysis by electrical analogy. Proc. Soc. Exp. Stress Anal. 12:13-22. 1954. (422D) (472AB)

Rippel, H. C. See Loeb, A. M. (222A)

Robertson, A. F., and Gross, D. An electrical-analog method for transient heat-flow analysis. J. Res. Nat. Bur. Stands. 61:105-115. 1958. (313A)

Robson, P. N. See Hartill, E. R., and McQueen, J. G. (Appendix II D1)

Roechmann, L. F. Mechanical model of a.c. transmission. Electrician 142: 1491-1493. 1949. (621)

Roeterink, T. M. Eine theoretische und experimentelle Untersuchung des Nutenfeldes einer unbelasteten Maschinen. Arch. fur Elekt. 7:292-319. 1919. (631)

Rogers, R. Some simple electrical analogies. Product Eng. 18:126-130. 1947. (621)

Romani, L., and Revuz, J. Sur la determination des systemes portants optima. Proc. 6th Int. Cong. Appl. Mech., London. 1946. (221AAA)

——. See Peres, J., and Malavard, L. (221AAA)

Roots, O. T. The electrical simulation of crossed beams. (In Russian.) Trudi Tallinsk. Politekhn. In-ta(A), 65:111-123. 1955. (422D)

Roscoe, R. The flow of viscous liquids around plane obstacles. Phil. Mag. 40:338-351. 1949. (222B)

Rose, A. Mechanical model for motion of electrons in magnetic field. J. Appl. Phys. 11:711-717. 1940. (651A)

Rosenthal, D. See Friedmann, N. E., and Yamamoto, Y. (441AD)

Rouse, H., and Abul-Fetough, Abdel-Hadi. Characteristics of irrotational flow through axially symmetric orifices. J. Appl. Mech. 17:A421. 1950. (221AAB)

——, and Hassan, M. M. Cavitation-free inlets and contractions; electrical analogy facilitates design problem. Mech. Eng. 71:213-216. 1949. (221AAC)

Rushton, K. R. See Redshaw, S. C. (457)

Russell, W. T., and MacNeal, R. H. Improved electrical analogy for analysis of beams in bending. Trans. ASME 75:349-354. 1953. Discussion 76:94-95. 1954. (433)

Ryan, J. J. The plate analogy as a means of stress analysis. Proc. Soc. Exp. Stress Anal. X(1):7-28. 1952. (451)

Ryasanov, G. A. Application of eddying electric fields when modelling plane circulation flows by the EGDA method. (In Russian.) Tr. Leningr. In-ta Inzh. Vod. Transp. 23:219-222. 1956. (221AAA)

Ryder, F. L. (A) Applications of electrical analogs of static structures. Proc. ASCE 85:7-32. 1959. (472AB)

———. (B) Electrical analogs of statically loaded structures. Proc. ASCE 79, Separate No. 376. 1953. (472AB)

———. (C) Electrical energy analogs of dynamically loaded framed structures. Report of Republic Aviation Corp. 1957. (472AB)

———. (D) Energy versus compatibility analogs in electrical simulators of structures. J. Aero/Space Sci. 26:108-116. 1959. (471B) (472AB) (472BD) (484)

———. (E) Structural simulator for static analyses. Consulting Engr. (St. Joseph, Mich.) 10:84-90. 1958. (472AB)

———. See Zaid, M. (472AB)

Sabarly, F. See Habib, P. (221AAC)

Sacchetto, E. Studio teorico sperimentale sulle capacita dei cavi brifase cordafi. L Elektro 8:533-544. 1921. (612)

Sadowsky, M. A. An extension of the sand-heap analogy in plastic torsion applicable to cross sections having one or more holes. Trans. ASME 63:A-166. 1941. (442A)

Sage, B. H. See Baer, D. H., Schlinger, W. G., and Berry, V. J. (313A)

———. See Jenkins, R., and Brough, H. W. (322C)

———. See Schlinger, W. G., Berry, V. J., and Mason, J. L. (322B)

Salet, G. (A) Amelioration de la methode de determination des pointes de tension dans les arbres de revolution soumis a torsion au moyen d'un modele electrique. Bul. de l'Assn. Tech. Mar. et Aero. 41:295-303. 1937. (441BB)

———. (B) Determination des pointes de tension dans les arbres de revolution soumis a torsion au moyen d'un modelle electrique. Bul. de l'Assn. Tech. Mar. et Aero. 40:341-351. 1936. (441BB)

Salvadore, M. G. See Weiner, J. H., and Paschkis, V. (441AC) (442B)

Samuel, A. L. Design of electron guns. Proc. IRE 33:233-240. 1945. (651D)

Sander, H. An experimental method for the determination of the torsional stiffness of arbitrarily shaped bars of simply connected cross section. (In German.) Bauingenieur 34:309-311. 1959. (441AG)

Sander, K. F. Electrolytic tank analogue. Beama J. 65:17-23. 1958. (Appendix II D1)

———, Oatley, C. W., and Yates, J. G. Factors affecting the design of an automatic electron-trajectory tracer. Proc. Instn. Elec. Engrs. 99(III):169-179. 1952. (651D)

———, and Yates, J. G. Accurate mapping of electric fields in electrolytic tank. Proc. Instn. Elec. Engrs. 100(2):167-175. 1953. Discussion, 176-183 and 569-570. (Appendix II D1)

Sanial, A. J. Loudspeaker design by electromechanical analogy. Tele-tech 6:38-43, 97-102, 104. 1947. (511A) (121BBB)

Sarjant, R. J. See Jackson, R., Wagstaff, J. B., Eyres, N. R., Hartree, D. R., and Ingham, J. (311A)

———. See Price, P. H. (311A)

Sauer, R. Relationship between the theory of deformation of surfaces and gas dynamics. (In German.) Arch. Math. 1:263-269. 1948-1949. (211C)

———. Theoretische einfuhrung in die gasdynamic. Berlin. (Sect. 17, par. 1) 1943. (211AC)

Saxton, H. L. Mechanical and electrical analogies of acoustical path. J. Acous. Soc. of Amer. 10:318-323. 1939. (511B) (121BBB)

REFERENCES

Say, M. G. Analogies. Proc. Instn. Elec. Engr. 97:21-22. 1950. (511A)

Scanlan, R. H. (A) Electrical resistance networks for beam and column problems. J. Aero. Sci. 21:787-789. 1954. (421B) (422D)

———. (B) Resistance network solution of some structural problems in deflection and stability. Proc. Soc. Exp. Stress Anal. 16(1):117-128. 1958. (421B) (422D) (462) (524)

———. (C) Steady-flow aeroelastic study by electrical analogy. J. Aero. Sci. 20:691-698. 1953. (221AAA)

Schaefer, H., and Stackowiack, R. Probe measurements in models for the determination of electric field distribution in multilayer dielectrics in a condenser field excited by ultrashort electric waves. Z. Techn. Phys. 21:367-372. 1940. (612)

Schlinger, W. G., Berry, V. J., Mason, J. L., and Sage, B. H. Prediction of temperature gradients in turbulent streams. Proc. Gen. Discussion on Heat Transfer. 150-153. 1951. (322B)

———. See Baer, D. H., Berry, V. J., and Sage, B. H. (313A)

Schlitt, H. Loesung einer Waermeleitungsaufgabe durch Analogiebetrachtungen. Archiv für Elektrotechnik 43:51-58. 1957. (313A)

Schmidl, H. (A) Field measurements of plane and rotationally symmetric electrostatic fields in different dielectric media by the electrolytic tank. (In German.) Tech. in Wissenschaftlichen Abhandlungen, Ausgabe E. 161. 1951; 10. 1952. (612)

———. (B) Principles of measurement for analog studies on grounding equipment. (In German.) Elektrotechnische Zeitschrift 75(17):557-559. 1954. (612)

Schmidt, W. See Batzel, S. (122BB)

Schneebeli, P. H. de la Marre. Nouvelles methodes de calcul pratique des ecoulements de filtration non permanents a surface libre. Houille Blanche 8: Special No. B. 760-772. 1953. (221AAA)

Schneider, P. J. (A) Conduction heat transfer. Addison-Wesley, Cambridge. 322. 1955. (312C)

———. (B) The prandtl membrane analogy for temperature fields with permanent heat sources or sinks. J. Aero. Sci. 19:644-645. 1952. (312A)

———, and Cambel, A. B. (A) Membrane apparatus for analogic experiments. Rev. Sci. Instr. 24:513-517. 1953. (Appendix II A 1b)

———, and Cambel, A. B. (B) Steady temperature fields in electrical coils by membrane analogy. ASME Preprint No. 53-SA-43. 1953. (312A)

Schoch, A. Lateral displacement of a totally-reflected beam of ultrasonic waves. Acustica 2:18-19. 1952. (121BBD)

Schofield, F. H. (A) The heat loss from a cylinder embedded in an insulating wall. Phil. Mag. 12(7):329. 1931. (313A)

———. (B) The heat loss from a plate embedded in an insulating wall. Phil. Mag. 10(7):480. 1930. (313A)

———. See Awberry, J. H. (313A)

Schoenfeld, J. C. (A) Analogy of hydraulic, mechanical, acoustic, and electric systems. Appl. Sci. Res. Sec. B 3:417-450. 1954. (111) (511B)

———. (B) Ein elektrische model voor het experimenteel bepalen van de vervorming van een statisch onbepaalde vlakke vak-of raamwerkconstructie. Ing. Grav. 59:0-41 to 0-43. 1947. (472BC)

———. (C) Propagation of tides and similar waves. Thesis, 's-Gravenhage, Staatsdrukkerij-en Uitgeverijbedrijf. 1951. (111)

Schott, See Spangenberg, and Walters. (641)

Schubert, F. See Baehr, H. D. (313A)

Schum, E. F. See Ellerbrock, H. H., Jr., and Nachtigall, A. J. (311A) (313A)

Schurig. See Hazen, and Gardner. (Appendix II E)
Schutrum, L. F. See Humphreys, C. M., Nottage, H. B., Franks, C. V., Huebscher, R. G., and Locklin, D. W. (313A)
Schwedoff. Poggendorff's Annalen, Ergänzungs 6:85. 1873. (Appendix II D2)
Scott, D. R. (A) Solving compressible flow problems by machine. Colliery Eng. 32:323-326. 1955. (122BB)
_____. (B) Solving ventilation-network problems by machine. Colliery Eng. 29:410-413, 423. 1952. (122BB)
_____. (C) Solving ventilation-network problems by machine. Colliery Eng. 32:196-199. 1955. (122BB)
_____, and Hinsley, F. B. Solution of ventilation-network problems. Trans. Instn. Min. Engr. 3:347-366. 1952. (122BB)
_____, and Hudson, R. F. Automatic analogue computer for solution of mine ventilation networks. J. Sci. Instr. 30:185-188. 1953. (122BB)
Scott, R. F. An hydraulic analogue computer for studying diffusion problems in soil. Geotechnique, London 7:55-72. 1957. (311B)
Searcy, D. F. See Meyer, H. I. (221AF)
Seban, R. A. Remarks on film condensation with turbulent flow. Trans. ASME 76:299-302. 1954. (322C)
Seidling, J. Berechnung von Ringleitungen. Gas Wasser Waerme 7:47-59. 1953. (122AB)
Selim, M. A. (A) Dams on porous media. Proc. ASCE 71:1518-1535. 1945. (221AAA)
_____. (B) Underflow and uplift pressure for dams and weirs on porous media by electric analogy. Ph. D. thesis, Univ. of Cal., Berkeley. 1941. (221AAA)
Sells, B. E. See Poritsky, H., and Danforth, C. E. (211AAC) (211AAD)
Servranckx, R. Resolution d'equations differentielles et d'equations aux derivees partielles du second ordre au moyen d'un reseau electrique. H. F. Electricite, Courants Faibles, Electronique 2(10):271-276. 1954. (Appendix II E)
Seymour, H. Heat transfer problems: new method for solving. Electrician 128:512. 1942. (311A)
Shapiro, A. H. (A) An appraisal of the hydraulic analogue to gas dynamics. MIT Meteor Report 34. 1949. (211AB)
_____. (B) Free surface water table. High Speed Aerodynamics and Jet Propulsion IX:309-321. Princeton University Press. 1954. (211AB)
Sharman, C. F. See Taylor, G. I. (211AAA) (211AAB)
Sharp, J. M., and Henderson, E. N. Compressor piping design for pulsation control and maximum compressor efficiency. Oil and Gas J. 54:113, 115-118. 1956. (121BAA)
Shearer, J. L. Electric network analog study of viscous flow normal to parallel, evenly spaced cylinders. Textile Research J. 29:467-476. 1959. (222B)
Shelton, G. L., Jr. See Borden, A., and Ball, W. E., Jr. (221AAA)
Shen, D. W. C., and Packer, J. S. Analysis of hunting phenomena in power stations by means of electrical analogues. Proc. Instn. Elec. Engrs. 101 (2, Power Eng.):21-34. 1954. (531A)
Shields, J. H., and MacNeal, R. H. The solution of elastic stability problems with the electric analog computer. Trans. ASME 81E:635-642. 1959. (421B) (422D) (461) (472AB)
Shinn, B. J. See Gelfand, R., and Tuteur, F. B. (Appendix II D1)
Shippy, D. J. An electrical analogy of the potential distribution in two-dimensional unsteady-flow fields. M. S. thesis, Iowa State University. 1954. (311A)
Shiu-Jen Kuh, E. Potential analog network synthesis for arbitrary loss functions. J. Appl. Phys. 24:897-902. 1953. (Appendix II E)
Shurley, L. A. See Dean, L. E. (313A)

REFERENCES

Siestrunck, R. (A) Potential flow in single screw machines. Nat. Etud. Rech. Aero. Rep. 32. 1949. (221AAA)
———. (B) Sur un mode de solution rheoelectrique des problemes de l'helice propulsive. C. R. Acad. Sci. Paris 219:411-413. 1944. (221AAA)
———. See Bernard, J. J. (221AAA)
———. See Malavard, L. (A), (221AAA); (B), (221AAA)
———. See Malavard, L., and Germain, P. (221AAA)
Sills, E. C. See Babister, A. W., Marshall, W. S. D., Lilley, G. M., and Deards, S. R. (221AAB)
Simmons, W. R. (A) Analog for solution of heat conduction problems. Preprint 233, AIChE, Nuclear Engineering and Science Congress, E.J.C., Cleveland, Ohio. 1955. (311A) (312D)
———. (B) Electrical geometrical analogue for two-directional steady state heat conduction with uniform internal heat generation. Mech. Engr. 76:762. 1954. (312D)
Simond, A. W., and Kron, L. C. A quantitative experimental method of mapping equipotential lines and its application to electric precipitator problems. Rev. Sci. Instr. 1:527-536. 1930. (612)
Simonyi, K. Calculation of stress and deformation waves in long rods by means of a method used in communication science. Acta Techn. Hung., Budapest 1(3):319-362. 1951. (512A)
Skilling, H. H. An electric analogue of friction for solution of mechanical systems. Trans. AIEE 50:1155-1158. 1931. (531A)
Skinner, S. Experiments on the flow of heat in metal sheets. Proc. Phys. Soc. 28:119-123. 1916. (611)
Skyrme, T. H. R. See Peierls, R. E. (633)
Slab analogy experiments. U. S. Bureau of Reclamation, Boulder Dam Project, Final Report, Pt. 5, Technical Investigations, Bul. 2. (451)
Smith, J. Stresses in ships. Eng. 82:436-439. 1906. (412) (454)
Smith, R. O. Heat transfer problems solved by electrical analogy. Aero. Digest 72:21-24. 1956. (313A)
Smith, W. R. Proc. Roy. Soc. Edin. 7:79-99. 1869-1870. (Appendix II D2)
Smits, H. G. Photoelastic determination of shrinkage stresses. Proc. ASCE 61:600. 1935. (452)
Smits, H. See Biot, M. A. (452)
Smoot, C. H. An experimental determination of air-gap reluctance. J. of Western Soc. of Engrs. 10:500-511. 1905. (631)
Sokolov, V. S. See Kostiuk, A. G. (313A)
Some applications of the method of hydraulic analogy. Publications Scientifiques et Techniques du Ministere de l'Air 144:43-77. 1939. (211AB)
Sonntag, R. Zur Torsion von runden Wellen mit verandlichem Durchmesser. Zeit. f. Angew. Math. u. Mech. 9:1-22. 1929. (441BB)
Soroka, W. W. (A) Bars in torsion; methods for analyzing torsional shear stress. Mach. Des. 22:122-123. 1950. (441AA)
———. (B) Mechanical and electrical analogies for vibration isolation problems. Prod. Eng. 21(1):104. 1950. (511A)
———. See Anderson, J. E. (531A)
———. See Gross, W. A. (A), (531A); (B), (455)
———. See Waner, N. S. (441AD)
Southwell, R. V. (A) On the analogues relating flexure and extension of flat plates. Quart. J. Mech. Appl. Math. 3:257-270. 1950. (451)
———. (B) Use of an analogue to resolve Stoke's paradox. Nature 181:1257-1258. 1958. (222B)
———. See Allen, Fox, and Motz. (643)
———. See Bradfield, K. N. E., and Hooker, S. G. (441AD)

REFERENCES

Spangenberg, Walters, and Schott. Electrical network analyzers for the solution of electromagnetic field problems. Proc. IRE 37:724, 866. 1949. (641)

Sponsler, G. C. (A) An automatic electron-trajectory tracer study of helix and band-type post deflection acceleration. Tech. Rep. 64, Lincoln Laboratory, M. I. T. 1954. (651D)

——. (B) Trajectory-tracer study of helix and band-type post deflection acceleration. J. Appl. Phys. 26:676-682. 1955. (651D)

Spooner, J. C. See Davey, P. E. (313A)

Stackowiack, R. See Schaefer, H. (612)

Stambaugh, R. B. Electrical analog method for studying elastomer behavior. Indus. and Eng. Chem. 44:1590-1594. 1952. (491)

Standart, G. See Prochazka, J., and Landau, J. (322AB)

Standrud, H. T. The design of non-linear resistances for an hydraulic calculating board. M. S. thesis, M. I. T. 1940. (122AB)

Stark, V. J. E. See Landahl, M. T. (221AAA)

Starke, G. O. See Stephenson, D. G. (Appendix II E)

Stenstrom, L. The SAAB gradient tank, an aid to aeroplane design. SAAB Sonics 12:18. 1950. (221AAC)

Stephenson, D. G., and Starke, G. O. Design of a π network for a heat-flow analog. Trans. ASME 81E:300-301. 1959. (Appendix II E)

Stephenson, E. A. See Miles, A. J. (211AD)

Stephenson, R. W., Eaton, J. R., and Duffy, F. L. Simplify analyzer adaptation; manipulative method coordinates network analyzer to solve distribution and flow pressure problems. Am. Gas Assn. Mo. 35:27-31, 44. 1953. (122BAA)

Stevens, K. N., Kasowski, S., and Fant, C. G. M. An electrical analog of the vocal tract. J. Acous. Soc. of Amer. 25:734-742. 1953. (121BBC)

Stevens, O. See Vreedenburgh, C. G. J. (A), (221AAA); (B), (221AAA); (C), (221AAA)

Stewart, H. D. Electrical model for investigation of crankshaft torsional vibrations in in-line engines. Trans. SAE 54:238. 1946. (531B)

Stodola, A. Dampf und gas turbinen. Berlin. 1924. (211AB)

Stoker, J. J. The formation of breakers and bores. Communications on Appl. Math. 1:1. 1948. (211AB)

Stokes, G. G. Mathematical proof of the identify of the stream lines obtained by means of a viscous film with those of a perfect fluid moving in two dimensions. Report of the 68th Meeting of the Brit. Assoc. for the Adv. of Sci. 143-144. 1898. (221AF)

Stokey, W. F., and Hughes, W. F. Tests of the conducting paper analogy for determining isopachic lines. Proc. Soc. Exp. Stress Anal. 12(2):77-82. 1955. (458)

Stone, M. The general torsion problem: solution by electric analogy. Proc. 3rd Int. Congr. Appl. Mech. Stockholm 2:167-170. 1930. (441AD)

Stout, J. E. See Laitone, E. V. (211AB)

Stress studies for Boulder Dam. U.S. Bureau of Reclamation, Boulder Dam Project, Final Reports (5), Technical Investigations. Bul. 4. (451)

Striegel, R. Mapping of potential fields in the electrolytic tank. (In German.) Arch. Tech. Messen V:312. 1943. (632)

Stroband, H. J. De voortplanting van het getij bepaald met behulp van de electrotechniek met inachtneming weerstandswet. Polytechnisch Tijdschrift, The Hague, Holland. 1948. (111)

Su, H. L. Electric analog for theory of adjustment and regression. J. Math. and Phys. 38:312-326. 1960. (Appendix II E)

Sunatani, C., Matuyama, T., and Hatamura, M. The solution of torsion problems by means of a liquid surface. Tohoko Univ. Tech. Rep. 12:374-396. 1938. (441AA) (441AB)

REFERENCES

Supper, H. L. Photoelasticimetrie et apsidometrie. Pub. Sci. et Tech. du. Min. de l'Air 106. 1937. (452)
———. See McGivern, J. G. (411)
Suryaprakasam, M. V., Reid, G. W., and Geyer, J. C. (B) Use of alternating-current network calculator in distribution system design. J. Am. Water Works Assn. 42:1154-1164. 1950. (122AA)
Sved, G. Electric resistance network analogue for solution of moment distribution problems. Australian J. Appl. Sci. 7:199-204. 1956. (472AB)
Svensson, N. L. An electric analog for the limit analysis of framed structures. Struct. Engr. 37:292-298. 1959. (471A) (472AB)
Swain, B. J. See Riesz, G. W. (422D) (472AB)
Swain, P. W. Electrical machine solves heat-transfer problems; device developed by V. Paschkis at Columbia University. Power 85:480-482. 1941. (311A)
Swanson, R. S., and Crandall, S. M. An electromagnetic-analogy method of solving lifting-surface-theory problems. U. S. NACA Adv. Restr. Rep. L5D23. 1945. (Wartime Rep. L-120) (221CA)
———, Crandall, S. M., and Miller, S. Lifting-surface-theory solution and tests of elliptic tail surfaces of aspect ratio 3 and 0.5 chord 0.85 span elevator. NACA Tech. Note 1275. 1947. (221CA)
———, and Priddy, E. L. Lifting-surface-theory values of the damping in roll and of the parameter used in estimating aileron stick forces. NACA Adv. Restr. Rep. L5F23. 1945. (Wartime Rep. L-53) (221CA)
Swearingen, J. S. Predicting wet-gas recovery in recycling operations. Oil Weekly 96:30, 32-35, 36, 38. 1939. (221AAA)
Swenson, G. W. (A) Analysis of nonuniform columns and beams by a simple d.c. network analyzer. J. Aero. Sci. 19:273-275. 1952. (421A)
———. (B) A direct current network analyzer for solving wave-equation boundary-value problems. Ph. D. thesis, Univ. of Wisconsin. 1951. (Appendix II E)
Swenson, G. W., Jr., and Higgins, T. J. A direct current network analyzer for wave equation boundary value problems. J. Appl. Phys. 23:126-131. 1952. (Appendix II E)
Swinton, R. S. Picture representation of membrane analogy for design in torque. Civil Eng. 8:346. 1938. (441AA) (442A)
Syogo, M. On the Mach number at the diffuser throat of an ejector according to the hydraulic analogy. J. Aero. Sci. 24:918-919. 1957. (211AB)
Szebehely, V. G. An analogy between open soap bubble and compressible flow. J. Aero. Sci. 17:809. 1950. (211AC)
———, and Pletta, D. H. The analogy between elastic solids and viscous liquids. Virginia Poly. Inst. Bul. Series 80. 1951. (454)
———, and Whicker, L. F. (A) Generalization and improvement of the hydraulic analogy. Proc. First U.S. Nat. Cong. of Appl. Mech. 893-901. (211AB)
———, and Whicker, L. F. (B) The exact analogy between free surface water flow and two-dimensional gas flow. Mech. Eng. 73:516. 1951. (211AB)
———, and Whicker, L. F. (C) The improved hydraulic analogy. Va. Acad. Sci. 1951. (211AB)
———. See Whicker, L. F. (211AB)
Tait, P. G. See Thomson, W. (441AF)
Tanabe, Y., and Yamada, S. Electrolytic tank analogue design and application of automatic control. Sci. Rep. Res. Inst. Tohoku Univ. Japan (A)10:133-174. 1958. (Appendix II D1)
Tanasawara, Y. See Nukiyama, S. (313A)
Tank, F. See Meyer, H. (441AD) (455)
Taylor, E. M. See Ram, G., and Vaidhianathan, V. I. (A), (221AAA); (B), (221AAA)
———. See Vaidhianathan, V. I., and Ram, G. (A), (221AAA); (B), (221AAA)

REFERENCES

Taylor, G. I. (A) Conditions at the surface of a hot body exposed to the wind. Gr. Brit. Adv. Comm. Aero. Rep. and Mem. 272. 2:423-429. 1916-1917. (322C)

———. (B) The flow round a body moving in a compressible fluid. Proc. Third Int. Cong. Appl. Mech., Stockholm I:263-275. 1930. (211AAA) (211AAB)

———, and Sharman, C. F. A mechanical method for solving problems of flow in compressible fluids. Proc. Roy. Soc. Lond. (A) 121:194-217. 1928. (211AAA) (211AAB)

———. See Griffith, A. A. (A), (432), (441AA); (B), (441AA); (C), (432); (D), (441AA); (E), (441AA), (441AB)

Taylor, M., and Vaidhianathan, V. I. Uplift pressure on weirs of simple and complex design. Proc. Int. Conf. Soil Mech. Foundation Eng. 1:146-149. 1936. (221AAA)

Tearnen, J. O. A water channel analogy to a problem in gaseous combustion. M. S. thesis, Univ. of Calif., Berkeley. 1951. (211AB)

Teasdale, R. D., and Higgins, T. J. Experimental determination of the capacitance of heavy-current busses comprised of solid or tubular rectangular conductors. Trans. AIEE 67(I):653-659. 1948. (612)

Teofilato, S. Contributo alla rappresentazione analitica di una corrente gassosa mediante una corrente idrica. Monogr. Sci. Aero. 5:1-22. 1947. (9. 1948.) (10. 1949.) (211AB)

———. A contribution to the study of similarity between hydro and gas dynamics. Monogr. Sci. Aero. 4. 1949. (211AB)

Tewari, S. J. See Cadambe, V. (431)

Theocaris, P. S. (A) Direct determination of stresses in plane elasticity problems based on the properties of isostatics. Trans. ASME 81E:227-234. 1959. (458)

———. (B) Electrical analogy method for separation of principal stresses along stress trajectories. Proc. Soc. Exp. Stress Anal. 14(2):11-20. 1957. (413) (458)

———. (C) Stress concentration produced in perforated strips under tension. Proc. Soc. Exp. Stress Anal. 16(1):129-136. 1958. (413) (458)

Theodorsen, Th. (A) The theory of propellers. 1--Determination of the circulation function and the mass coefficient for dual-rotating propellers. U.S. NACA Rep. 775. 1944. (221AAA)

———. (B) Theory of propellers, Chap. V. McGraw-Hill. 1948. (221CB)

Thiel, A. Photogrammetrisches Verfahren zur vesuchsmassigen Losung von Torsionsaufgaben (Nach einem Seifenhautgleischnis von L. Foppl). Ing. Arch. 5:417-419. 1934. (441AA) (441AB)

Thomson, W., and Tait, P. G. Natural philosophy. Cambridge University Press, London. 1886. (441AF)

Thornton, W. M. The distribution of magnetic induction and hysteresis loss in armature. J. Instn. E. E. 37:125-139. 1906. (631)

———, and Williams, O. J. The distribution of dielectric stress in three-phase cables. Engineering 88:297-299. 1909. (613)

Thum, A., and Bautz, W. (A) Die Ermittlung von Spannungsspitzen in verdrehbeanspruchten Wellen durch ein elektrisches Modell. Zeit. Verein. Deut. Ing. 78:17-19. 1934. (441BB)

———, and Bautz, W. (B) Ermittlung der Verdrehspannungen in gekerbten Kinstruktionsteillen durch Modellversuche. Arch. f. Tech. Messen 4:V132-11, T113-15. 1934. (441BB)

———, and Bautz, W. (C) Kerbempfindlichkeit von Stahlen. Schweiz. Bauz. 106:25. 1935. (441BB)

———, and Bautz, W. (D) Zur Frage der Formziffer. Zeit. Verein Deut. Ing. 79:1303-1306. 1935. (441BB)

REFERENCES

Thwaites, A. L. See Woodcock, A. H., and Breckenridge, J. R. (311A) (321A) (330)

Timmis, A. C. Negative feed-back amplifier. J. Post Office Elec. Engr. 29:71-72. 1936. (121BBB)

Timoshenko, S. A membrane analogy to flexure. Proc. Lond. Math. Soc. (2)20: 398. 1922. (431)

———, and Goodier, J. N. Theory of elasticity. McGraw-Hill, New York. 1934. (432) (441AG)

Tipler, W. An electrical analogue to the heat regenerator. Proc. Seventh Int. Cong. Appl. Mech. 3:196-210. 1948. (321A)

Tirroloy. See Balachowsky, G. (612)

Tomita, Y. (A) A study of high-speed gas flow by the hydraulic analogy. (2nd Report) (In Japanese.) Trans. Japan Soc. Mech. Engrs. 21:22-29. 1955. (211AB)

———. (B) A study of high speed gas flow by the hydraulic analogy. (4th Report, Flow around an airfoil). Bul. Japan Soc. Mech. Engrs. 2:663-669. 1959. (211AB)

———. See Itaya, S. (211AB)

Touloukian, Y. S. See Klein, E. O. P., and Eaton, J. R. (311A)

Torgeson, W. L., Kitchar, A. F., and Hill, B. F. Study of the application of an electrolytic tank to 3-dimensional asymmetrical bodies (as applicable to aircraft icing). WADC TR 55-354 (PB111792). 1958. (313A)

Toupin, R. A. A variational principle for the mesh-type analysis of a mechanical system. J. Appl. Mech. 19:151-152. 1952. (511B)

Tracing electron paths. Gen. Elec. R. 61:39. 1958. (651C)

Trayer, G. W., and March, H. Torsion of members having sections common in aircraft construction. U.S. NACA Rept. 334. 1930. (441AA)

Trent, H. M. (A) An alternative formulation of the laws of mechanics. J. Appl. Mech. 19:147-150. 1952. (511B)

———. (B) Equivalent circuit for vibrating beam which includes shear motions. J. Acous. Soc. Am. 22:355-356. 1950. (523)

———. (C) Practical aspects of electro-mechanical analogies. N.R.L. Engineering Colloquium. 1947. (511B)

Tribus, M. Intermittent heating for aircraft ice protection with application to propellers and jet engines. Trans. ASME 73:1117-1118, 1128-1130. 1951. (311A)

Trofimenko, A. T. See Potseluiko, V. A. (311A)

Trudsø, E. An electric model of beams and plates. Bygnsstat. Medd. 26:1-22. 1955. (422D) (462)

Truesdell, C. The analogy between irrotational gas flow and minimal surfaces. J. Aero. Sci. 18:502. 1951. (211AC)

Trunk, E. F. Laclede applies McIlroy network analyzer to design calculations for distribution system. Am. Gas. J. 179:14-15. 1953. (122BAB)

Tumura, T. On the laminar flow of viscous fluid through a straight pipe. Trans. Soc. Mech. Eng. Tokyo, 2(6):59-61. 1936. (441AG)

Tuteur, F. B. See Gelfand, R., and Shinn, B. J. (Appendix II D1)

Uhrig, R. E. Analogies using non-identical equations. Proc. ASCE 81(775) 1955. (422D)

Umanski, E. S. See Kvitka, A. L., and Agarev, V. A. (492)

Vaidhianathan, V. I., and Ram, G. (A) Electrical method for determining pressure distribution under hydraulic works. Proc. Punjab Eng. Cong. Paper 190. 1936. (221AAA)

———, and Ram, G. (B) On the electrical method of investigating the uplift pressure under dams and weirs. Res. Pub. Punjab Irrig. Res. Inst. 5(4). 1935. (221AAA)

REFERENCES

Vaidhianathan, V. I., Ram, G., and Taylor, E. M. (A) Pressure under weirs — depressed floor with and without sheet piles. Res. Pub. Punjab Irrig. Res. Inst. 5(6). 1935. (221AAA)

———, Ram, G., and Taylor, E. M. (B) Uplift pressure on weirs. A floor with a line of sheet piles. Proc. Ind. Acad. Sci. 2:646-655, Sect. A (6). 1935. (221AAA)

———. See Luthra, H. R. (221AAA)

———. See Ram, G. (A), (221AAA); (B), (221AAA)

———. See Ram, G., and Taylor, E. M. (A), (221AAA); (B), (221AAA)

———. See Taylor, M. (221AAA)

Vainberg, D. V. Analogy between problems for plane stress and bending of a circular plate of varying thickness under asymmetrical load. (In Russian.) Prikl. Mat. Mekh. 16:749-752. 1952. (451)

Van Der Meer, S. See Brouwer, G. (472AB)

Vandrey, F. Investigations concerning the treatment of plane subsonic flow with the aid of electric analogy. (In German.) Gottingen. 1944. (211C)

Van Haver, V. Solving acoustical problems by means of electrical models. (In Dutch.) Tech. Wet. Tijdschr. 22:141-150. 1953. (121BBB)

Van Veen, J. (A) Analogy between tides and a.c. electricity. Eng. 184:498-500; 520-521; 544-545. 1947. (111)

———. (B) Electrische nabootsing van getijden. De Ingenieur. 3. 1946. (111)

———. (C) Getijstroomberekeningen met behulp van wetten analoog aan die van ohm and kirchhoff. De Ingenieur, 3. 1937. (111)

Vartia, K. O. Analog circuit solves open channel problem. Eng. News-Rec. 149:44-45. 1952. (112B)

Vaughan, D. C. Relaxation methods, a three-dimensional mechanical analogy. Quart. J. Mech. and Appl. Math. 5:462-465. 1952. (472AF)

Vening-Meinesz, F. De verdeeling der spanningen in een lichaam dat zich volgens de wet van hooke gedraagt. De Ingenieur. (26-3):180-185. 1911. (432)

Vernon, M. Les analyserus electriques a resistances et capacites. Technique Moderne 42:97-104, 173-180:245-252. 1950. (313A)

Vidal, J. L'analyse des champs thermiques par analogie électrique. Revue Universelle des Mines 13:32-50. 1957. (313A)

Vierling, D. E. See Perry, H. A., Jr., and Kohler, R. W. (122AA)

Vincent, J. L'écoulement inverse en analogie rhéoélectrique. ONERA NT 3/1101.A. 1953. (221AAA)

Vineyard, G. H. Simulation of trajectories of charged particles in magnetic fields. J. Appl. Phys. 23:35-39. 1952. (651B)

Vogel, L. C., and Krueger, R. F. An analog computer for studying heat transfer during a thermal recovery process. J. Petr. Technol. (Petr. Trans. AIME 204) 7:205-212. 1955. (311A)

Voigt, W. Eine neue methode zur untersuchunz der varmeleitung in krystallen. Ann. der Phys. 60:350-367. 1897. (611)

Voissel, P. Resonanzerscheinungen in der saugleitung von kompressoren und gasmotoren. Mitteilungen über Forschungsarbeiten auf dem Gebiete des Ingenieurwesens. (121BAA)

Voith, R. J. See Douglas, J. F. H. (632)

Volterra, V. On the temperature in the interior of a mountain. (In Italian.) Nuovo Cim. 4, series 6:111-126. 1912. (313A)

von Bruck, J. See Barkhausen, H. (612)

von Karman, T. (A) The analogy between fluid friction and heat transfer. Trans. ASME 61:705-710. 1939. (322C)

———. (B) Eine praktische anwendung der analogie zwischen uberschallstromung

REFERENCES

in gasen und uberkritesder stromung in offenen gerinnen. Zeits. f. Angew. Math. Mech. 18:49-56. 1938. (211AB)

———, and Biot, M. A. Mathematical methods in engineering. McGraw-Hill, New York. 1940. (511A)

Vreedenburgh, C. G. J., and Stevens, O. (A) Electric investigations of underground water flow nets. Proc. Int. Conf. Soil Mech. Foundation Eng. 1:219-222. 1936. (221AAA)

———, and Stevens, O. (B) Electrisch onderzoek van potentiaalstroomingen in vloeistoffen in het byzonder toegepast op vlakke grondwater-stroomingen. De Ingenieur 32. 1933. (221AAA)

———, and Stevens, O. (C) Elektrodynamische untersuchungen von potentialstromungen in flussigkeiten. Premier Cong. Grands Barrages, Stockholm IV:217-226. 1933. (221AAA)

Vyalov, S. S., and Kogan, L. G. Solution of some heat problems by methods of electric analogies. Izv. Akad. Nauk SSR Otd. tech. Nauk. 1:39-52. 1951. (311A)

Wagner, C. F. Mechanical demonstrator of travelling waves. Elec. Eng. 58:414-420. 1939. (621)

Wagstaff, J. B. See Jackson, R., Sarjant, R. J., Eyres, N. R., Hartree, D. R., and Ingham, J. (311A)

Walker, G. E. Electrical method of plotting streamlines. Mech. World 133:294-295. 1953. (221AB)

Wall, T. F. Generator surging; equivalent mechanical model. Electrical Rev. (London), 150:65-66. 1952. (621)

Wallis, R. A. See Pizer, H. I., and Warden, M. C. (221AAA)

Walters. See Spangenberg, and Schott. (641)

Walther, A. Ein elektrisches Naherungsgleichnic fur die Rotationssymmetrische Potentialsstromung. Zeit. f. Tech. Phys. 16:19-22. 1935. (441BB)

Waner, N. S., and Soroka, W. W. Stress concentration for structural angles in torsion by conducting sheet analogy. Proc. Soc. Exp. Stress Anal. 11(1):19-34. 1953. (441AD)

Warden, M. C. See Pizer, H. I., and Wallis, R. A. (221AAA)

Warner, P. C. See Hagg, A. C. (524)

Warren, C. E. See Criner, H. E., and McCann, G. D. (511A) (511B)

———. See McCann, G. D., and Criner, H. E. (531A)

Warshavsky, L. A., and Fedorovich, V. Electro-mechanical analogies. Isvestiya Electro-Promish-Iennosti Slavovo Toka. 3:51-63. 1936. (551A)

Weber, E. See Kopf, E. (441AA)

Wegner, V. Über den Zusammenbang von Stromungs und Spannungsproblemen. Ing. Arch. 5:449-469. 1934. (454)

Weibel, E. E. (A) Application of soap-film studies to photoelastic stress determination. Assoc. Int. des Ponts et Charpentes — Memoires 3:421-438. 1935. (452)

———. (B) Studies in photoelastic stress determination. Trans. ASME 56:637. 1934. (452)

Weiner, F. (A) Further remarks on intermittent heating for aircraft ice protection. Trans. ASME 73:1131-1136. 1951. (311A)

———. (B) The thermal analyzer. California Engr., Univ. of Calif., Berkeley. 1950. (313A)

Weiner, J. H. Resistor-network solution for the torsion of hollow sections. Trans. ASME (J. Appl. Mech.) 75:562-564. 1953. (441AC)

———, Salvadori, M. G., and Paschkis, V. Resistor-network solution for elastoplastic torsion problem. Proc. ASCE 81(671):1-10. 1955. Discussion. 82(EM2, 946):5-7. 1956. (441AC) (442B)

REFERENCES

Weissenbach, B. Berechnung von Waermeschaltplaenen mit Hilfe der kirchhoffschen Saetye. Brennstoffe-Waerme-Kraft 10:505-507. 1958. (322AA)

Welch, W. P. A proposed new shock measuring instrument. Exp. Stress Anal. 5(1):39-51. 1947. (511A)

Wells, A. A. On solution of beam-on-elastic foundation problems by means of mechanical analogue. Proc. Inst. Mech. Engr. (War Emergency Proc. 63) 163:307-310, 316-327. 1950. (422B) (465)

Werner, B. L., and Moore, R. C. McIlroy network analyzer solves Baltimore distribution problems. Water Works Eng. 109:343, 383-385. 1956. (122AB)

Werner, K. H. Ein waermeleitungsmodell des tiefseetelegraphenkabels. Frequenz 9:280-281. 1955. (313A)

Wesser, U. See Blass, E. (313A)

Westergaard, H. M. (A) Deflection of beams by the conjugate beam method. J. Western Soc. Engr. 26:369. 1921. (422C)

———. (B) Graphostatics of stress functions. Trans. ASME 56:141. 1934. (451)

———. (C) Memorandum to chief design engineer, U. S. Bureau of Reclamation. 1931. (451)

Weston, M. K. Some factors determining the performance of solid insulation. Inst. Elec. Engrs. Students Quart. J. 1955. (Digest: J. Instn. Elec. Engrs. 1:464-465. 1955.) (612)

Whicker, L. F., and Szebehely, V. G. Determination of aerodynamic forces on high speed aircraft by means of an improved hydraulic analogy. Va. Poly. Inst. Engr. Exp. Sta. Rep. 1951. (211AB)

———. See Szebehely, V. G. (A), (211AB); (B), (211AB); (C), (211AB)

Whinnery, J. R., Concordia, C., Kron, G., and Ridgway, W. Network analyzer studies of electromagnetic cavity resonators. Proc. IRE 32:360-367. 1944. (641)

———, and Ramo, S. A new approach to the solution of high-frequency field problems. Proc. IRE 32:284-288. 1944. (641)

Whitehead, S. Mathematical methods applicable to linear phenomena. J. Sci. Instr. 21:73-80. 1944. (511B)

Whiting, S. E. See Kennelly, A. E. (612)

Wianko, T. See Biot, M. (524) (532C)

Wieghardt, K. Über ein neues Verfahren, verwickelte Spannungsverteilungen in elastischen Korpern auf experimentellem Wege zu finden. Mitteilungen über Forschungsarbeiten auf dem Gebiete des Ingenieurwesens 49:15-30. 1908. (451)

Wiggins, A. M. Electro-mechanical analogy in acoustic design. Radio 30:28-29. 1946. (121BBB)

Wiles, G. G., and Grave, D. F. H. Heat flow studies using electrolytic tank. J. Chem., Met. and Min. Soc. S. Africa 55:149-153. 1954. Discussion 56:255-257. 1955. (313A)

Willers, F. A. Die Torsion eines Rotationskorper um seine Achse. Zeit. f. Math. u. Phys. 55:225-263. 1907. (441BB)

Willey, F. G. Electrical solution of thermal problems. Electronics 19:190. 1946. (311A)

Williams, J. C. Solution of temperature distribution on problems by computation and electric analogy. Brit. J. Appl. Phys. 3:197-199. 1952. (311A)

Williams, M. L. Some thermal stress design data for rocket grains. J. ARS 29:260-267. 1959. (313A)

Williams, O. J. See Thornton, W. M. (613)

Wilson, F. R. See Farr, H. K. (Appendix II D1)

Wilson, L. H., and Miles, A. J. Application of membrane analogy to solution of heat conduction problems. J. Appl. Phys. 21:532-535. 1950. (313B)

Wilts, C. H. Investigation of methods for computing flutter characteristics of supersonic delta wings and comparison with experimental data. NASA TN D-5. 1959. (524) (532C)

_____. See Green, L., Jr. (212B)
_____. See McCann, G. D. (211AAD) (313A)
_____. See McCann, G. D., and Locanthe, B. N. (Appendix II E)
_____. See MacNeal, R. H., and McCann, G. D. (521) (532C)
Wolf, A. Use of electrical models in study of secondary recovery projects. Oil and Gas J. 46:94, 96-98. 1948. (221AAA)
Wolfenson, L. B. See Reid, G. W. (122AA)
Woodcock, A. H., Thwaites, A. L., and Breckenridge, J. R. Electrical analog for studying heat transfer in dynamic situations. Mech. Engr. 81:71. 1959. (311A) (321A) (330)
Wright, K. R. Model approach to a groundwater problem. Proc. ASCE 84(IR4, 1862):1-9. 1958. (221AAA)
Wright, W. Column analogy solution of rotational yield of beam and frame supports. Civ. Eng. (London) 54:745-746. 1959. (422A)
Wyckoff, R. D., and Botset, H. G. An experimental study of the motion of particles in systems of complex potential distribution. Physics 5:265. 1934. (221AAA)
_____, Botset, H. G., and Muskat, M. The mechanics of porous flow applied to water flooding problems. Trans. AIME (Petroleum Division) 103:219. 1933. (221AAA)
_____, and Reed, D. W. Electrical-conduction models for the solution of water-seepage problems. Physics 6:395-401. 1935. (221AAA)
_____. See Muskat, M. (221AAA)
Wyszomirski, A. Stromlinein und Spannungslinien. Ein Versuch, Probleme der Elastizitatslehre nirt Hilfe Hydraulischer Analogien Experiment ell zu Lasen. Dresden. 1914. (441BB)
Yamada, S. See Tanabe, Y. (Appendix II D1)
Yamamoto, Y. See Friedmann, N. E., and Rosenthal, D. (441AD)
Yamauchi, T. The experimental analysis for rigid frame structure by an analogous circuit consisting of resistors. (In Japanese.) Trans. Japan Soc. Civ. Engrs. 60:30-36. 1959. (472AB)
Yarnold, G. D. Mechanical analogy to motion of electrons in gases. Phil. Mag. 29:47-51. 1940. (653)
Yates, J. G. See Sander, K. F. (Appendix II D1)
_____. See Sander, K. F., and Oatley, C. W. (651D)
Yeung, Y. See Le Corbeiller, P. (511A) (511B)
Yu, Y.-Y. On the complex representation of the general extensional and flexural problems of thin plates and their analogies. J. Franklin Inst. 260:269-282. 1955. (451)
Zachrisson, L. E. On the membrane analogy of torsion and its use in a simple apparatus. Trans. Roy. Inst. Tech., Stockholm. 44. 1951. (441AA)
Zaid, M., and Ryder, F. L. Electrical analogue to a celled tube subjected to flexure and torsion. Aircraft Eng. 31:94-100. 1959. (472AB)
Zangar, C. N., and Haefeli, R. J. Electric analog indicates effect of horizontal earthquakes shock on dams. Civ. Eng. 22:54-55. 1952. (221D)
Zee, C.-H. The use of combined electrical and membrane analogies to investigate unconfined flow into wells. Ph. D. thesis, Utah State Agric. Coll. 1952. (221AC)
Zhoukovsky, M. I. Application of hydro-gas analogy for an approximate investigation into supersonic flow in the profile gratings. (In Russian.) Teploenergetika 2:19-33. 1959. (211AB)
Zlotowski, I., and Prebus, A. F. Computation of electron trajectories in electrostatic lenses of rotational and plane symmetry. Phys. Rev. 67:203. 1945. (Abstract) (652)
_____. See Prebus, A. F., and Kron, G. (652)
Zschaage, W. Nachahmung des elecktrischen Feldes von Leitungen in elektrolytischen Trog. Elekt. Zeits. 46:1215-1219. 1925. (612)
Zupnik, T. See Landes, F. (313A)